AF823783

Springer Theses

Recognizing Outstanding Ph.D. Research

Aims and Scope

The series "Springer Theses" brings together a selection of the very best Ph.D. theses from around the world and across the physical sciences. Nominated and endorsed by two recognized specialists, each published volume has been selected for its scientific excellence and the high impact of its contents for the pertinent field of research. For greater accessibility to non-specialists, the published versions include an extended introduction, as well as a foreword by the student's supervisor explaining the special relevance of the work for the field. As a whole, the series will provide a valuable resource both for newcomers to the research fields described, and for other scientists seeking detailed background information on special questions. Finally, it provides an accredited documentation of the valuable contributions made by today's younger generation of scientists.

Theses are accepted into the series by invited nomination only and must fulfill all of the following criteria

- They must be written in good English.
- The topic should fall within the confines of Chemistry, Physics, Earth Sciences, Engineering and related interdisciplinary fields such as Materials, Nanoscience, Chemical Engineering, Complex Systems and Biophysics.
- The work reported in the thesis must represent a significant scientific advance.
- If the thesis includes previously published material, permission to reproduce this must be gained from the respective copyright holder.
- They must have been examined and passed during the 12 months prior to nomination.
- Each thesis should include a foreword by the supervisor outlining the significance of its content.
- The theses should have a clearly defined structure including an introduction accessible to scientists not expert in that particular field.

More information about this series at http://www.springer.com/series/8790

Giordon Stark

The Search for Supersymmetry in Hadronic Final States Using Boosted Object Reconstruction

Doctoral Thesis accepted by the
University of Chicago, Illinois, USA

Springer

Giordon Stark
University of Chicago
Chicago, IL, USA

ISSN 2190-5053 ISSN 2190-5061 (electronic)
Springer Theses
ISBN 978-3-030-34547-1 ISBN 978-3-030-34548-8 (eBook)
https://doi.org/10.1007/978-3-030-34548-8

© Springer Nature Switzerland AG 2020
This work is subject to copyright. All rights are reserved by the Publisher, whether the whole or part of the material is concerned, specifically the rights of translation, reprinting, reuse of illustrations, recitation, broadcasting, reproduction on microfilms or in any other physical way, and transmission or information storage and retrieval, electronic adaptation, computer software, or by similar or dissimilar methodology now known or hereafter developed.
The use of general descriptive names, registered names, trademarks, service marks, etc. in this publication does not imply, even in the absence of a specific statement, that such names are exempt from the relevant protective laws and regulations and therefore free for general use.
The publisher, the authors, and the editors are safe to assume that the advice and information in this book are believed to be true and accurate at the date of publication. Neither the publisher nor the authors or the editors give a warranty, expressed or implied, with respect to the material contained herein or for any errors or omissions that may have been made. The publisher remains neutral with regard to jurisdictional claims in published maps and institutional affiliations.

This Springer imprint is published by the registered company Springer Nature Switzerland AG.
The registered company address is: Gewerbestrasse 11, 6330 Cham, Switzerland

The story so far: In the beginning the Universe was created. This has made a lot of people very angry and been widely regarded as a bad move.

—Douglas Adams,
The Restaurant at the End of the Universe

And I urge you to please notice when you are happy, and exclaim or murmur or think at some point, 'If this isn't nice, I don't know what is'.

—Kurt Vonnegut,
A Man Without a Country

This is me.

—Keala Settle

Supervisor's Foreword

It is an honor to introduce the work of my former graduate student, Dr. Giordon Stark, whose work delves deeply into the study of the patterns created in the highest energy human-made collisions ever at the Large Hadron Collider (LHC).

A modern particle physics experiment is an observatory of the subatomic structure of nature. We conduct measurements millions of times per second using a multitude of technologies, wavelengths, time domains, and methodologies. We study processes that take place in less than a trillionth of a second and search for the existence of never-before-seen phenomena. The extreme energies of the LHC yield an unprecedented resolving power for these measurements and create dramatic challenges in searching for signals that may be hiding inside the patterns that the debris created by these intense collisions leaves behind in our instruments. These patterns, and in particular those formed by hadronic systems, may reveal the answers to several deep mysteries of physics, including the origins of dark matter, particle mass and family hierarchies, and more.

The work performed by Dr. Giordon Stark in the process of producing this thesis represents one of the most comprehensive and in-depth studies of the use of Lorentz-boosted hadronic final state systems in the search for signals of supersymmetry conducted to date at the LHC. A thorough assessment is performed of the observables that provide enhanced sensitivity to new physics signals otherwise hidden under an enormous background of top quark pairs produced by Standard Model processes. This is complemented by an ingenious analysis optimization procedure that allowed for extending the reach of this analysis by hundreds of GeV in mass of these hypothetical new particles. Lastly, the combination of deep, thoughtful physics analysis with the development of high-speed electronics for identifying and selecting these same objects is not only unique but also revolutionary. The Global Feature Extraction system that Dr. Giordon Stark played a critical role in bringing to fruition represents the first dedicated hardware device for selecting these Lorentz-boosted hadronic systems in real time using state-of-the-art processing chips and embedded systems. Therefore, not only does the work described in this thesis represent a detailed study of a potential revolutionary signature of new physics, but it also builds a platform—quite literally—for searching for such signatures well into the future of the LHC and the ATLAS Experiment. It is a great honor for Giordon Stark that his thesis is being published by Springer.

Chicago, IL, USA David W. Miller

Acknowledgments

Even now that I am writing the acknowledgments of this thesis, it does not seem to quite feel real just yet. In any case, I want to take the time to reflect back on my not-quite-so-straight path that eventually led to this cohesive set of writings on the research I have done with David Miller and the UChicago ATLAS group over the past 5 years. I still recall when I sat in David's office in the old HEP building wondering if he would be crazy enough to let me work for him. I started with the hardware project, using what I have learned from Rai Weiss and Sam Waldman at LIGO. Funnily enough, this is the second "David Miller" I have had the pleasure to work with, the first at Caltech when I was an undergrad working on the submillimeter wave observatory. Altogether, it has been a lot of fun, and hopefully we will still be able to keep working together even after I leave UChicago for a postdoc.

I have many other people to thank, but most notably my parents: Suzane and Henry Stark, mom and dad. I am grateful for all of the love and support, especially through the many battles over the years just to make sure I had the same level of access as anyone else, so that I could (and did) succeed. Even though you probably do not understand most of this thesis, you guys have always been there, thank you. The rest of my family has also been incredibly supportive and never stopped believing in me: my brothers Michael, Aron, and Harrison; my sister Rachael; my uncles Morty, Philip, and Brian; and my (fake) sister Kathy Buckley. They have all been helpful when I needed a break from my studies from time to time and head down to Florida or over to Long Island.

I would like to also thank my committee members for bearing with me over this past year—Mel Shochet, Dave Schmitz, LianTao Wang, Heinrich Jaeger. They have had a lot of fantastic questions and comments that helped me learn how to explain and frame my research in the context of the bigger picture, and their help has been much appreciated. As well, all my professors and teachers over the years who have, in some way or form, guided me along this path of physics including Scotty Howard, Marilyn Pedek, Brian Murray, Mary Murray (no, they are not related), Beth Bobay, Jeannie Cornwell, and William Mech. Additionally, Harvey Newman is an excellent advisor and I was lucky to have him for the 4 years I was an undergrad at Caltech.

Even though I never did anything remotely related to high energy physics as an undergrad, it was probably not a surprise that I ended up running into Harvey again in Switzerland as a grad student.

The UChicago ATLAS HEP group has been an awesome place to work with a ton of brilliant people. I have enjoyed working closely with my fellow grad students Joakim, Miles, Kevin, Jeff, Karol, Patrick, Lesya, and Todd; and I have learned a lot from the postdocs Max, Reina, Gabriel, Jamie, and John. Max, Reina, Joakim, Miles, and I were David's group of misfits doing all kinds of amazing research, and it seems to be interesting timing that (apart from Miles who graduated last year) all of us are moving on around the same time. I guess things just kind of work out like that in life. Joakim, thanks for all the late-night chats when I was in the midst of writing a thesis and for just being there. Max, how the hell did you stand me? It has been a lot of fun and we have definitely leaned on each other a lot for things and you are definitely one of the best postdocs I have known. Reina, we have had quite a few adventures together, including climbing all over Chamonix in France and I am glad we got the chance to work together on gFEX. Lesya, thanks for being a fantastic friend and going to all of those broadway shows with me and being here, except for when you moved to Switzerland and now you are "over there." I also want to thank the folks of the electronics design group here: Mary Heintz (manager) and Fukun Tang (engineer); never saying "no" to any of my crazy ideas and whose support helped make the hardware project a reality.

There are a lot of people to thank still, especially because I have found a home in the Chicago community here. Most especially, thanks to Simone, Anni, and Lilly. Simone, I do not know if you have ever watched Grey's Anatomy (I have) but I like to think of us as Meredith and Cristina on that show. We have had a ridiculous amount of fun, especially all of those random late nights when we find ourselves in the middle of the city, hungry, and cannot find a place to eat that is open so late. It would be funnier had it not happen dozens of times, we really need to find more late-night places. Anni, I have always tried to support you as much as you have supported me and I still feel like it is never enough. It is funny that we are both graduating at the same time and I am proud of you and you deserve every opportunity you are going to get because of all the work and passion you put in as an interpreter. And to Lilly (really, Eliza), it is funny that we both messed up each other's names the first time we met in that Starbucks in Old Town. You have been there when I was struggling from time to time over the past 4 years we have known each other, and you have taught me a lot about seeing the world differently. I will not ever forget that. And to the rest of the Chicago Deafies: Evelyn, Sandy, JEF, Nick, Jason, Melissa, Jesse, Yehonatan, Cicely, Matthew, Crom, Peter, and many, many others; as well as the friends I have made along the way: McKenna, Helena, Caitlin, Umi, Boo, Alithea, Leah, and easily a hundred other people that is just too many to name in an already-too-long thesis. And finally, I would like to especially acknowledge those who have enabled access to my research, Rivka Hozinsky and the many interpreters I have used along the way both here in America and around the world. Thank you for handling all of my requests over the years, no matter how short notice or crazy they might have been. And finally, those who I have met as a result of working at

CERN, the highly international collaboration of scientists including many Matts, Davids, and Bens, as well as Michael Begel, Revital Kopeliansky, Trisha, Chiara, Ian, Elliot, Brianna, Tuna, Lukas, Jon, Attila, Nils, Chris, Peter, Micha (whom I roomed with and bonded over our shared injuries due to the evil mandoline slicer), and the many other colleagues from across the ring. The thousands of people that make up CERN form a wonderful group to discuss techniques and foster ideas in the hundreds of meetings and conferences and workshops and plenaries and "weeks."

I have experienced a crazy, eventful 5 years all over the world, and it still has not quite hit me yet how amazing this scientific collaboration is. I still cannot believe David was crazy enough to take me on as a graduate student. To all I have named and those I have not, this work is dedicated to you. Merci beaucoup á tous.

Contents

Chapter 1
Introduction

The Large Hadron Collider (LHC) operates at the highest energy scales ever artificially created in particle collision experiments with a center-of-mass energy $\sqrt{s} = 13$ TeV. In addition, the high luminosity allows the unique opportunity to probe the Standard Model at the electroweak scale and explore for rare signs of new physics beyond the Standard Model. The coupling of the third-generation top quark to the Higgs boson introduces large, quadratic, radiative corrections to the Higgs mass, requiring a significant amount of fine-tuning that results in a nearly perfect correction of the Higgs mass from the Planck scale to the observable electroweak scale. A possible solution to the naturalness problem proposes a collection of supersymmetric partners to the Standard Model particles with the mass of lightest particles at the electroweak scale: the gluino, the stop squarks, and the lightest supersymmetric particle.

All matter interacts via the four fundamental forces: gravitational, electromagnetic, weak, and strong; at least up to the scale of the weak interactions. Gravity is very well-described by Einstein's theory of general relativity. The remaining three forces are described by a group of theories that describe fundamental particle physics and the interactions of all known elementary particles, the Standard Model. Formulated over the last century, this theory was kickstarted by Glashow's [170] discovery of combining electromagnetic and weak interactions in 1961. The Standard Model has stood up to rigorous testing by many experiments and shown to be robust. However, this is not a complete model given the success so far, as certain assumptions are still made that need to be reconciled. Chapter 2 introduces the theories of the Standard Model and motivates the search for new physics. Chapter 3 introduces the Large Hadron Collider and the ATLAS detector, and its role in enabling searches for physics beyond the Standard Model. Chapter 4 discusses the current limitations in recording all of the data produced by the collider and the ATLAS detector's solution to managing this enormous influx of data. This chapter finishes off with an introduction of the necessary instrumentation upgrades for the ATLAS detector in parallel with the upgrades to the collider. At this point

© Springer Nature Switzerland AG 2020

G. Stark, *The Search for Supersymmetry in Hadronic Final States Using Boosted Object Reconstruction*, Springer Theses,
https://doi.org/10.1007/978-3-030-34548-8_1

the reader will have a broad understanding of the fundamental interactions of particles, designing a massive and complex hardware system that enables us to probe for new physics, and being able to record all of the raw data of proton–proton collisions. But now, we must be like Sherlock Holmes, using the footprints of the collision data to look for patterns to reconstruct a picture of the original collision and what happened. Chapter 5 explores the many tried-and-proven techniques used by the ATLAS collaboration to reconstruct many of these fundamental particles. The energy scale of the proton–proton collisions at the ATLAS detector produces showers of Lorentz-boosted partons that form massive hadrons with interesting substructure, a tell-tale signature of many beyond the Standard Model theories, necessitating specialized reconstruction techniques described in Chap. 6. Chapter 7 applies these boosted reconstruction techniques to a particular search for gluinos, a new theoretical particle whose existence at the electroweak scale could help answer some questions about the Standard Model. The results of the search for gluinos are detailed in Chap. 8. This search, and many others, can benefit from the future upgrades of the ATLAS detector whose physics impact is being studied in Chap. 9 to enhance the detector's sensitivity to these boosted objects that are copiously produced. Finally, Chap. 10 provides some concluding remarks about the search for new physics and the exciting outlook of the LHC physics program.

Chapter 2
Standard Model (and Beyond!)

The Standard Model (SM) is a set of theories that describe fundamental particle physics and the interactions of all known elementary particles, except gravity.[1] Kickstarted by Glashow's [170] discovery of combining electromagnetic and weak interactions in 1961, it has evolved since then into its current form that we know today. Many precision analyses have been performed at many particle physics experiments such as AGS, E288, PETRA, UA1, D0, DONUT, and the experiments at the LHC (see Table 2.2). All of these experiments have measured the cross-section for various processes and show good agreement to the predictions of the Standard Model. For example, Fig. 2.1 describes the production cross-section measurements measured by the ATLAS detector compared to theoretical expectations for common decay processes. The data/theory ratio is shown on the right side of the figure and serves to show how successful the Standard Model has been. However, it had posed a few problems such as requiring spontaneous symmetry breaking in order to explain the heavy masses of the bosons that mediate the weak interactions (see Sect. 2.1.2). The Higgs boson, discovered in 2012 [7], explained this missing piece.

Section 2.1 will provide background information about the Standard Model, the theories, and its particles. There are still many other puzzles of the Standard Model that need to be reconciled and will be discussed in Sect. 2.2, thanks in large part due to [207, 244].

[1] As far as we know, gravity is too weak to play any significant role in ordinary particle processes.

© Springer Nature Switzerland AG 2020

G. Stark, *The Search for Supersymmetry in Hadronic Final States Using Boosted Object Reconstruction*, Springer Theses,
https://doi.org/10.1007/978-3-030-34548-8_2

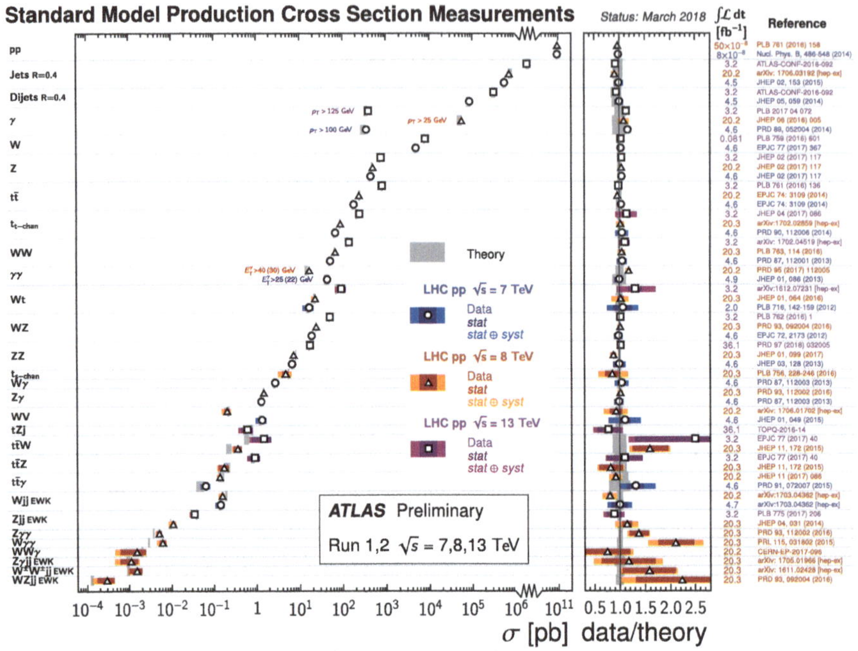

Fig. 2.1 Summary of several Standard Model total and fiducial production cross-section measurements, corrected for leptonic branching fractions, compared to the corresponding theoretical expectations [54]. All theoretical expectations were calculated at next-to-leading-order (NLO) or higher. The dark-color error bar represents the statistical uncertainty. The lighter-color error bar represents the full uncertainty, including systematics and luminosity uncertainties. The data/theory ratio, luminosity used, and reference for each measurement are also shown. Uncertainties for the theoretical predictions are quoted from the original ATLAS papers. They were not always evaluated using the same prescriptions for PDFs and scales. The Wgamma and Zgamma theoretical cross-sections have non-perturbative corrections applied to the next-to-next-to-leading-order (NNLO) fixed order calculations [36]. Not all measurements are statistically significant yet

2.1 The Standard Model

The SM is the most comprehensive quantum field theory of particle physics today. It encompasses a single, concise model made of up two theories: the Glashow–Weinberg–Salam theory of QED (Sect. 2.1.2) which describes the electromagnetic and weak nuclear forces and QCD (Sect. 2.1.3) which describes the strong nuclear force, with two classes of particles: fermions and bosons. These two theories form the symmetry group of the Standard Model [170, 231, 235, 256]

$$\underbrace{\mathrm{SU_C}(3)}_{\mathrm{QCD}} \otimes \underbrace{\mathrm{SU_L}(2) \otimes \mathrm{U_Y}(1)}_{\mathrm{QED}} . \tag{2.1}$$

$SU_C(3)$ is denoted with a subscript to ensure it is not confused with the non-gauge theory flavor SU(3), $SU_L(2)$ represents the weak gauge vectors in the theory, and $SU_Y(1)$ denotes the gauge group of weak hypercharge. As the SM is a quantum field theory, the fundamental objects are quantum fields. These are as follows:

- fermionic fields ψ^L, ψ^R (for left/right chirality),
- electroweak boson fields W_1, W_2, W_3, B,
- gluon field G,
- and the Higgs field ϕ.

For example, the massless electroweak boson fields are given mass due to the Higgs mechanism through mixing, to create physically observable particles [147]

$$Z = \cos\theta_W W_3 - \sin\theta_W B, \tag{2.2}$$

$$A = \sin\theta_W W_3 + \cos\theta_W B, \tag{2.3}$$

$$W^{\pm} = \frac{1}{\sqrt{2}}(W_1 \mp i W_2). \tag{2.4}$$

In Eq. (2.2), θ_W is the Weinberg angle or weak mixing angle $\sin^2\theta_W = 0.2223(21)$ [212].

Figure 2.2 summarizes the fermions and bosons known today with Table 2.2 providing a brief timeline of the discoveries. Table 2.1 summarizes the fundamental forces and how they interact with the different particles of the SM (Table 2.2).

Fermions are spin-$\frac{1}{2}$ particles and follow Fermi–Dirac statistics, **bosons** are integer spin and follow Bose–Einstein statistics. These particles are the result of enforcing the symmetry in Eq. (2.1) by introducing fields and interactions as mentioned in Table 2.1. Fermions and bosons also have anti-particles of the same mass but opposite quantum charge. The **photon** is a mediator of the electromagnetic force and couples to all fermions with a non-zero electromagnetic charge; itself being massless, neutrally charged, and with spin 1. The electrically neutral **gluon** is the mediator of the strong force and couples to all fermions with a color[2] charge. The gluon also carries color charge, color/anti-color pair, so it also participates in strong interactions[3] unlike the photon. The color flavor is SU(3) which means given the three colors red-green-blue, there are actually nine possible combinations of color/anti-color but only eight gluons in reality. The ninth possibility is a colorless singlet that is unobservable via strong interaction $\left(r\bar{r} + g\bar{g} + b\bar{b}\right)/\sqrt{3}$ and does not exist. Gluons are massless with spin 1. The other nice thing about the color terminology is that all naturally occurring particles are colorless.[4] It is a nice rule

[2]Color does not actually mean "color" as if a quark actually appears red. Physicists would say that a quark has one unit of red-ness, for example.

[3]Quarks and gluons have different strong coupling strengths, with quark-gluon color factor $C_F = 4/3$ and gluon-gluon color factor $C_A = 3$ [209].

[4]Total amount of each color is zero or all three colors are present in equal amounts.

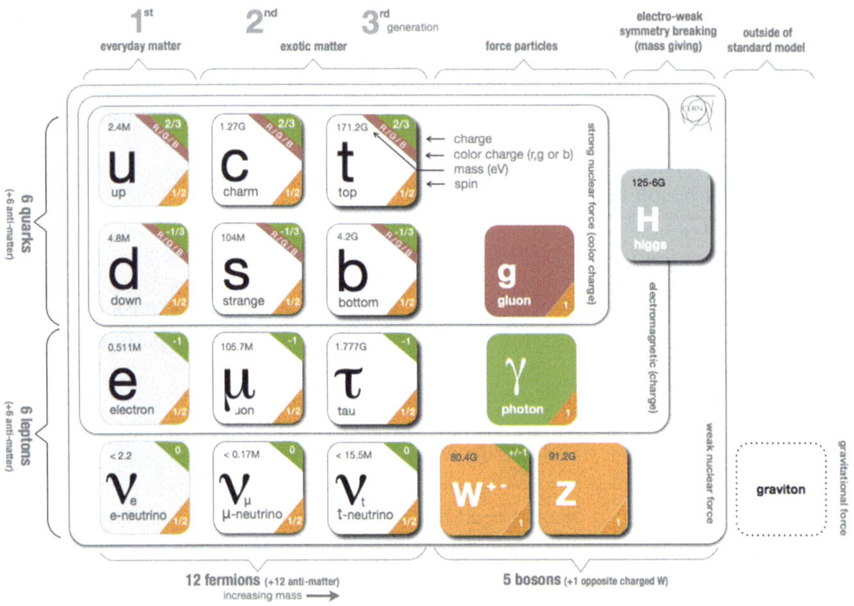

Fig. 2.2 A diagram of the Standard Model of particles [225]. Shown are three generations of 12 fermions (quarks and leptons), all with spin $\frac{1}{2}$. The five force carriers (bosons) are shown: gluon, photon, W/Z bosons, and the Higgs boson. Also depicted is the graviton, a theoretical mediator of the gravitational force which is not currently in the Standard Model. All gauge bosons, except for the Higgs boson

Table 2.1 The strengths of the interactions (forces) are shown relative to the strength of the electromagnetic force for two u quarks separated by 10^{-18} m, the scale of quarks [146]

	Interaction			
Property	Gravitational	Weak	Electromagnetic	Strong
Acts on	Mass–Energy	Flavor	Electric charge	Color charge
Particles experiencing	All	Quarks, Leptons	Charged	Quarks, Gluons
Particles mediating	Graviton	W/Z bosons	Photons	Gluons
Strength at 10^{-18}m	10^{-41}	0.8	1	25

that helps to explain why you cannot make a particle out of two quarks qq or four quarks $qqqq$, but instead see particles like mesons $q\bar{q}$, baryons qqq, and the antibaryons $\bar{q}\bar{q}\bar{q}$. The **W/Z bosons** are mediators of the weak force and couple to all fermions. The W bosons have electromagnetic charges of ± 1 while the Z boson is electromagnetically neutral, all with spin 1.

Finally, each force has an associated radiation where a real or virtual particle can be emitted. A photon can be radiated through the electromagnetic force, and this is

Table 2.2 The abridged timeline of particle physics discoveries of the fermions and bosons that make up the Standard Model known today

What	When	Who	Paper
Photon	1895	Wilhelm Röntgen	[217]
Electron	1897	J.J. Thomson	[250]
Proton	1919	Ernest Rutherford	[233]
Neutron	1932	James Chadwick	[133]
Muon	1937	Seh Neddermeyer, Carl Anderson	[215]
Electron neutrino	1956	Clyde Cowan, Frederick Reines	[229]
Muon neutrino	1962	BNL (AGS)	[154]
Up Quark			
Down Quark	1969	SLAC	[118, 120]
Strange Quark			
Charm Quark	1974	SLAC and MIT	[96, 98]
Tau	1975	SLAC-LBL	[219]
Bottom Quark	1977	Fermilab (E288)	[179]
Gluon	1979	DESY (PETRA)	[105]
W/Z Bosons	1983	CERN (UA1)	[97, 131]
Top Quark	1995	Fermilab (D0, CDF)	[8, 9]
Tau Neutrino	2000	Fermilab (DONUT)	[194]
Higgs Boson	2012	CERN LHC (ATLAS, CMS)	[7, 134]

known as bremsstrahlung. A quark can radiate a gluon[5] through the strong force. A similar process can also occur through the weak force where a quark can radiate a W/Z boson [262]. The search presented in this thesis focuses primarily on the strong interaction.

2.1.1 Spontaneous Symmetry Breaking

Spontaneous Symmetry Breaking (SSB) is the process in which a symmetry of a theory is not realized[6] in the lowest energy configuration (the vacuum expectation value, v.e.v. or vev). The classical example of describing such a situation is to imagine a pencil standing straight up on a table. The pencil is in a state of maximum energy with infinitely many ground states when it is lying horizontal on the table. The high energy state has a symmetry of rotation about the z-axis, but none of the ground states has this symmetry! So a physicist will say that when the pencil falls over, the rotational symmetry about the z-axis is "spontaneously broken".

[5] Gluon showers.

[6] I say realized, and not "broken," because I believe the phrase "broken" confuses people. There is nothing that is broken, but simply "transformed."

To explain this with a toy model [147], consider a complex scalar field $\Phi = (\phi_1 + i\phi_2)/\sqrt{2}$. The Lagrangian density for this is

$$\mathcal{L} = \partial_\mu \Phi^\dagger \partial^\mu \Phi - m^2 \Phi^\dagger \Phi. \tag{2.5}$$

If Φ is constant, independent of space and time, only the $m^2\Phi^\dagger\Phi$ term contributes to the energy. Since the mass, m, is real, m^2 is positive and the energy is a minimum with the trivial solution $\phi_1 = \phi_2 = 0$. So $\Phi = 0$ is the ground state. Now, take the same equation but flip the sign in front of m^2 and now the Lagrangian is unstable as it is not bounded from below. One can make this stable again by introducing a term $(m^2/2\phi_0^2)(\Phi^\dagger\Phi)^2$, and then the Lagrangian density is

$$\mathcal{L} = \partial_\mu \Phi^\dagger \partial^\mu \Phi - V(\Phi^\dagger \Phi), \qquad V(\Phi^\dagger \Phi) = \frac{1}{2\phi_0^2} m^2 \left[\Phi^\dagger \Phi - \phi_0\right]^2 + \text{constant}. \tag{2.6}$$

Just like in Eqs. (2.5) and (2.6) has minimum energy when Φ is constant (independent of space and time) where $\Phi^\dagger\Phi = \phi_0$. Instead of a unique field Φ, there is an infinite number of vacuum states described by $|\Phi|^2 = \phi_0$. In Eq. (2.6), there is a global U(1) symmetry $\Phi \to \Phi' = e^{-i\theta}\Phi$ such that $\mathcal{L} \to \mathcal{L}' = \mathcal{L}$. If one picks out a particular direction in (ϕ_1, ϕ_2) space for which Φ is real, and take the vacuum state to be $(\phi_0, 0)$, the U(1) symmetry is lost. That is, the Lagrangian has some "global" symmetry that appears to have been lost when a ground state is picked out for the field.

So what does SSB give us? Well, you need to reinterpret the new fields after the loss of the symmetry. To expand about the ground state, the procedure is to put in $\Phi = \phi_0 + (\chi + i\phi)/\sqrt{2}$ for two real scalar fields χ, ψ, so the Lagrangian is now written in two terms: $\mathcal{L} = \mathcal{L}_{\text{free}} + \mathcal{L}_{\text{int}}$ with a free component and an interacting component corresponding to interactions between the free particles. Here

$$\mathcal{L}_{\text{free}} = \frac{1}{2}\partial_\mu \chi \partial^\mu \chi - m^2\chi^2 + \frac{1}{2}\partial_\mu \psi \partial^\mu \psi \tag{2.7}$$

represents the free particle fields and contains terms quadratic in the fields. Notice that in Eq. (2.7), there is a $-m^2\chi^2$ term which implies that the χ field has a scalar spin-zero particle of mass $m\sqrt{2}$. For the ψ field, there is no corresponding term so it is a massless, scalar, spin-zero particle. ψ is known as a Nambu–Goldstone[7] boson which are massless particles that always arise as a result of the loss of a global symmetry [214].

[7]Yoichiro Nambu was a professor here at the University of Chicago.

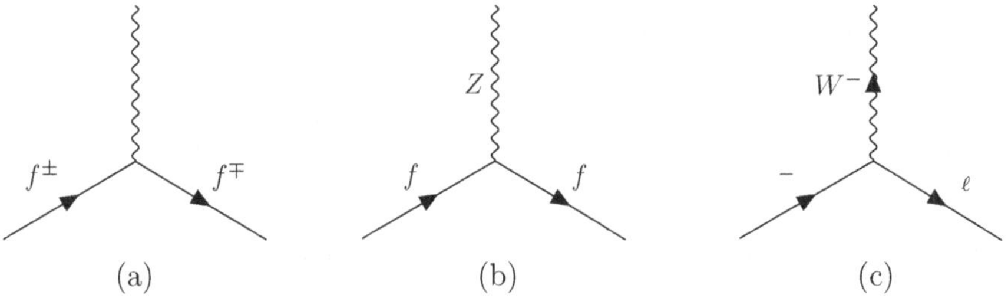

Fig. 2.3 The elementary processes of quantum electrodynamics. Note that time is horizontal (a convention in ATLAS). (**a**) A charged particle, f, enters, emits (or absorbs) a photon, γ, and exits, electromagnetic. (**b**) The Z boson mediates such processes, weak neutral. (**c**) A lepton converts into corresponding neutrino with emission or absorption of $W^{\pm}$, weak charged. These diagrams were made with TikZ-Feynman [162]

2.1.2 *Quantum Electrodynamics (QED)*

QED is the oldest and perhaps simplest of the SM theories and has influenced the design of other theories. The QED theory corresponds to the $SU_W(2) \otimes U_Y(1)$ symmetry that is spontaneously broken by the Higgs mechanism providing mass eigenstates corresponding to the Z^0, $W^{\pm}$ bosons, and the photon. All electromagnetic and weak phenomena are reducible to fundamental processes in Fig. 2.3. To describe more complicated processes, you simply combine two or more replicas of this vertex. Each vertex introduces a factor of $\alpha = 1/137$ which is a small number, so only needs to sum over a smaller number of Feynman diagrams to get a reasonable approximation of the probability amplitude.

In order to describe QED, it will be sufficient to describe the process by which the masses of the electroweak bosons arise through the loss of global symmetries

$$SU(2)_L \times U(1)_Y \rightarrow U(1)_{EM}. \tag{2.8}$$

This idea was explored by Yang and Mills in 1954 [263] and will be re-explored here. First, introduce a two-component field $\Phi = (\Phi_A, \Phi_B)$ where $\Phi_A = \phi_1 + i\phi_2$ and $\Phi_B = \phi_3 + i\phi_4$. In this case, a simple Lagrangian density that has global $U(1) \otimes SU(2)$ symmetry is described by Eq. (2.6). If $V(\Phi^{\dagger}\Phi) = m^2\Phi^{\dagger}\Phi$, this Lagrangian density would correspond to four independent free scalar fields with the same mass m. In the SM, we need to describe the local symmetries from the global symmetries. Defining τ^k as the generators of SU(2), which are identical to the Pauli spin matrices in Eq. (2.9)

$$\tau^0 = \begin{pmatrix} 1 & 0 \\ 0 & 1 \end{pmatrix}, \tau^1 = \begin{pmatrix} 0 & 1 \\ 1 & 0 \end{pmatrix}, \tau^2 = \begin{pmatrix} 0 & -i \\ i & 0 \end{pmatrix}, \tau^3 = \begin{pmatrix} 1 & 0 \\ 0 & -1 \end{pmatrix}. \tag{2.9}$$

The U(1) transformation, $\Phi \rightarrow \Phi' = e^{-i\theta\tau^0}\Phi$ requires the introduction of a vector gauge field $B_\mu(x)\tau^0$ to become a local symmetry

$$B_\mu(x) \rightarrow B'_\mu(x)0 = B_\mu(x) + \frac{2}{g_1}\partial_\mu\theta, \qquad i\partial_\mu \rightarrow i\partial_\mu - \frac{g_1}{2}B_\mu, \tag{2.10}$$

where g_1 is a dimensionless parameter of the theory. For SU(2) where $U = e^{-i\alpha^k\tau^k}$ for three real numbers, α^k and τ^k are the generators in Eq. (2.9), a vector gauge field $W^k_\mu(x)$ is introduced:

$$\begin{aligned} \boldsymbol{W}_\mu(x) = W^k_\mu(x)\tau^k, \quad \boldsymbol{W}_\mu(x) \rightarrow \boldsymbol{W}'_\mu(x) &= \boldsymbol{U}(x)\boldsymbol{W}_\mu(x)\boldsymbol{U}^\dagger(x) \\ &+ \frac{2i}{g_2}(\partial_\mu \boldsymbol{U}(x))\boldsymbol{U}^\dagger(x), \end{aligned} \tag{2.11}$$

where g_2 is another dimensionless parameter of the theory. Finally, one needs to define the covariant derivative D_μ as

$$D_\mu\Phi = \left[\partial_\mu + \frac{ig_1}{2}B_\mu + \frac{ig_2}{2}\boldsymbol{W}_\mu\right]\Phi, \qquad D'_\mu\Phi' = e^{-i\theta}\boldsymbol{U}D_\mu\Phi. \tag{2.12}$$

So the locally gauge invariant Lagrangian density corresponding to Eq. (2.6) is

$$\mathcal{L}_\Phi = (D_\mu\Phi)^\dagger D^\mu\Phi - V(\Phi^\dagger\Phi). \tag{2.13}$$

So now we are ready to write out the dynamical contribution to the Lagrangian density associated with the gauge fields:

$$\mathcal{L}_{\text{dyn}} = -\frac{1}{4}B_{\mu\nu}B^{\mu\nu} - \frac{1}{8}\text{Tr}\left(\boldsymbol{W}_{\mu\nu}\boldsymbol{W}^{\mu\nu}\right), \tag{2.14}$$

with the field strength tensor for $B_\mu(x)$ straightforward to write out. As the SU(2) group is non-abelian, $\boldsymbol{W}_\mu(x)$ is trickier

$$B_{\mu\nu} = \partial_\mu B_\nu - \partial_\nu B_\mu, \tag{2.15a}$$

$$\boldsymbol{W}_{\mu\nu} = \left[\partial_\mu + \frac{ig_2}{2}\boldsymbol{W}_\mu\right]\boldsymbol{W}_\nu - \text{same, but } \mu \leftrightarrow \nu. \tag{2.15b}$$

Now, because of the nice features of the Pauli matrices, specifically that $\text{Tr}(\tau^i)^2 = 2$ and $\text{Tr}(\tau^i\tau^j) = 0, i \neq j$, Eq. (2.14) can be written more simply as

$$\mathcal{L}_{\text{dyn}} = -\frac{1}{4}B_{\mu\nu}B^{\mu\nu} - \frac{1}{4}W^3_{\mu\nu}W^{3\mu\nu} - \frac{1}{2}W^-_{\mu\nu}W^{+\mu\nu}, \tag{2.16}$$

where the W field has defined complex mixing for convenience

$$W_\mu^\pm = \frac{1}{\sqrt{2}}\left(W_\mu^1 - iW_\mu^2\right), \qquad W_{\mu\nu}^\pm \text{ written similarly, and} \tag{2.17a}$$

$$W_{\mu\nu}^3 = \partial_\mu W_\nu^3 - \partial_\nu W_\mu^3 - ig_2\left(W_\mu^- W_\nu^+ - W_\nu^- W_\mu^+\right). \tag{2.17b}$$

Now, we are at the point with Eq. (2.16) to apply the methodology of losing the symmetry as described previously in Sect. 2.1.1. Since there are three real parameters $\alpha^k(x)$ in SU(2), a gauge is chosen such that $\Phi_A = 0$ (two conditions) and $\Phi_B = \phi_0$ are real (one condition). The ground and excited states are then of the form

$$\Phi_{\text{ground}} = \begin{pmatrix} 0 \\ \phi_0 \end{pmatrix}, \tag{2.18a}$$

$$\Phi_{\text{excited}} = \begin{pmatrix} 0 \\ \phi_0 + h(x)/\sqrt{2} \end{pmatrix}, \qquad h(x) \text{ is real} \tag{2.18b}$$

so plugging this into Eq. (2.13), one obtains

$$\mathcal{L}_\Phi = \frac{1}{2}\partial_\mu h \partial^\mu h + \frac{g_2^2}{2} W_\mu^- W^{+\mu}\left(\phi_0 + \frac{h}{\sqrt{2}}\right)^2 + \underbrace{\frac{1}{4}\left(g_1^2 + g_2^2\right) Z_\mu Z^\mu \left(\phi_0 + \frac{h}{\sqrt{2}}\right)^2}_{Z_\mu = W_\mu^3 \cos\theta_w - B_\mu \sin\theta_w}$$
$$\underbrace{- m^2 h^2 + \frac{m^2}{\phi_0\sqrt{2}} h^3 + \frac{m^2}{8\phi_0^2} h^4}_{V(h)}. \tag{2.19}$$

$V(h)$ here is the Higgs potential which takes on the shape of a Mexican hat; a local maxima at the origin and the potential drops off before rising up again with a local minima along a circle around the origin. $A_\mu = W_\mu^3 \sin\theta_w + B_\mu \cos\theta_w$ (the orthogonal complement to Z_μ) with

$$\cos\theta_w = \frac{g_2}{\sqrt{g_1^2 + g_2^2}} \qquad \text{and} \tag{2.20a}$$

$$\sin\theta_w = \frac{g_1}{\sqrt{g_1^2 + g_2^2}}, \tag{2.20b}$$

where w stands for the Weinberg angle. So we have $\mathcal{L} = \mathcal{L}_{\text{dyn}} + \mathcal{L}_{\Phi}$ from Eqs. (2.16) and (2.19). Putting it all together and rewriting a little bit[8]

$$\begin{aligned}\mathcal{L} =& \frac{1}{2}\partial_\mu h \partial^\mu h - m^2 h^2 \\ &- \frac{1}{4} Z_{\mu\nu} Z^{\mu\nu} + \frac{1}{4}\phi_0^2 \left(g_1^2 + g_2^2\right) Z_\mu Z^\mu \\ &- \frac{1}{4} A_{\mu\nu} A^{\mu\nu} \\ &- \frac{1}{2}\left[\left(D_\mu W_\nu^+\right)^* - \left(D_\nu W_\mu^+\right)^*\right]\left[D^\mu W^{+\nu} - D^\nu W^{+\mu}\right] + \frac{1}{2} g_2^2 \phi_0^2 W_\mu^- W^{+\mu} \\ &+ \mathcal{L}_{\text{int}},\end{aligned} \tag{2.21}$$

where $Z_{\mu\nu} = \partial_\mu Z_\nu - \partial_\nu Z_\mu$ ($A_{\mu\nu}$ is written similarly) and $D_\mu W_\nu^+ = (\partial_\mu i g_2 \sin\theta_w A_\mu) W_\nu^+$. Looking at this, one can extract out the masses of the particles associated with the fields:

$$m_A = 0, \tag{2.22a}$$

$$m_W = \phi_0 \frac{g_2}{\sqrt{2}} = 80.385 \pm 0.015\,\text{GeV}, \tag{2.22b}$$

$$m_Z = \phi_0 \sqrt{\frac{g_1^2 + g_2^2}{2}} = 91.1876 \pm 0.0021\,\text{GeV}, \tag{2.22c}$$

$$m_h = m\sqrt{2} = 125.09 \pm 0.24\,\text{GeV}. \tag{2.22d}$$

From experimental observations [218], we know all of these masses experimentally, including the mass of the Higgs boson found on July 4th, 2012 [7, 134] by the ATLAS and CMS collaborations. Finally, notice that $\cos\theta_w = m_W/m_Z$ is a reported ratio in PDG [218] as well. So what we have seen from basic principles of QED is that starting with a two-component complex field (composed of four real fields), one can find the global symmetry of SU(2)⊗SU(1), lose that symmetry locally as in Eq. (2.8), trigger the Higgs mechanism, and find a Nambu–Goldstone boson instead. The real, initially massless fields now gain mass[9] through their interaction with $h(x)$ and we can write out the interacting portion of the Lagrangian $\mathcal{L}_{\text{int}}$.

[8]It helps to have a really, *really* big chalkboard here.

[9]Well, except for the massless photon of course. Technically, the photon interacts with the "Higgs doublet" but this is not the component of the Higgs field whose excitations are the Higgs bosons.

2.1.3 Quantum Chromodynamics (QCD)

QCD is a quantum field theory describing the strong force, governed by the symmetry $SU_C(3)$ [175, 222]. I will state the Lagrangian density for this theory to illuminate how this compares to Eq. (2.21) but the procedure is very similar to QED. In QCD, there are three fields[10] for each flavor of quark and are put into color triplets. The top quark, for example, looks like Eq. (2.23).

$$\boldsymbol{t} = \begin{pmatrix} t_r \\ t_g \\ t_b \end{pmatrix}, \tag{2.23}$$

where t_c, $\{c \mid r, g, b\}$ represents the four-component Dirac spinors. You state a local SU(3) transformation under which the theory is invariant $\boldsymbol{q} \to \boldsymbol{q}' = U\boldsymbol{q}$. This lets us write down Eq. (2.24) where the gluon gauge fields, $\boldsymbol{G}_\mu$, are similar to the weak gauge fields, $\boldsymbol{W}_\mu$, and the covariant derivative $D_\mu = \partial_\mu + ig\boldsymbol{G}_\mu$ as by Yang–Mills construction [263].

$$\mathcal{L}_{\text{QCD}} = \underbrace{-\frac{1}{4}\sum_{a=1}^{8} G^a_{\mu\nu}G^{a\mu\nu}}_{\mathcal{L}_{\text{gluon}}} + \underbrace{\sum_{f=1}^{6}\left[\bar{\boldsymbol{q}}_f i\gamma^\mu \left(\partial_\mu i g \boldsymbol{G}_\mu\right)\boldsymbol{q}_f - m_f \bar{\boldsymbol{q}}_f \boldsymbol{q}_f\right]}_{\mathcal{L}_{\text{quark}}} \tag{2.24}$$

$\mathcal{L}_{\text{gluon}}$ has a sum over the eight gluons of $SU_C(3)$ and provides the kinetic terms for gluons and their self-interactions as in Fig. 2.4c. $\mathcal{L}_{\text{quark}}$ has a sum over the six flavors of quarks with m_f the "true" masses given to the quarks by coupling to the Higgs field; and provides the kinetic terms for the quarks and their interactions with gluons.

Any number of interactions may follow from a single initial state, but the probability of a final state occurring decreases as the complexity of the final state increases. A set of Feynman diagrams representing basic strong force interactions is shown in Fig. 2.4. The probability of a given Feynman diagram is determined by many factors, including the probabilities of each interaction point, all proportional to the strong coupling constant.

All lowest order QCD diagrams are on the order of $O(\alpha_S^2)$, such as for example, the Feynman diagram representing the strong force binding two quarks together to make hadrons such as neutrons and protons is shown in Fig. 2.5.

So it seems that at least up to this point, QCD looks suspiciously like QED, and that is not an accident. There are some important differences, such as the size of the coupling constants where QED introduces a factor $\alpha_{\text{QED}} = 1/137$

[10] Read: color.

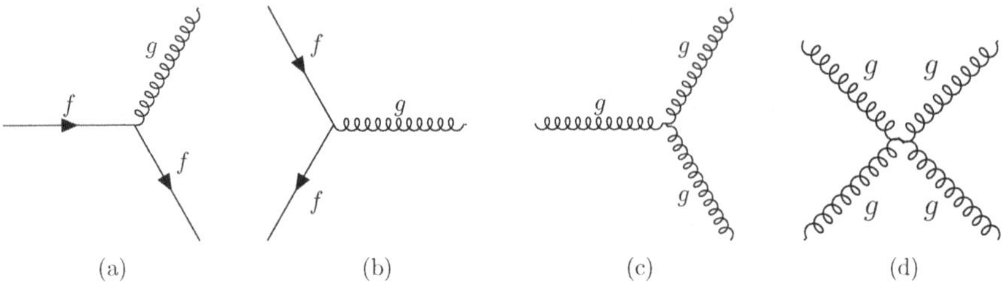

Fig. 2.4 A few Feynman diagrams of basic QCD interactions to lowest order, including (**a**) gluon radiation, $q \rightarrow qg$, (**b**) quark/anti-quark annihilation, $q\bar{q} \rightarrow g$, (**c**) gluon splitting, $g \rightarrow gg$, and (**d**) gluon self-coupling, $gg \rightarrow gg$. These diagrams were made with TikZ-Feynman [162]

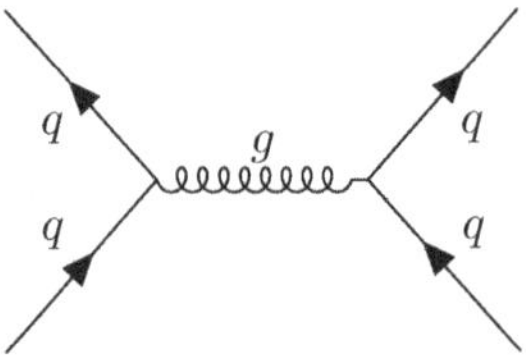

Fig. 2.5 A Feynman diagram at leading order (LO) with probability amplitude proportional to the square of the strong coupling constant. This particular Feynman diagram represents the interaction between quarks that, for example, binds them into hadrons. These diagrams were made with TikZ-Feynman [162]

but for QCD, the factor $\alpha_{\text{QCD}} > 1$ is larger.[11] This was initially a headache as calculations[12] beyond NLO would contribute just as equally, if not more. It was found that the strong coupling constant decreases at higher energy scales (or probing smaller distances) and is called a "running coupling constant."[13] This discovery by Gross, Wilczek, and Politzer won the Nobel Prize in 2004 [174–176, 223]. This phenomenon is known as **asymptotic freedom**, and allows the Feynman diagrams as a legitimate tool for QCD calculations in the high-energy regime. As the energy scale goes up, the strength of the strong force goes down to zero, allowing for perturbative calculations. In the other direction, as the energy scale goes down, a non-perturbative approach needs to be taken as the coupling constant blows up [144]. This will be discussed briefly later in this section. As mentioned in [175, 176], there is a kind of competition between the quark loops and gluon loops in the Feynman diagrams that determines whether the effective coupling constant increases or decreases at short distances. It turns out, comparing the coupling constants as a function of the energy scale between QED and QCD, it is clear to see why the running coupling is different [135] in Eq. (2.25).

[11] Hence the theories are said to be strongly coupled or weakly coupled.

[12] These calculations would involve infinitely more loops.

[13] This also happens for α_{QED} too.

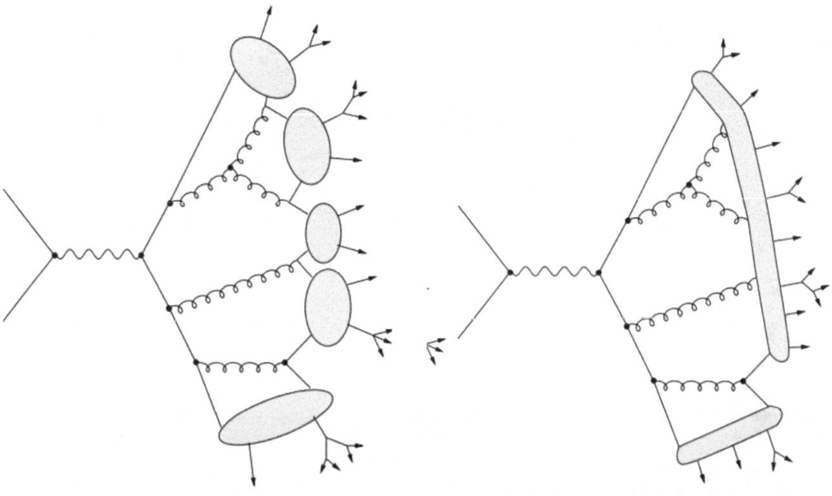

Fig. 2.6 Cartoon of the (**a**) cluster hadronization model which treats individual color singlets separately and the (**b**) Lund string hadronization model which propagates field lines of color flux [254]. These two leading models approximate the non-perturbative process of hadronization to map colored partons onto stable, colorless hadrons

$$\alpha_{\text{QED}}\left(Q^2\right) = \frac{e^2}{4\pi - \frac{e^2}{3\pi}\ln\left(\frac{Q^2}{4m^2}\right)} \quad \text{and} \tag{2.25a}$$

$$\alpha_{\text{QCD}}\left(Q^2\right) = \frac{g^2}{4\pi - \frac{g^2}{4\pi}\frac{1}{n_c}\left[2n_f - 11n_c\right]\ln\left(\frac{Q^2}{\lambda^2}\right)}. \tag{2.25b}$$

The special piece to notice is in the denominator of α_{QCD} in Eq. (2.25) where n_c, n_f are for the number of colors and number of flavors in the theory. If this piece is negative, that is, $f(n_f, n_c) \equiv 2n_f - 11n_c < 0$, then the α_{QCD} decreases at short distances (large Q^2). For SM QCD there are 6 flavors (quarks) and 3 colors, so $f(n_f, n_c) < 0$ and this is the basis of asymptotic freedom where color-charged particles barely interact with each other at small distances.

On the opposite end of the energy scale, for low Q^2 and large distances on the order of femtometers, a non-perturbative approach needs to be taken to evaluate the interactions. Physics simulators, which try to approximate the non-perturbative behavior of QCD, pick from two different popular options shown in Fig. 2.6. The cluster model starts with gluon splitting into $q\bar{q}$ to form clusters that are used to predict final state hadrons. The Lund string model, on the other hand, uses the $q\bar{q}$ pair to estimate the intensity of the color flux string and generates gluons and hadrons based on kinks in this flux.

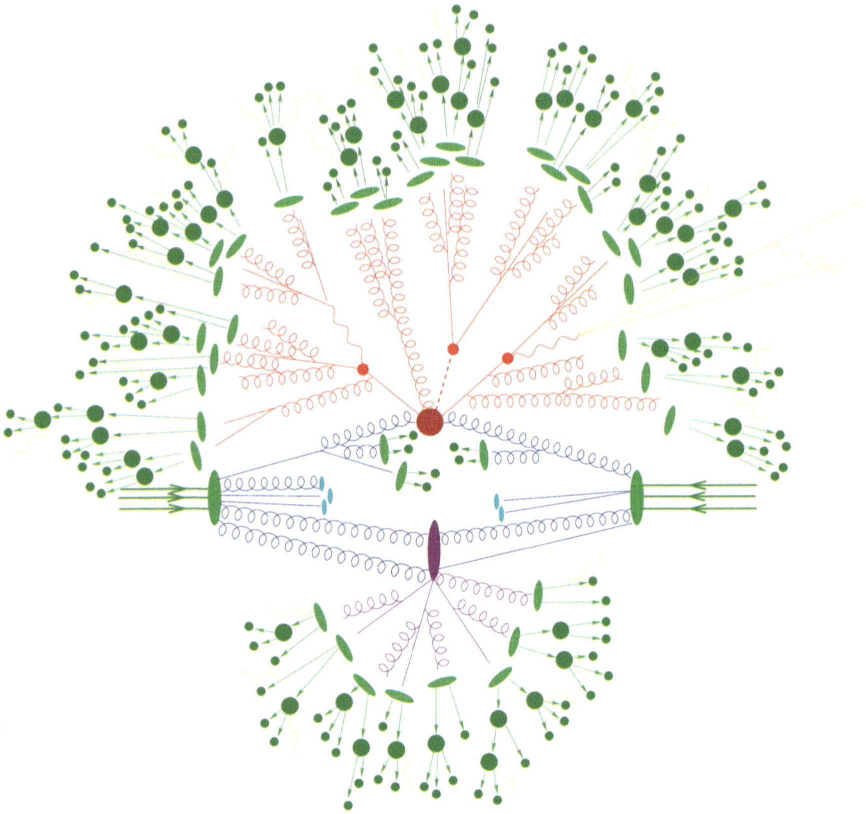

Fig. 2.7 Sketch of a hadron–hadron collision as simulated by a Monte Carlo event generator [181]. The red blob in the center represents the hard collision, surrounded by a tree-like structure representing bremsstrahlung as simulated by parton showers. The purple blob indicates a secondary hard scattering event. Parton-to-hadron transitions are represented by light green blobs, dark green blobs indicate hadron decays, while yellow lines signal soft photon radiation

There currently is no analytic proof of this behavior (or the transition to this behavior) known as **color confinement**, but it can be observed experimentally at a particle detector. To describe it in a qualitative manner, as quarks and gluons separate, the strong force increases in strength. At a certain point, it is energetically favorable to produce a quark/anti-quark pair rather than put in more work to separate the two particles. In other words, separating two particles with color will produce bound states which are colorless. What this means for detectors like ATLAS is that non-colorless particles cannot be directly detected due to color confinement. All physically and directly observable particles are colorless.

The boosted partons (quarks and gluons) that come flying out of the proton–proton collisions with large amounts of energy will create colorless bound states. This process is called **hadronization** and refers to the transition of colored partons to colorless hadrons. Partons can also radiate collinear gluons which in turn radiate $q\bar{q}$ collimated pairs, through a process known as **showering**. These steps are shown in Fig. 2.7, a partonic representation of the process of a single colored parton generating multiple, colorless, measurable hadron showers. The green arrows on either side of the event are the proton bunches which have gluons radiating which form two different groups of interactions. The hard scatter[14] (large red circle) of a proton–proton collision is the highest energy interaction in the event. The secondary interactions (purple blob) form the underlying event, involve smaller momentum transfers. From the hard scatter, the high energy partons shower according to perturbative QCD (red showers). At a low enough energy level where perturbation theory becomes invalid and color confinement takes over, the partons hadronize (green blobs) into various colorless hadrons.

2.1.4 Parton Distribution Function

The name "parton" was proposed by Richard Feynman in 1969 [120] as a generic description for any particle constituent within the proton, neutron, and other hadrons. At first, the hadrons were thought to consist of doublets and triplets of quarks ($q\bar{q}$ and qqq). However, through high-energy proton–proton collisions at the LHC and the interaction processes, these "valence" quarks and gluons can also produce an arbitrary number of lower-energy virtual partons, "sea"[15] quarks, and gluons. These were first observed by James Bjorken and Emmanuel Paschos in 1969 [117]. Now, we know that protons (neutrons) are made up of two (one) up quarks, u, and one (two) down quark, d, along with the gluons, g, that hold them together.

While the Standard Model enables the calculations of cross-sections as a function of the energies of colliding partons (quarks and gluons), the LHC is a proton–proton collider. It is also useful to know the cross-sections for a given proton energy, the parton distribution function (PDF) [160]. This is because the collisions at the LHC are really between partons inside the protons. The PDF is a function that provides the probability density of finding the given parton in the given hadron with the given momentum. PDFs are parameterized by Q^2 and "Bjorken x" (or just x). The Q^2 corresponds to the energy scale of the collision[16] process and x represents the momentum fraction of the proton that the interacting parton holds. For proton–proton collider experiments like LHC, proton PDFs are the most interesting, reconstructed using data from proton scattering experiments. Figure 2.8

---

[14] At the large energies of the LHC, the "core process" here are gluon–gluon scattering.

[15] Personally, as a deaf person, I propose that we call them "ocean" or "plum" quarks, to reduce confusion with the existing "c" quarks.

[16] The center-of-mass energy at a proton–proton collider, $\sqrt{s}$, is not related to Q^2, but is instead the upper bound on Q^2.

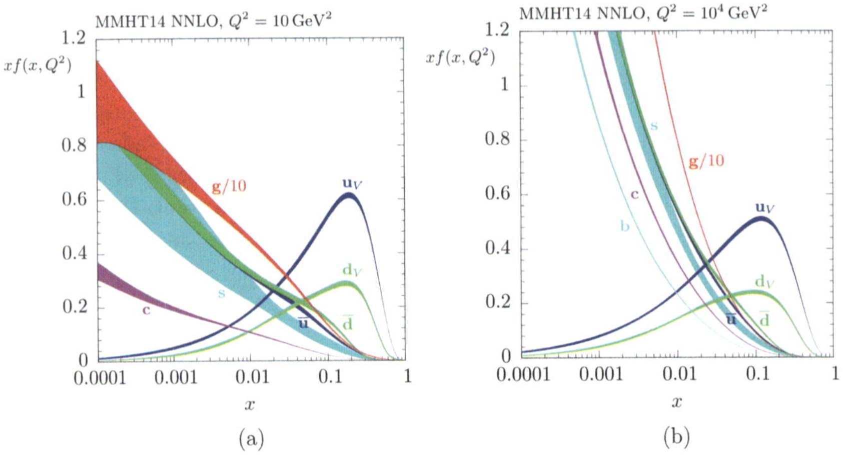

Fig. 2.8 MMHT2014 NNLO PDFs at, (**a**) $Q^2 = 10\,\mathrm{GeV}^2$ and (**b**) $Q^2 = 10^4\mathrm{GeV}^2$ with associated 68% confidence-level uncertainty bands [177]. The function $xf(x, Q^2)$ is plotted versus x for different flavors: $u, \bar{u}, d, \bar{d}, s = \bar{s}$, and g

shows one such example of a PDF: the u, d quark, and the g gluon generally dominate at low energies (low Q^2), while other virtual partons are more likely to participate in the interaction processes at high energies (high Q^2). Many other PDFs exist [103, 104, 177, 184], and for LHC Run 2, the global PDFs NNPDF3.0, MMHT14, and CT14 are the latest used which use the data from LHC Run 1 for further constraints. There is some uncertainty in these PDFs which contribute to uncertainties in the predicted proton–proton cross-sections and are often one of the dominant sources of uncertainty for many important searches and analyses at the LHC, especially for precision cross-section measurements.

Factorization [145] is a concept that was implicit in the discussion about hadronization and PDFs. In particular, what factorization allows us to do is define a cutoff scale Q_F above which collinear radiation is directly treated and below which it is absorbed into the PDF definition. Effectively, this allows us to separate the calculation of phenomena which are perturbatively calculable from phenomena which are not. The total cross-section σ for a collision process [221], $ab \to n$ may be derived by integrating over all possible initial state momentums for partons a and b, hadron h, the parton p_{T} fraction x_a^h and x_b^h, and weighting them by their PDF f_a^h and f_b^h can be written as shown in Eq. (2.26).

$$\sigma(Q_F, Q_R) = \sum_{a,b} \int_0^1 \mathrm{d}x_a \mathrm{d}x_b \int_{\text{hadrons}} f_a^h\left(x_a, Q^2\right) f_b^h\left(x_b, Q^2\right) \mathrm{d}\sigma_{ab\to n}. \tag{2.26}$$

The other scale involved is the renormalization scale Q_R [251, 257] which accounts for the logarithmically divergent contributions of the Lagrangian through the process of renormalization. Unlike Q_F which represents the scale at which the

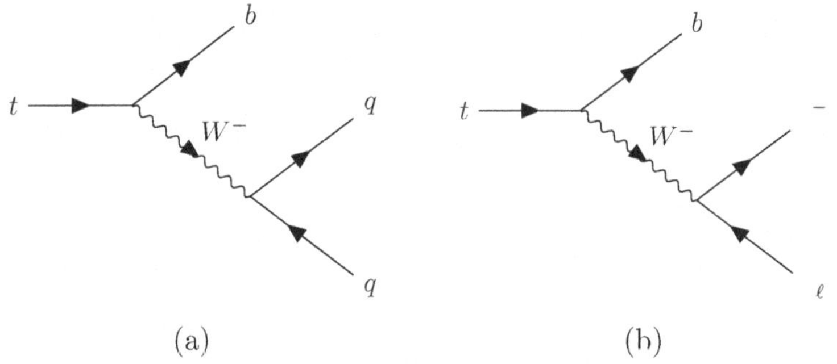

Fig. 2.9 Feynman diagrams showing the top quark decays for (**a**) hadronic and (**b**) leptonic. These diagrams were made with TikZ-Feynman [162]

hadron is being probed, Q_R is a non-physical effect that accounts for the limited knowledge used in lower order calculations.

2.1.5 Top Quark Decays

The top quark is a 3rd generation parton and is fundamental to this thesis analysis as we search for four Lorentz-boosted top quarks in the final state. As such, it is appropriate to provide a little bit more detail about the top quark and its decay. The timescale for strong force interactions is on the order of 10^{-24} s. The top quark has a lifetime of 10^{-25} s which is due to its large mass. Thus, the top quark is a unique parton in that it decays before it can hadronize, allowing physicists to measure the "bare mass" of the top quark [226]. Figure 2.9 shows the two dominant decay modes of a top quark, through the weak interaction, producing a W-boson and a down-type quark (down, strange, or bottom).

The W-boson branching ratios are listed in Eq. (2.27) [218]. The top quark will decay hadronically (to two quarks) approximately 70% of the time and leptonically approximately 30% of the time. As well as top quark decays, I also discuss $t\bar{t}$[17] which has three different kinds of decays: hadronic (both W bosons decay hadronically), semi-leptonic (one W boson decays leptonically), and fully leptonic (both W bosons decay leptonically). $t\bar{t}$ decays hadronically about 50% of the time, semi-leptonically about 40% of the time, and full-leptonically about 10% of the time.

$$\mathrm{BR}(W \rightarrow e\bar{\nu}_e) = 0.1046 \pm 0.0042(\mathrm{stat}) \pm 0.0014(\mathrm{syst}), \tag{2.27a}$$

$$\mathrm{BR}(W \rightarrow \mu\bar{\nu}_\mu) = 0.1050 \pm 0.0041(\mathrm{stat}) \pm 0.0012(\mathrm{syst}), \tag{2.27b}$$

[17] Colloquially "tee-tee-bar" or written "ttbar."

$$\mathrm{BR}(W \to \tau\bar{\nu}_\tau) = 0.1075 \pm 0.0052(\mathrm{stat}) \pm 0.0021(\mathrm{syst}), \tag{2.27c}$$

$$\mathrm{BR}(W \to q\bar{q}) = 0.6832 \pm 0.0061(\mathrm{stat}) \pm 0.0028(\mathrm{syst}). \tag{2.27d}$$

2.2 Beyond the Standard Model

The Standard Model has been tested over the last few decades by many experiments and shown to be robust. The fermion fields of leptons and quarks interact through the mediation of vector bosons. The renormalizability of the SM requires that the vector boson fields be introduced through the requirement of local gauge symmetry as in Sects. 2.1.1–2.1.3. However, we know this is not a complete model given the success so far, as certain assumptions are still made that need to be reconciled, motivated by naturalness, such as:

- the matter/anti-matter asymmetry not observed in the detector [127],
- the fine-tuning required to the quantum corrections to keep the Higgs mass around the electroweak scale [100],
- the lack of inclusion of gravity, and the lack of dark matter candidates [116] even though it is largely agreed upon that dark matter exists [10],
- the scale difference between the Planck scale and the electroweak scale (the so-called Hierarchy problem) [252],
- and many more [23, 210, 227, 253]

Many Beyond the Standard Model (BSM) theories have been proposed, all with a variety of testable signatures. Attempts have been made to carry unification further, by combining the electroweak and strong interactions in a higher, unified symmetry, which could only manifest at extremely high energies of order 10e13 TeV. The Higgs boson interacts with all other gauge fields in QED and QCD so that quantum loops in the Feynman diagrams are created to correct the Higgs mass. It seems absurd, and incredibly coincidental, that the Higgs mass, which was expected to be around the Planck scale, receives gigantic corrections on the order of 10^{17} to be on the electroweak scale. In addition, the coupling of the Higgs to some quarks covers two orders of magnitudes which does not seem *natural*. Hence, **naturalness** is a strong motivation for many physicists, myself included.

If you look at SM and understand the corrections to the Higgs mass, the Feynman diagram in Fig. 2.10 shows an example of the loop correction to the Higgs mass that requires such precise fine-tuning that it does not seem natural for the Higgs mass to be as light as observed in 2012. The top mass has the largest coupling, and therefore the largest correction to the Higgs mass, which means this quantum-level correction is roughly described by Eq. (2.28), where λ_t is the Yukawa coupling of the top and Λ_{UV} is the ultraviolet momentum cutoff used to regulate the loop integral of the theory [207], which for the SM is the Planck mass. This naturalness motivation is the strongest motivation for trying to find supersymmetry.

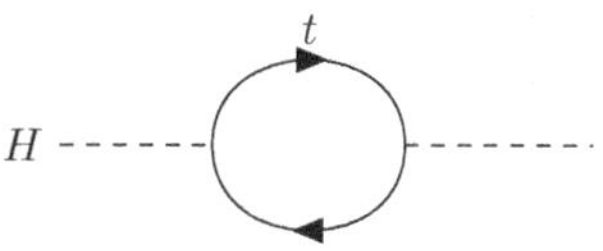

Fig. 2.10 An example of a loop diagram which corrects the Higgs mass [207]

$$\Delta m_H^2 = -\frac{\lambda_t|^2}{8\pi^2}\Lambda_{\mathrm{UV}}^2 + \ldots \tag{2.28}$$

For the rest of this section, I discuss the theoretical framework of supersymmetry, the supersymmetric particles, and introduce the simplified models that I studied.

2.2.1 *Supersymmetry*

Supersymmetry (SUSY) [168, 173, 237, 253, 259, 260] is a generalization of space-time symmetries that predicts new bosonic partners for the fermions and new fermionic partners for the bosons of the SM. If R-parity[18] is conserved [166], SUSY particles are produced in pairs and the lightest supersymmetric particle (LSP) is stable. The scalar partners of the left-handed and right-handed quarks, the squarks $\tilde{q}_{\mathrm{L}}$ and $\tilde{q}_{\mathrm{R}}$, can mix to form two mass eigenstates $\tilde{q}_1$ and $\tilde{q}_2$, ordered by increasing mass. SUSY can solve the hierarchy problem [156, 157, 186, 234] reducing unnatural tuning in the Higgs sector by orders of magnitude, provided that the superpartners of the top quark have masses not too far above the weak scale. The large top Yukawa coupling results in significant $\tilde{t}_{\mathrm{L}}$–$\tilde{t}_{\mathrm{R}}$ mixing so that the mass eigenstate $\tilde{t}_1$ is typically lighter than the other squarks [163, 187].

If supersymmetry exists, it should contain SSB. From a theoretical perspective, there should be a Lagrangian density that is invariant under supersymmetry but a ground state that is not. This is analogous to what has been discussed before in Sect. 2.1.1. On top of this, the theory should be renormalizable to compensate for the effects of self-interactions, infinities arising in calculated quantities, and the differences in descriptions between small-distance-scale physics and large-distance-scale physics [207]. In a supersymmetric extension of the SM [166, 167], each of the known fermions (bosons) is therefore either in a chiral or gauge supermultiplet and must have a superpartner boson (fermion) with spin differing by $\frac{1}{2}$ unit. The names, while appearing somewhat humorous, serve to make the connection from the superpartner to their physical SM partner more obvious are generated as follows:

- the names for the spin-0 partners of the quarks and leptons are constructed by prepending an "s" for scalar (or superpartner) to be called squarks, sleptons, and sfermions,

---

[18] Also known as matter parity. All SM particles and Higgs boson have even R-parity $P_R = +1$ while the squarks, sleptons, gauginos, and higgsinos have odd R-parity $P_R = -1$.

- the symbols for the squarks and sleptons are the same as for the corresponding fermion, but with a tilde added such as $\tilde{e}_L$, $\tilde{e}_R$,[19]
- the supersymmetric fermions take the name of their superpartner bosons, but with an "ino" appended, such as "wino," "gluino."

Given this fermion-boson symmetry, as well as R-parity, the Lagrangian density for an unbroken symmetric theory can be written down. Using a similar mechanism to break this symmetry, gauge fields are introduced, and mass states arise as a mixing of the gauge states. Unlike electroweak which is slightly easier to break, there are a few gotchas this time:

- In SM, there is one Higgs boson; in SUSY, there are two complex Higgs doublets. The reason for this is that the fermionic partner of a Higgs must be able to cancel gauge anomalies which are usually the traces of hypercharge matrices. In the SM this works out because $Y = 0$, but for SUSY, $Y = \pm 1$. So there must be two complex Higgs doublets to account for each hypercharge variation. This is a heuristic motivation.
- In the SM, the Yukawa couplings between the Higgs and the 3rd generation fermions (t, b, τ) are much larger than the first and second generations. Normally, it is not very easy to diagonalize the gauge eigenstates for the fermions, however in the minimal supersymmetric model (MSSM) that is being considered, an approximate can be made to treat the Yukawa couplings for first and second generation as negligible. An example is shown in Eq. (2.29) [207].
- The higgsinos and electroweak gauginos mix with each other because of the effects of electroweak symmetry breaking. The neutral higgsinos and electroweak gauginos mix to form **neutral**inos,[20] and the charged versions mix to form **chargi**nos.

$$\boldsymbol{M}_{\chi^0} = \begin{pmatrix} M_1 & 0 & -m_Z s_W c_\beta & m_Z s_W s_\beta \\ 0 & M_2 & m_Z c_W c_\beta & -m_Z c_W s_\beta \\ -m_Z c_\beta s_W & m_Z c_\beta c_W & 0 & -\mu \\ m_Z s_\beta s_W & -m_Z s_\beta c_W & -\mu & 0 \end{pmatrix},$$
$$s_k \equiv \sin\theta_k,\ c_k \equiv \cos\theta_k \tag{2.29}$$

In Eq. (2.29), the M_i terms come from the soft component of the supersymmetric Lagrangian [207, eq. 6.3.1]. Diagonalizing this matrix allows us to form the neutralinos χ_k^0 as a mixture of the wino, bino, and higgsinos gauge eigenstates. Different mixtures of these gauge eigenstates correspond to different decay products in the final state. A similar procedure exists for the charginos $\chi_k^\pm$. Both neutralinos and charginos are conventionally ordered (and labeled) in k in terms of increasing

[19]The leptons and quarks have left/right handedness and superpartners for each version, as the superpartners are spin-0. SM neutrinos ν_ℓ are always left-handed, so superpartners are just $\tilde{\nu}_\ell$.

[20]Not, as I sometimes mistakenly think, the superpartners of the neutrinos which are the sneutrinos.

Table 2.3 The undiscovered particles in the minimal supersymmetric Standard Model (with sfermion mixing for the first two families assumed to be negligible) [207]

Names	Spin	P_R	Gauge eigenstates	Mass eigenstates
Higgs bosons	0	+1	$H_u^0, H_d^0, H_u^+, H_d^-$	$h^0, H^0, A^0, H^\pm$
Squarks	0	−1	$\tilde{u}_L, \tilde{u}_R, \tilde{d}_L, \tilde{d}_R$	(same)
			$\tilde{s}_L, \tilde{s}_R, \tilde{c}_L, \tilde{c}_R$	(same)
			$\tilde{t}_L, \tilde{t}_R, \tilde{b}_L, \tilde{b}_R$	$\tilde{t}_1, \tilde{t}_2, \tilde{b}_1, \tilde{b}_2$
Sleptons	0	−1	$\tilde{e}_L, \tilde{e}_R, \tilde{\nu}_e$	(same)
			$\tilde{\mu}_L, \tilde{\mu}_R, \tilde{\nu}_\mu$	(same)
			$\tilde{\tau}_L, \tilde{\tau}_R, \tilde{\nu}_\tau$	$\tilde{\tau}_1, \tilde{\tau}_2, \tilde{\nu}_\tau$
Neutralinos	$\frac{1}{2}$	−1	$\tilde{B}^0, \tilde{W}^0, \tilde{H}_u^0, \tilde{H}_d^0$	$\tilde{\chi}_1^0\ \tilde{\chi}_2^0\ \tilde{\chi}_3^0\ \tilde{\chi}_4^0$
Charginos	$\frac{1}{2}$	−1	$\tilde{W}^\pm, \tilde{H}_u^+, \tilde{H}_d^-$	$\tilde{\chi}_1^\pm, \tilde{\chi}_2^\pm$
Gluino	$\frac{1}{2}$	−1	$\tilde{g}$	(same)
Goldstino (gravitino)	$\frac{1}{2}$ $\frac{3}{2}$	−1	$\tilde{G}$	(same)

mass, such that $\tilde{\chi}_1^\pm < \tilde{\chi}_2^\pm$ and $\tilde{\chi}_1^0 < \tilde{\chi}_2^0 < \tilde{\chi}_3^0 < \tilde{\chi}_4^0$. Table 2.3 shows a summary of the various SUSY particles and their mass eigenstates.

Now, as shown in Fig. 2.11 for SUSY compared to Fig. 2.10 for SM, the stop squark is the bosonic superpartner to the top quark which provides an equal and opposite contribution[21] to the correction of the Higgs mass. In the limit of top squark masses $m_{\tilde{t}_k} \gg m_t$ much greater than the top quark mass, the largest **finite** correction to the Higgs mass m_{h^0} is [207] shown in Eq. (2.30), where $\Delta_{\text{threshold}}$ is a small correction based on the top squark mixing angle and the Higgs quartic coupling, α is a mixing angle of the Higgs couplets, and λ_t is the top Yukawa coupling. So in addition to having a light stop [156, 190, 261], there is also a strong motivation[22] for a light gluino [25], as the gluino couples to the stop squark and pulls the stop mass up. And finally, since the Higgsinos also contribute, and the Higgs and Higgsinos should have similar masses, and the Higgsinos mix with the wino and bino to form neutralinos, we can motivate a light neutralino particle.

$$\Delta\left(m_{h^0}^2\right) = \frac{3}{4\pi^2}\cos^2\alpha\lambda_t^2 m_t^2\left[\ln\left(m_{\tilde{t}_1} m_{\tilde{t}_2}/m_t^2\right) + \Delta_{\text{threshold}}\right]. \tag{2.30}$$

Figure 2.12 shows the theoretical cross-sections of the supersymmetric particles at the LHC assuming a center-of-mass collision energy $\sqrt{s} = 13$ TeV. Compared to Run-I, the rate of gluino production has increased by a factor of 50. Since

[21] Spin-statistics theorem states that fermions have a negative contribution and bosons have a positive contribution.

[22] Warning: slightly heuristic argument ahead.

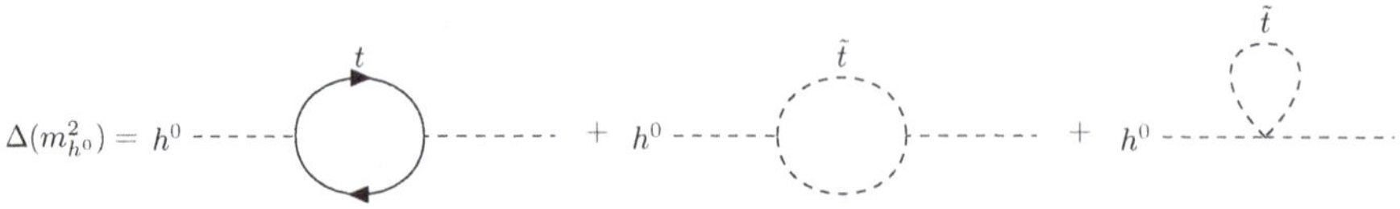

Fig. 2.11 An updated version of Fig. 2.10 with the inclusion of the stop squark, top quark one-loop diagrams [207]. The stop squark is a bosonic superpartner of the fermionic top quark, provides equal and opposite contribution to the top quark loop, cancelling out the contribution. There are two loops because there are two bosonic partners for the top quark, a fermion with spin

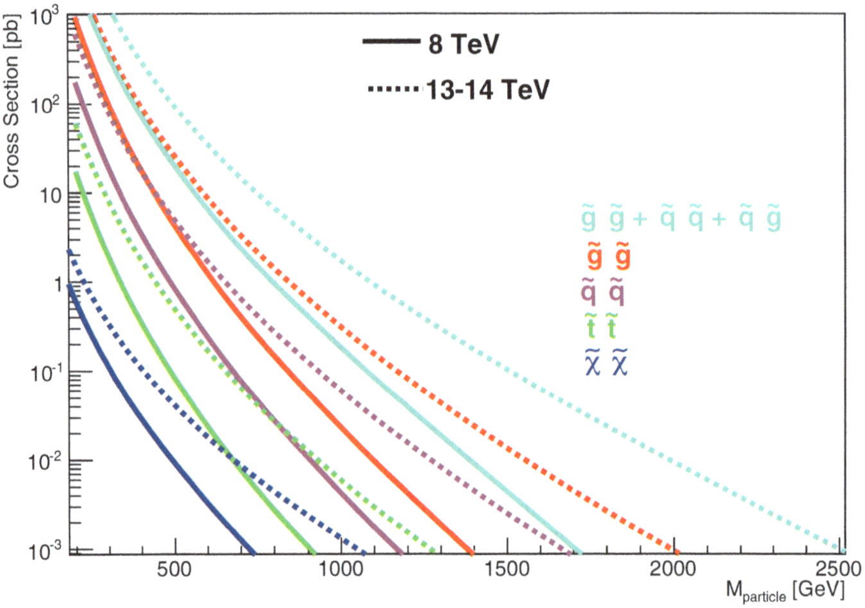

Fig. 2.12 Theoretical cross-sections of gluino pair production are shown in the model of equal degenerate squark masses, as a function of gluino mass at $\sqrt{s} = 13\,\text{TeV}$ [119, 195]. Gluinos, because of their strong color coupling, have the highest theoretical cross-section of the sparticles that could be found at the LHC

naturalness [107] is a strong motivator for the gluinos ($\tilde{g}$) to have a mass around the TeV scale in order to limit their contributions to the radiative corrections to the top squark masses, also at the TeV scale, and the lightest supersymmetric partner $\tilde{\chi}_1^0$ is also motivated to be light as well, one expects these particles to be produced copiously during Run 2 operation of the LHC at 13 TeV. For these reasons, the search for gluino production with decays via top squarks is a highly motivated search to perform. In Sect. 2.2.2, I introduce the simplified SUSY model that is the crux of the search presented in this thesis.

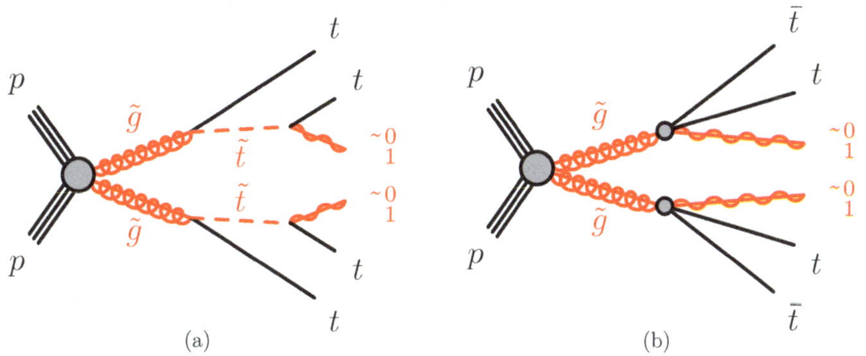

Fig. 2.13 The decay topology of the simplified model for $\tilde{g} \to \bar{t}\tilde{t}_1 \to t\bar{t}\tilde{\chi}_1^0$ for both (**a**) on-shell and (**b**) off-shell stops, $\tilde{t}_1$. The difference between the two diagrams is that in the off-shell diagram, the stops are integrated out of the simplified model by setting the mass of the stop to 5 TeV

2.2.2 Searching for New Physics Using Simplified Models

A model of new physics is defined by a TeV-scale effective Lagrangian describing its particles and their interactions. The efforts so far have focused around motivating the lightness of stops, gluinos, and neutralinos; but have we neglected the other sparticles? For the sake of experimental physicists, a simplified model is generally a limit of a more general new-physics scenario with all but a few particles integrated out [15, 16, 143] by setting them to very large mass scales. Simplified models are useful to provide topology-based limits on searches to identify the boundaries of search sensitivity and derive limits on more general models by reinterpreting [150] the limits in the context of a different signal topology. Another particular reason a simplified model helps in the search for new physics is to understand the sensitivity of the detector technology and analysis design. Experimentalists and theorists alike can identify kinematic ranges for which existing searches are not efficient or sensitive, and then define new search strategies to attempt to cover the gaps in the exploration of phase space. An example of two simplified models is shown in Fig. 2.13 for gluino production with final states consisting of four top quarks and a large missing transverse energy[23] from the neutralinos $\tilde{\chi}_1^0$. The 1st and 2nd generation squarks are assumed to be much larger than the gluino mass. The gluino can produce $t\bar{t} + \tilde{\chi}_1^0$ by decaying either off-shell through a heavier stop squark or on-shell through a lighter stop squark.

These simplified models can be parameterized allowing for projected views in phase space. The cross-section of gluinos (Fig. 2.12), $\sigma(pp \to \tilde{g}\tilde{g} \to X)$ is one

[23]This is an assumption, for R-parity conserving (RPC) scenarios such that $\tilde{\chi}_1^0$ is stable, does not decay, and escapes the detector unseen. In R-parity violating (RPV) scenarios [106, 152, 159, 167], the lightest supersymmetric particle is unstable and decays to SM particles.

parameter. In both the on-shell and off-shell models in Fig. 2.13, there are two parameters for the gluino mass $m_{\tilde{g}}$ and the neutralino mass $m_{\tilde{\chi}_1^0}$. In the on-shell model in Fig. 2.13a, there is an extra parameter for the mass of the stop squark $m_{\tilde{t}_1}$, but the off-shell model does not have this parameter, setting the mass of the stop squark $m_{\tilde{t}_1} = 5\,\text{TeV}$. Finally, the branching ratio for $\tilde{g}$ to decay to $\tilde{t}_1\,\bar{t}$ is assumed to be 100% in this simplified model to reduce the number of parameters. This is clearly not physical (if we find SUSY) but models with multiple decay modes can be studied by taking linear combinations of the results of simplified models for 100% branching ratios.

At the end of the day, one needs to remember that all of these are theories and we, as experimentalists, make many assumptions to simplify the theories into a set of reduced observables to search for. In particular, the search presented in this analysis has assumed that the $\tilde{t}$ have higher masses than the rest of the squarks, but it could be possible that the $\tilde{t}$ has a lower mass. If there is indeed a sign of new physics observed through the search designed around a simplified model, further studies and analysis reinterpretations need to be performed to determine what that new physics is. In the next chapter, I will discuss how we can leverage the world's most powerful collider to search for new physics.

Chapter 3
The Large Hadron Collider and the ATLAS Detector

This chapter provides a brief introduction to the world's largest and most powerful particle accelerator, the LHC. I will describe the LHC in the context of this thesis and how the ATLAS detector fits into the picture. More information about the design, construction, and operation of the LHC can be found in [164].

The rest of the chapter is organized as follows: Sect. 3.1 describes the LHC and the experiments located along the collider; Sect. 3.3 describes the operating schedule of the LHC; and Sects. 3.4–3.8 describe the ATLAS detector instrumentation used in this thesis whose successful operation provided the datasets that allowed me to perform my analysis.

3.1 Overview

The Large Hadron Collider (LHC) [164] at the European Organization for Nuclear Research (European Organization for Nuclear Research (CERN)) is a 27 km super-conducting particle accelerator located approximately 100 m underground. A diagram of the LHC is shown in Fig. 3.1.

The LHC's role is to collide beams of protons in opposite directions at four locations along the ring of the machine. Each of these four locations contains an excavated cavern that houses one of the four LHC experiments: ATLAS [28], CMS [248], LHCb [249], and ALICE [245]. The beams of protons are guided around the accelerator ring using 1232, 15 m long super-conducting dipole magnets [232] which provide a strong 8.3 T magnetic field for bending the proton trajectories. 392 main quadrupole magnets [232], each 5–7 m long, are used to help keep the proton bunches in a tight beam with four magnetic poles arranged

© Springer Nature Switzerland AG 2020

G. Stark, *The Search for Supersymmetry in Hadronic Final States Using Boosted Object Reconstruction*, Springer Theses,
https://doi.org/10.1007/978-3-030-34548-8_3

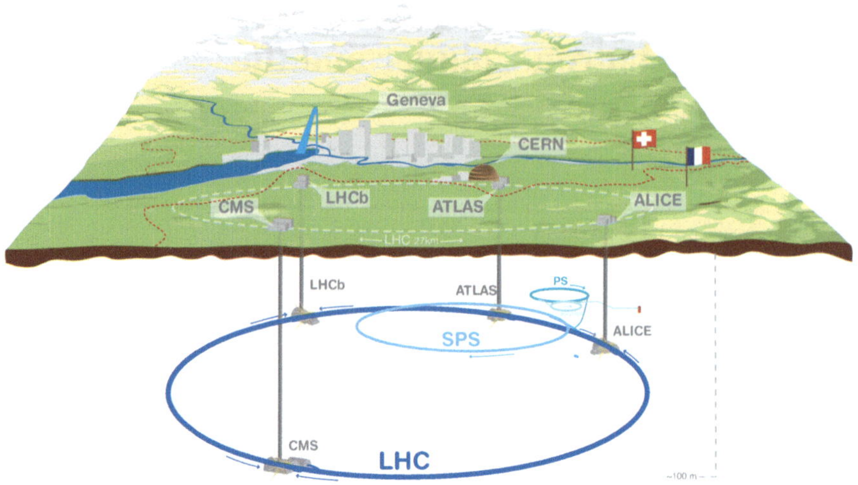

Fig. 3.1 A diagram of the LHC which sits on the border between Switzerland and France, near the city of Geneva [132]. There are four main experiments located here: ALICE, ATLAS, CMS, and LHCb. All of the experimental apparatus are located around 100 m underground where the stable rock is located. Also shown here is the Super Proton Synchrotron which is the second-to-last stage of the injector chain before the particles enter the LHC tunnel

symmetrically around the beam pipe to squeeze the beam either horizontally or vertically. The aim of the LHC is to reveal the physics beyond the Standard Model with center-of-mass collision[1] energies of up to $\sqrt{s} = 14$ TeV.

However, it should be noted that the LHC is only the last step in the injector chain, where protons are accelerated from low energies in defined steps to their final energy as illustrated in Fig. 3.2. Their journey starts at a linear accelerator aptly called Linac2[2] which accelerated protons to 50 MeV. The protons are injected in to the Proton Synchotron Booster (PSB), which accelerates them to 1.4 GeV. After the PSB, the protons are sent to the Proton Synchotron (PS) to reach 25 GeV of energy. They are then sent to the Super Proton Synchotron (SPS) where the protons are accelerated up to 450 GeV of energy. Finally, they are injected into the LHC. Under nominal operating conditions, where the LHC can run for many hours,[3] each proton beam can have 2808 bunches and on the order of 10^{11} protons per bunch.

[1]Technically, it is bunches of protons colliding with bunches of protons, rather than a single proton with a single proton; each proton has $\sqrt{s}/2$ energy.

[2]Linac1 was retired in early 1990s.

[3]Current record in 2015–2016 data run was 37 h with fill #5045.

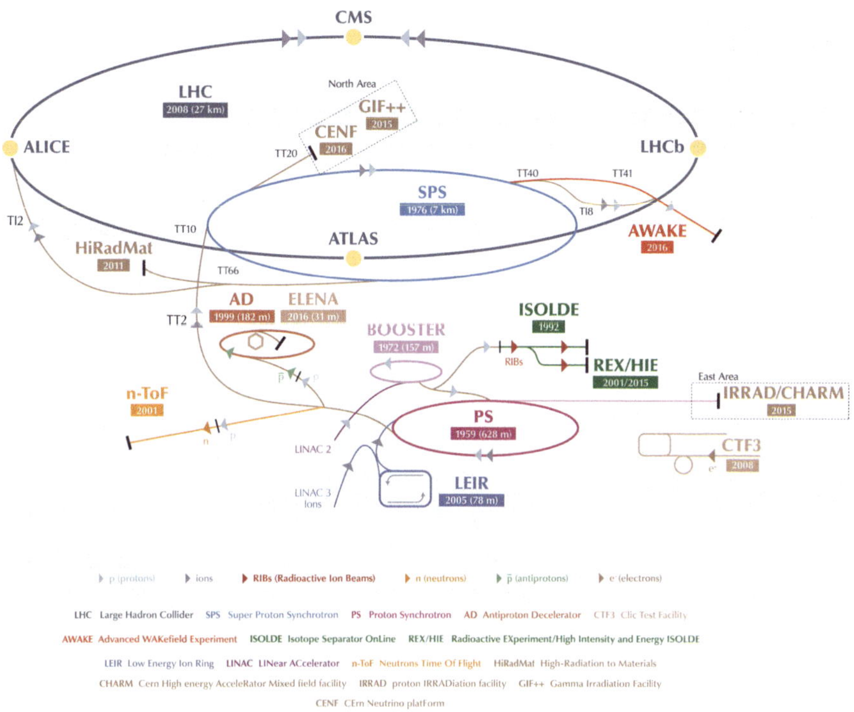

Fig. 3.2 A diagram of LHC injector complex [164, 211]. Protons are accelerated in the following chain: Linac2→Proton Synchrotron Booster→Proton Synchrotron→Super Proton Synchrotron→Large Hadron Collider

3.2 LHC Upgrades

After 2019, the statistical gain in running the accelerator without a considerable luminosity increase beyond its design value will become marginal. The running time at a constant luminosity and $\sqrt{s}$ necessary to half the statistical error in the measurements will be more than 10 years at the end of 2019. Therefore to maintain scientific progress and to explore its full capacity, the LHC will need to have a decisive increase of its luminosity. The CERN Council has updated the LHC upgrade plan taking this into consideration. The schedule for the upgrades and operation of the LHC accelerator complex, leading to the High Luminosity LHC (HL-LHC) [180], is illustrated in Fig. 3.3.

3.3 Operation of the LHC in Run 2

The last thing I want to cover in discussing the LHC is about the data it provides from a physics point of view. The center-of-mass energy and integrated luminosity

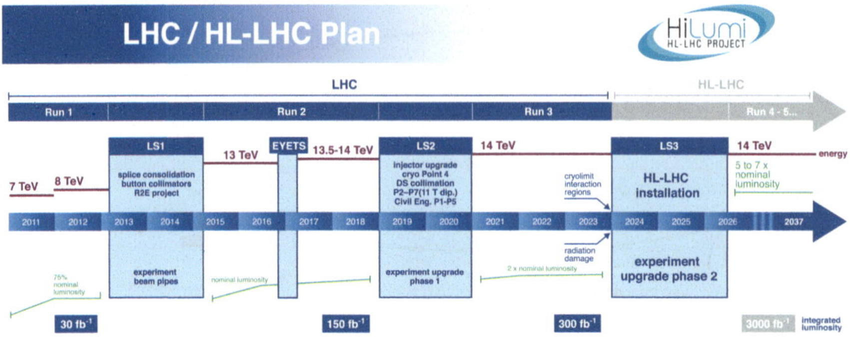

Fig. 3.3 LHC and High-Luminosity LHC plan [180]

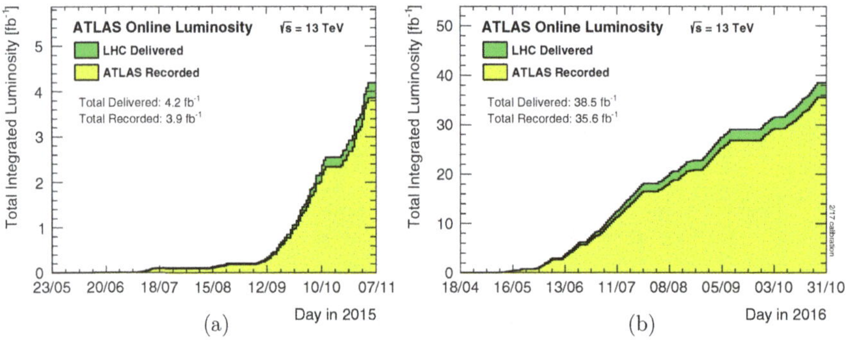

Fig. 3.4 Cumulative luminosity versus time delivered to (green) and recorded by ATLAS (yellow) during stable beams for proton–proton collisions at $\sqrt{s}$ = 13 TeV center-of-mass energy in (**a**) 2015 and (**b**) 2016 [57]. The difference between delivered and recorded luminosity reflects the small inefficiency of the data acquisition in ATLAS. The uncertainty in the combined 2015 + 2016 integrated luminosity is 2.1%. It is derived, following a methodology similar to that detailed in [72], from a calibration of the luminosity scale using x–y beam-separation scans performed in August 2015 and May 2016

are two of the most important characteristics of the dataset. For the 2015–2016 data-taking run which this thesis is written on, the center-of-mass energy is $\sqrt{s} = 13$ TeV with a total integrated luminosity of 36.1 fb^{-1} as seen in Fig. 3.4. The LHC will be shut down in 2018–2012 for a series of repairs and upgrades, after which it is expected to be run at $\sqrt{s} = 14$ TeV.

Luminosity can be a little confusing to understand as physicists often have two different related terms. The integrated luminosity in Fig. 3.4 is proportional to the total number of collisions (or events) recorded while the instantaneous luminosity in Fig. 3.5 is proportional to the bunch crossing rate and represents the potential number of collisions per second. The integrated luminosity, L_{int} is then meant to be

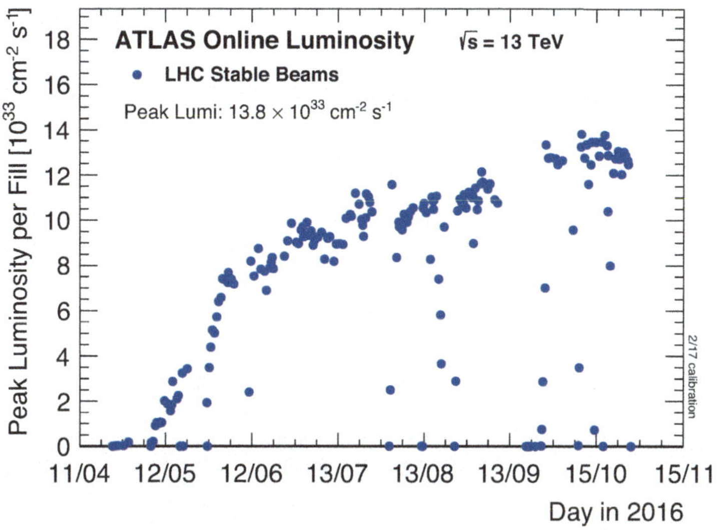

Fig. 3.5 The peak instantaneous luminosity delivered to ATLAS during stable beams for proton–proton collisions at $\sqrt{s} = 13$ TeV center-of-mass energy is shown for each LHC fill as a function of time in 2016 [57]. The luminosity is determined using counting rates measured by the luminosity detectors

understood as the integral of the instantaneous luminosity $L_{\text{inst.}} \equiv L$[4] over the data collection period. That is,

$$L_{\text{int}} = \int L\text{d}t = 36.1\ \text{fb}^{-1} \tag{3.1}$$

and the machine luminosity depends only on the beam characteristics [164]. For a Gaussian beam, this can be written as

$$L = \frac{N_b^2 n_b f_{\text{rev}} \gamma_r}{4\pi \varepsilon_n \beta^*} F, \tag{3.2}$$

where N_b is the number of particles per bunch,[5] n_b is the number of bunches per beam, f_{rev} is the revolution frequency, γ_r is the relativistic gamma factor, ε_n is the normalized transverse beam emittance, β^* is the beta function[6] at the collision point, F is the geometric luminosity reduction factor due to the crossing angle at the interaction point:

[4]This is often referred to as the machine luminosity as well, or just L.

[5]Assuming that each beam has the same number of bunches N_b.

[6]Small β^* corresponds to a narrower beam, related to the transverse size of the particle beam at the interaction point. The narrower the beam, the more "squeezed" it is, and so this also corresponds to a smaller geometric factor, F.

Table 3.1 A summary of the LHC typical parameters for Run 2 operating and data taking as taken from [164]

Parameter	Run 2 LHC [164]
Circumference	26,659 m
Beam energy in collision	7 TeV
Protons per bunch N_b	1.15e11
Bunches per beam n_b	2808
Radiofrequency cavity frequency	400 MHz
Proton speed β	0.9999999991
Lorentz factor γ_r	7460.523
Revolution frequency f_{rev}	11.2455 kHz
Beam current	0.58 A
Crossing angle θ_c	285 μrad
Beta function at collision point β^*	0.55 m
Transverse beam emittance ε_n	3.75 μm
RMS bunch length σ_z	7.55×10^{-2}m
Transverse RMS beam size σ^*	16.6 μm
Expected peak luminosity	$10\,\text{nb}^{-1}/\text{s}$

This design is based on the 25 ns bunch crossing separation. At full power, the LHC beam intensity is given above. Other parameters, such as beam amplitude parameters are typical values which are kept small to achieve high luminosity

$$F = \left(1 + \left(\frac{\theta_c \sigma_z}{2\sigma^*}\right)^2\right)^{-1/2}, \tag{3.3}$$

where θ_c is the full crossing angle at the interaction point, σ_z is the RMS bunch length, and σ^* is the transverse RMS beam size at the interaction point. Using the nominal LHC parameters in [164] and summarized in Table 3.1, the expected peak luminosity is $L = 1\text{e}34\,\text{cm}^2/\text{s} = 0.36\,\text{fb}^{-1}/\text{h}$ for both ATLAS and CMS, which are the high-luminosity experiments at the LHC. A classical route to increase the luminosity is to reduce the β^* with stronger and larger aperture quadrupole magnets. This requires a larger crossing angle which reduces the geometrical factor, but is compensated for with crab cavities to generate transverse electric fields [108].

The larger the integrated luminosity, the larger the dataset which enables the study of physics beyond the Standard Model and precision measurements of rare processes.

3.3.1 Pile-Up at the LHC

Due to the cross-section for interaction and the total number of protons per bunch, the probability to observe multiple proton–proton interactions per bunch crossing

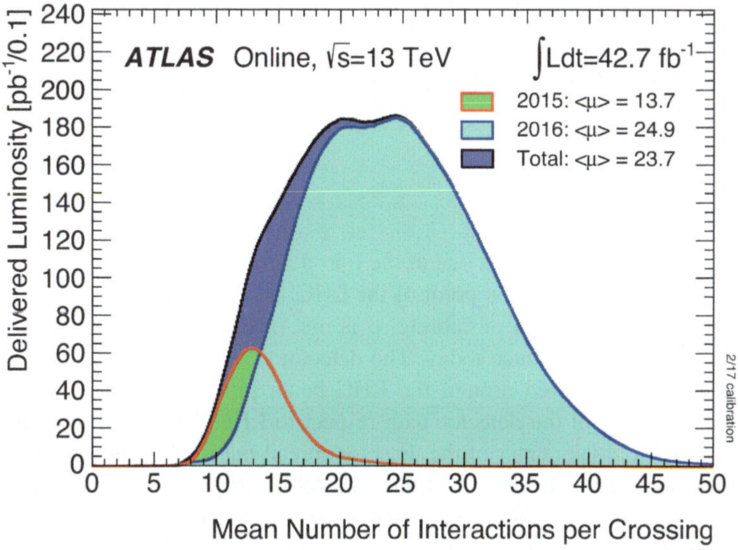

Fig. 3.6 The distribution of the mean number of interactions per crossing for the 2015–2016 proton–proton collision data at $\sqrt{s} = 13\,\text{TeV}$ center-of-mass energy [57]. All data delivered to ATLAS during stable beams is shown, but not necessarily all of this data is suitable for an analysis

increases as the luminosity increases. The multiple proton–proton interactions is referred to as **pile-up** but comes in two main forms:

1. **in-time** pile-up refers to the multiple proton–proton interactions that in the same bunch crossing that is currently being recorded and
2. **out-of-time** pile-up refers to the effect of seeing multiple proton–proton interactions outside of the bunch crossing that is currently being recorded.

Out-of-time pile-up is primarily an electronic effect due to the long integration times of various detector components. The 2015–2016 data run had up to 50 proton–proton interactions per bunch crossing as seen in Fig. 3.6. Pile-up is often referred to as μ, and the time-average pile-up is reported as $\langle\mu\rangle$. The actual number of interactions per bunch crossing can fluctuate with Poisson statistics. The wide variation seen in the figure is due to two primary effects. During a run of proton–proton collisions, the number of protons in a bunch will decrease over time (as does luminosity) and so μ will also decrease. The peak μ is often seen at the start of a run, with the peak luminosity. The other main source of the fluctuation is due to the tweaking of the LHC beams, such as changing β^* to get a narrower or wider beam. These sorts of large-scale changes to the beam properties are often either due to optimizing the beam for maximum physics impact or to respond to issues with subsystems of the accelerator complex such as power issues or magnet issues.

Pile-up is very important for future upgrades as the LHC will have an increased luminosity. Further studies of pile-up in the context of the instrumentation upgrades I work on are described in Sect. 9.4.1.

3.4 ATLAS Overview

The ATLAS detector Fig. 3.7 is one of the four main experiments at the LHC and is centered at Point 1, a collision point of the LHC. With over 100 million electronic channels and over 3000 km of cabling, it is one of the largest and most complex particle detectors in existence today. The detector is located approximately 100 m underground and centered around the LHC beam pipe. See Sect. 3.5 for details about the geometry of the detector with respect to LHC. Particles produced at the interaction point at the center of the detector spread out in all directions, hence the encompassing cylindrical design of the ATLAS detector. The ATLAS detector is built up of three main sub-detector pieces: tracking system, calorimetry, and muon tracking system.

The rest of the sections are ordered as seen in Fig. 3.7 from the inside-out, providing an overview of each sub-detector and its role in studying high energy particle physics collisions. Section 3.6 discusses the Inner Detector (ID), a tracking system that uses ionization to measure the trajectory of charged particles with help of the enclosing 2 T solenoid magnet. Section 3.7 describes the calorimetry

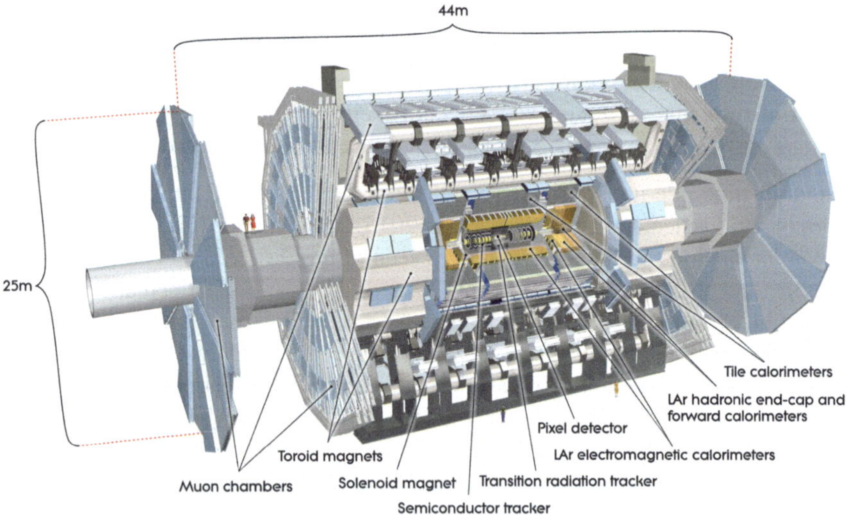

Fig. 3.7 A cut-away view of the ATLAS detector [28]. The dimensions of the detector are 25 m in height and 44 m in length. The overall weight of the detector is approximately 7000 t. This figure groups up major instrumentation components of the detector. Two people in red are shown for scale just to the right of the muon chambers on the left side of the figure

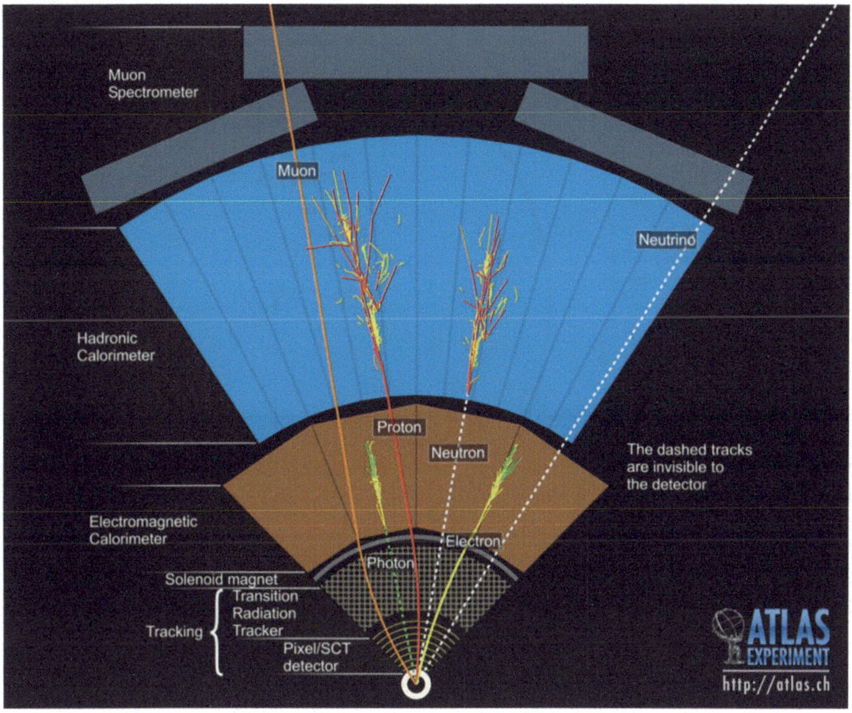

Fig. 3.8 A slice of the ATLAS detector depicting the various particle interactions with each component of the detector [28]. Dashed tracks in this figure are invisible to the detector component that the line is overlaid on top of. Muon track (orange) and neutrinos (dashed, white) pass through the entire detector. Electrons (yellow/green), photons (yellow/green), and hadrons (red/yellow) are fully absorbed by the calorimeter system. Charged particles like protons, electrons, and muons are curved by the solenoid magnet within the tracking system

system surrounding the solenoid magnet, composed of the electromagnetic and hadronic calorimeters that use scintillation and ionization to measure the energy of electrons, photons, and hadrons. The muon system, described in Sect. 3.8, surrounds the calorimeters and contains toroid magnets that uses ionization to measure the trajectories of deflected muons leaving the detector. Neutrinos are the only other Standard Model particles that leave the detector, but undetected. This is neatly depicted in Fig. 3.8 which shows a cartoon diagram of a slice of the detector with the various particle interactions.

3.5 ATLAS Geometry

ATLAS uses a right-handed coordinate system with its origin at the nominal interaction point in the center of the detector. The positive x-axis is defined by the

Fig. 3.9 A top-down cartoon of the LHC, the SPS, and the four experiments at the LHC: ATLAS, CMS, LHCb, and ALICE. A common coordinate system is used in ATLAS. The positive x-axis points from the interaction point at the center of the ATLAS detector to the center of the LHC ring. the positive y-axis points from the interaction point upward to the surface of the earth. The z-axis runs along the beam line, with the detector half at positive z-values referred to as the "A-side" (Geneva side) and the detector half at negative z-values referred to as the "C-side" (Jura side)

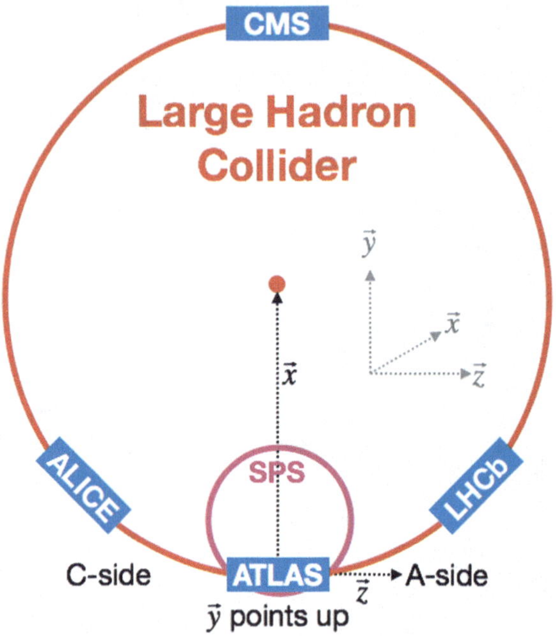

direction from the interaction point to the center of the LHC ring, with the positive y-axis pointing upwards, while the beam direction defines the z-axis as seen in Fig. 3.9. The x-y plane is perpendicular to the beam line and is referred to as the transverse plane. Cylindrical coordinates (r, ϕ) are used in the transverse plane, ϕ being the azimuthal angle around the z-axis. The objects measured in the ATLAS detector have momenta that can be described using two quantities $\vec{p} = (p_{\mathrm{T}}, p_z)$ with p_{T} the momentum of the particle in the transverse plane and p_z the momentum of the particle along the beam axis.

The pseudorapidity η in Fig. 3.10 is defined in terms of the polar angle θ by

$$\eta = -\ln\tan\left(\frac{\theta}{2}\right), \tag{3.4}$$

where $\eta = 0$ is perpendicular to the beam axis and large values of $|\eta|$ are close to the beam axis. Positive η is in the positive z-side and negative η is on the negative z-side.

Rapidity y is defined as

$$y = \frac{1}{2}\ln\left[\frac{E + p_z}{E - p_z}\right], \tag{3.5}$$

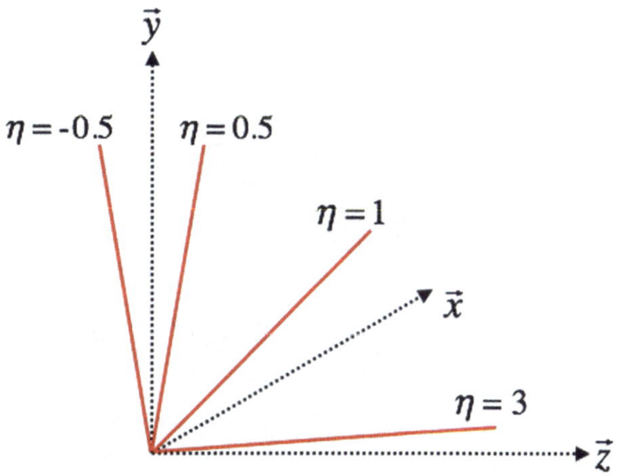

Fig. 3.10 A cartoon representation of selected pseudorapidity (η) values overlaid on Cartesian geometry axes (dashed black lines). Red lines are drawn for $\eta = \pm 0.5, 1.0, 3.0$

where E denotes the energy and p_z is the component of the momentum along the beam direction. The position of an object is often described in terms of (η, ϕ). The distance ΔR between objects in η-ϕ space is

$$\Delta R = \sqrt{(\Delta\eta)^2 + (\Delta\phi)^2}. \tag{3.6}$$

The choice of geometry for pseudorapidity (or rapidity) and ϕ is because differences in rapidity are Lorentz-invariant under boosts along the beam axis. If you boost Eq. (3.5) along the z-axis

$$\begin{aligned} y' &= \frac{1}{2} \ln \left[\frac{\gamma E - \beta\gamma p_z + \gamma p_z - \beta\gamma E}{\gamma E - \beta\gamma p_z - \gamma p_z + \beta\gamma E} \right] \\ &= y + \frac{1}{2} \ln \left[\frac{1-\beta}{1+\beta} \right] \\ &= y + \tanh^{-1} \beta. \end{aligned} \tag{3.7}$$

So the difference in rapidities is Lorentz-invariant. Since $\Delta\phi$ is measured in the x-y plane, it is also invariant under Lorentz boosts along the beam axis. This means that ΔR is also Lorentz-invariant.

3.6 Tracking in the Inner Detector

The inner detector [91, 178] is the first part of ATLAS to see the outgoing particles of the proton–proton collisions. It is built around the beam pipe with cylindrical geometry as shown in Fig. 3.11 within a 2 T solenoid magnet. It is designed to be compact with excellent momentum resolution of charged particle tracks above $p_{\mathrm{T}} \geq$

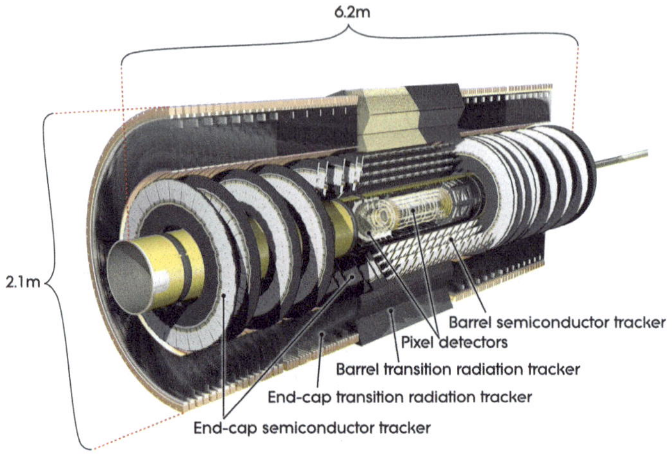

Fig. 3.11 A longitudinal view of the ID compromising of the three main pieces: pixel detector, semiconductor tracker, and transition radiation tracker [28]. It is 6.2 m in length and 2.1 m in height

500 MeV[7] for $|\eta| < 2.5$ (Fig. 3.12). The inner detector is made of three specific subcomponents as seen in Fig. 3.13.

The pixel detector [258] is the closest to the beam pipe and has 80 million pixels (or readout channels) covering $1.7\,\text{m}^2$ for $|\eta| \leq 2.5$. Each pixel has an area[8] of $20{,}000\,\mu\text{m}^2$ with a position resolution of[9] $14\,\mu\text{m}$ in ϕ and $115\,\mu\text{m}$ in z directions.[10] The three barrel layers have 1456 pixel modules, each with 46,080 readout channels. The three pixel disks in each end-cap have 144 modules, 6.6 million readout channels. The pixel detector provides one measurement per barrel layer for each charged particle track and full pattern recognition capability to reconstruct tracks at nominal LHC parameters. It is also crucial to the identification and reconstruction of both primary and secondary vertices; the latter which is seen in the decay of particles containing a b-quark or for b-tagging of jets, necessary to perform the analysis search in Chap. 7. This is highlighted in green in Fig. 3.13.

The semiconductor tracker [27] surrounds the pixel detectors. This is a silicon microstrip tracker that consists of 4088 two-sided modules with over 6 million

[7]The solenoid magnetic field strength of 2 T means that charged particles need $p_T \geq 500\,\text{MeV}$ at 2 T to escape the ID and reach the calorimeters. The magnetic field from the solenoid is not perfect and fluctuates down to 0.5 T on the ends of the detector. A minimum p_T threshold is applied to reduce the rate of fake tracks.

[8]$50\,\mu\text{m}$ in ϕ direction and $400\,\mu\text{m}$ in z direction, along beam axis.

[9]$\text{area}/\sqrt{12}$ is the upper limit in resolution for a digital readout [115]. Modern pixel detectors can achieve better resolution by using charge measurement to determine which pixel a charged particle was closer to.

[10]Better resolution in the ϕ direction as this is enclosed in a solenoid magnet so that charged particles will bend along the ϕ direction.

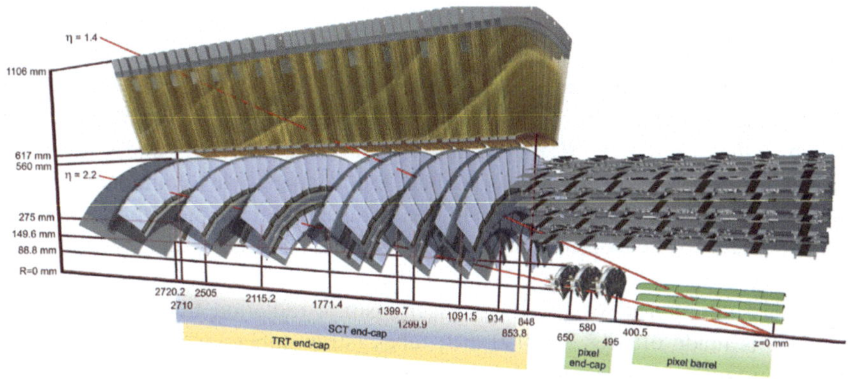

Fig. 3.12 Drawing showing the sensors and structural elements traversed by two charged tracks of $p_T = 10\,\text{GeV}$ in the end-cap ID at $\eta = 1.4$ and $\eta = 2.2$ [28]. The track at $\eta = 1.4$ passes through the beam pipe, the three pixel layers, four SCT disks with double layers, and approximately 40 straws TRT end-cap. A particle at $\eta = 2.2$ passes through the beam, only the first pixel layer, two of the end-cap pixel disks, and the last four end-cap SCT disks, but does not pass through the TRT straws which covers $|\eta| \leq 2.0$

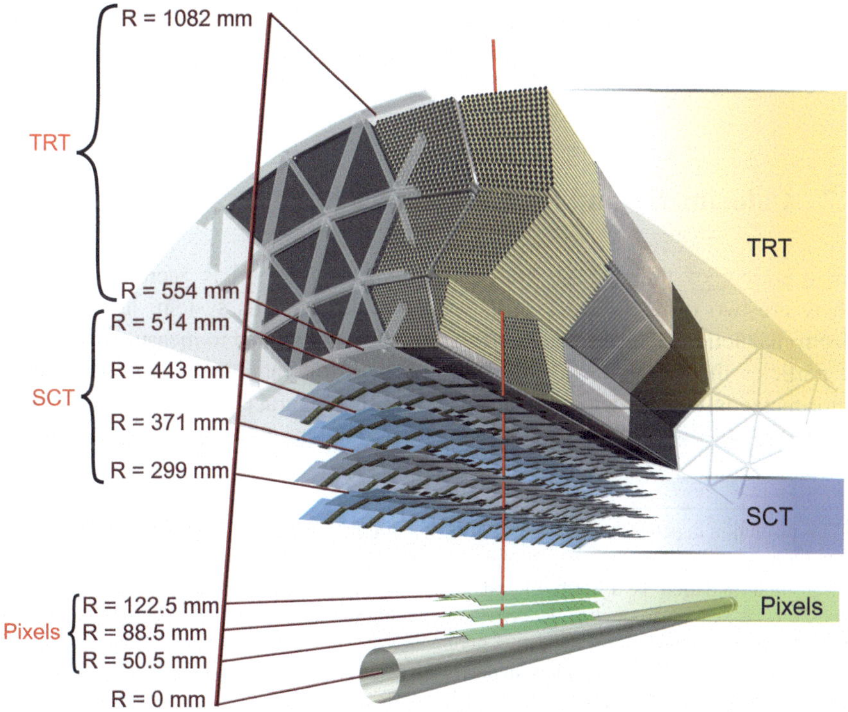

Fig. 3.13 A radial view of the ID with the detector elements crossed by a charged particle $p_T = 10\,\text{GeV}$ [28]. The track passes through the beam pipe, three pixel layers, four SCT disks with double layers, and approximately TRT 40 straws

strips (or readout channels) covering 63 m^2 for $|\eta| \leq 2.5$. All of the modules are distributed over four barrel layers and 9 disks in each end-cap (18 end-cap disks total). The readout strips are placed 80 μm and rotated by 50 mrad with respect to each other, provide a position resolution of 17 μm in the transverse plane and 580 μm in the z-axis. The SCT is designed to provide between 4 and 9 precision measurements per track in the intermediate radial range. This component, along with the pixel detector, contributes to the measurement of momentum, impact parameter, and vertex identification of a charged particle track. This is highlighted in blue in Fig. 3.13.

The last component of the ID is the transition radiation tracker [246, 247]. The TRT is made of over 350,000 drift tubes (straw tubes, or readout channels) covering 12 m^3 of volume for $|\eta| < 2.0$. The basic detector element, straws, is 4 mm in diameter, 144 cm (37 cm) long in the barrel (end-cap) providing a position resolution of 130 μm [230][11] in ϕ. In the barrel (end-cap), there are 52,544 (245,760) straws over 73 layers (160 straw planes) which provide transition radiation[12] tracking for charged particle identification. All the charged particle tracks will traverse through at least 36 straws.[13] The charge collection time in the straw is an important parameter for tracking performance. At a fixed transverse momentum for a charged particle, a light-mass charged particle will emit more transition radiation photons than a heavier-mass charged particle.[14] Therefore, the TRT is an important component for discrimination between electrons and charged hadrons for $|\eta| < 2.0$.

3.7 Calorimetry and the Calorimeter System

An overview of the ATLAS calorimetry system [93, 95] is seen in Fig. 3.14. There are two different types of calorimeters used in this system: hadronic and electromagnetic. An electromagnetic calorimeter is designed to measure the energy of particles that interact via the electromagnetic interaction,[15] while a hadronic calorimeter is designed to measure particles that interact via the strong nuclear force. ATLAS uses both of these types of calorimeters, as a sampling calorimeter.

---

[11]This resolution depends strongly on the drift-time (or drift-distance). See studies in Figs. 20–23 from [246].

[12]Transition radiation is a form of radiation when a charged particle passes through the 70%/27%/3% xenon/carbon dioxide/oxygen mixture in the straw tubes. The energy of the photon emitted is proportional to the relativistic Lorentz factor.

[13]Only for $|\eta| < 2.0$. The exception is in the region between barrel and end-cap where this number goes down to at least 22 straws.

[14]Recall that these charged particles are travelling curved trajectories in the solenoid's magnetic field and heavier particles bend less than lighter particles. Lighter particles spend more time in the drift tubes and thus emit more radiation.

[15]Such as bremsstrahlung, pair production.

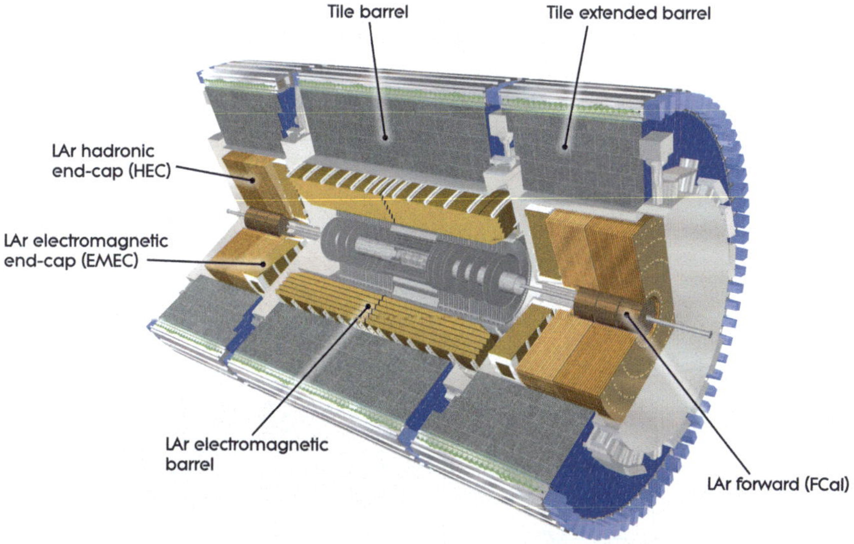

Fig. 3.14 A longitudinal, cut-away view of the ATLAS calorimeter system showing the tile calorimeter barrel, the tile calorimeter extended barrel, the liquid argon electromagnetic barrel, the liquid argon electromagnetic end-cap, the liquid argon hadronic end-cap, and the forward calorimeter [28]

A sampling calorimeter is one in which the "active" material that provides the detectable signal is different from the dense "absorber" material that reduces particle energy. Because the dense material is chosen to absorb a lot of the particle energy, only a fraction of the energy is measurable by the detector sensors. This requires a calibration to the measured calorimeter energy by studying the calorimeter response, and will be discussed in more detail in Chap. 5. The calorimeters have a large responsibility in providing coverage for the full $-4.9 < \eta < 4.9$ range, while having sufficient granularity for precision measurements, and providing containment for both electromagnetic and hadronic showers from electrons, photons, and hadrons (see Fig. 3.8). This containment is important for punch-through, where energy leaks outside the calorimeters to the muon spectrometers, but also to ensure a good $E_{\mathrm{T}}^{\mathrm{miss}}$ measurement, which is crucial for many physics programs, in particular supersymmetry searches like mine (see Chap. 7).

The LAr electromagnetic barrel (EMB) and EMEC are lead/liquid-argon detectors with "accordion geometry" as seen in Fig. 3.15 covering $|\eta| < 3.2$ for precision electromagnetic shower measurements. This specialized geometry provides complete and uniform coverage over ϕ without any cracks, while allowing low latency readout of the data. This geometry has three radial layers. The first sampling layer, known as "strips," is finely segmented in $\Delta\eta = 0.0031$[16] with 8 strips in front of

[16] We would like to resolve two photons coming from the Higgs decay, versus other decays such as a pion.

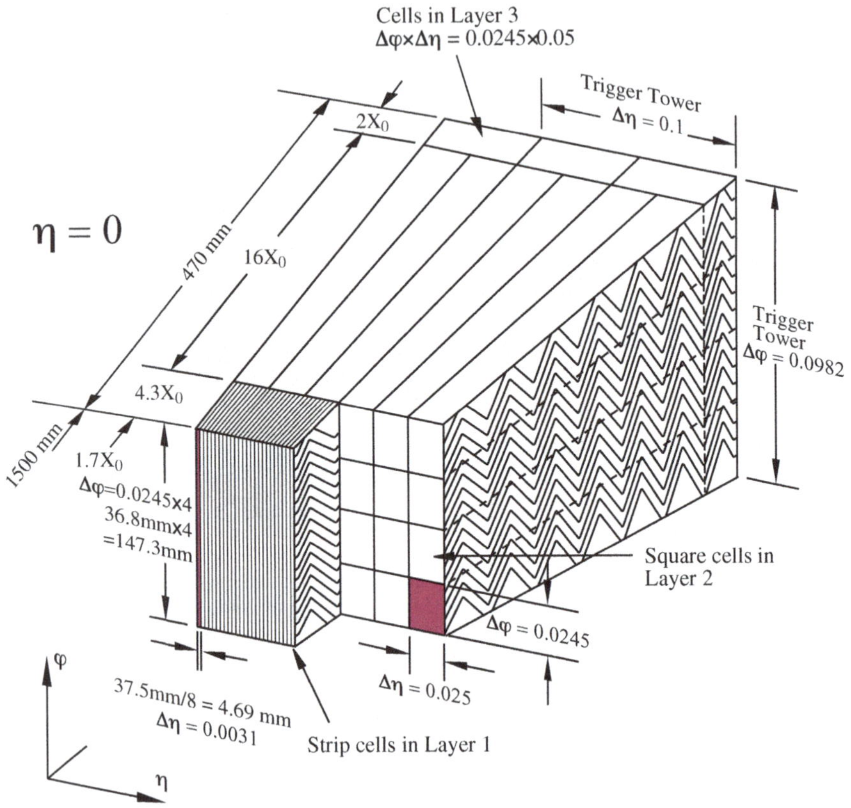

Fig. 3.15 This is a sketch of a LAr accordion module where the different layers are visible in ϕ which is pointing up in this figure [28]. The granularity in η and ϕ of the calorimeter cells for each of the three sampling layers and of the trigger towers ($\eta \times \phi = 0.1 \times 0.1$) is also shown. These trigger towers will be discussed more in Chap. 4

each cell. The second sampling layer, which collects the largest fraction of energy of the electromagnetic shower, has fine segmentation of $\Delta\eta = 0.025$ and $\Delta\phi = 0.0245$. The last layer collects the tail end of the electromagnetic shower, and thus can have a coarser segmentation of $\Delta\eta = 0.05$. The fine "strips" in the first layer allow for discrimination of electromagnetic showers from electrons/photons versus energetic pions. For example, a neutral pion can shower to photons ($\pi^0 \rightarrow \gamma\gamma$) and the angular distance between the two photons can be small,[17] the fine "strips" allow

[17] See [165], a π^0 with $E \sim 50$ GeV will have two decay photons with $\Delta R < 1$ cm at 150 cm from the interaction point. If no sufficient resolution, this looks like a single photon which can also be faked by hadronic showers. For a large background like multijet which has a cross-section of 10e8 larger than $H \rightarrow \gamma\gamma$ cross-section, rejecting jets faking photons like these is important for physics impact.

for discrimination of photon showers from pion showers. The EMB is composed of two half-barrels and covers $|\eta| < 1.475$. The LAr EMEC is composed of two wheels and covers the region $1.375 < |\eta|3.2$. An additional thin LAr presampler covering $|\eta| < 1.8$ allows corrections for energy losses upstream of the electromagnetic calorimeter (EMCal).

To measure the energy of hadrons, the hadronic calorimeters Tile, HEC, and forward calorimeter (FCal) cover $0 < |\eta| < 4.9$. The tile calorimeter is a sampling calorimeter using steel as the absorber and scintillator as the active medium covering the region $|\eta| < 1.7$. As seen in Fig. 3.14, it is located behind the EMB and EMEC and divided into a long (central) barrel that is 5.8 m in length covering $|\eta| < 1.0$ and two extended barrels on each side of the detector each 2.6 m in length covering $0.8 < |\eta| < 1.7$ with radius $r = 2.28$–4.25 m. Each barrel consists of 64 modules (or wedges) as seen in Fig. 3.16. The HEC uses LAr with a copper/liquid-argon sampling calorimeter which covers $1.5 < |\eta| < 3.2$. There are two wheels on

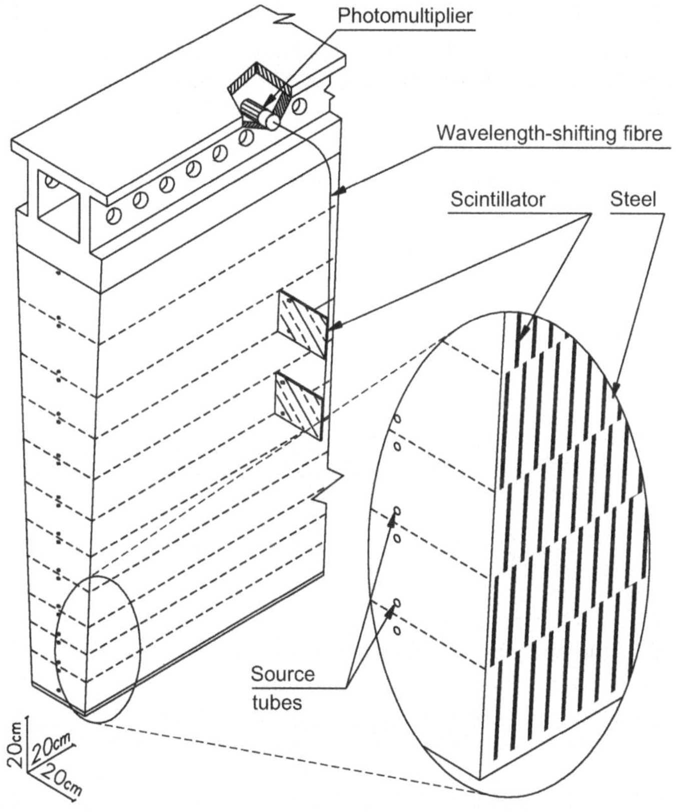

Fig. 3.16 This is a sketch of a Tile calorimeter (Tile) module showing how the mechanical assembly and the optical readout are integrated together [28]. The various components of the optical readout are shown: the tiles, the fibers, and the photomultiplier tubes. Each wedge is approximately $\Delta\phi = 0.1$ which is around 20 cm

Table 3.2 Summary of the sampling calorimeters in the calorimetry section, their coverage in η, and the 179,166 readout channels

		Layers			
Detector	Shower	Absorber	Active	Coverage	Number of channels
EMB	EM	Lead	Liquid-Argon	$\|\eta\| < 1.475$	99,712
EMEC	EM	Lead	Liquid-Argon	$1.375 < \|\eta\| < 3.2$	62,208
Tile	Had	Steel	Scintillator	$\|\eta\| < 1.0$	5760
	Had	Steel	Scintillator	$0.8 < \|\eta\| < 1.7$	4092
HEC	Had	Copper	Liquid-Argon	$1.5 < \|\eta\| < 3.2$	5632
FCal	EM	Copper	Liquid-Argon	$3.2 < \|\eta\| < 4.9$	1008
	Had	Tungsten	Liquid-Argon		754
Total					179,166

Here, "EM" means the calorimeter component measures an electromagnetic shower, while "Had" means the calorimeter component measures a hadronic shower

each side of the detector, with each wheel consisting of 32 wedge-shaped modules. Finally, the FCal extends the hadronic calorimeter sampling range by providing coverage over $3.1 < |\eta| < 4.9$, with much coarser granularity. There are 3 modules on each side for the FCal, one electromagnetic module (copper/liquid-argon) and two hadronic modules (tungsten/liquid-argon) (Table 3.2).

To wrap up this section, I want to briefly discuss an important characteristic of calorimeters: energy resolution. A natural feature that comes out of the calorimeter is the improved energy resolution as the energy increases.[18] Luckily, the upgrades at the LHC mean that even more highly energetic particles will be measured by the calorimeters and that comes with improved resolution at no cost.[19] In most cases, the calorimeter energy resolution improves with energy as $1/\sqrt{E}$, where E is the energy of the incident particle. For practical purposes, the resolution is reported [165] as a number with three components as in Eq. (3.8)

$$\frac{\sigma(E)}{E} = \frac{a}{\sqrt{E}} \oplus \frac{b}{E} \oplus c, \tag{3.8}$$

where the symbol "$\oplus$" indicates a quadratic sum. The first term a represents the "stochastic term," the second term b represents the "noise term," and the third term represents the "constant term." Each of these terms is understood by their dependency on the energy of the incident particle E. The "stochastic" term arises out of the calorimeter response being proportional to the number of track segments in the shower and a statistical argument can be made to show that this depends on $\sqrt{E}$. This is usually the dominant term that limits the resolution of a calorimeter

[18]Another feature is that the electromagnetic showers grow in size as well, which is why deeper calorimeters are needed at higher energies.

[19]There are downsides such as much harder radiation impacting the instrumentation means more repairs, maintenance, and upgrades.

at low energies. The "noise" represents the electronic noise in the readout chain. The "constant" term includes contributions that do not depend on the energy of the particle such as how the calorimeter shapes the particle impact point or nonuniformity of the detector geometry. At high energies, calorimeter resolution is limited by the "constant" term. For ATLAS, the quoted energy resolution [218, Chapter 33. Particle detectors] for the EMCal is

$$\frac{\sigma(E)}{E} = \frac{10\%}{\sqrt{E}} \oplus \frac{0.3}{E} \oplus 0.4\%. \tag{3.9}$$

To interpret this correctly, a 100 GeV electron will have $\frac{\sigma(E)}{E} = 1\% \oplus 0.003 \oplus 0.4\% = 1.1\%$ while a 10 GeV electron will have $\frac{\sigma(E)}{E} = 40\%$. At this low energy, the tracking Sect. 3.6 will help improve this measurement from the calorimeter. In [42], ATLAS measured the jet energy resolution in Run 1 to be from 20% to 10% for jets within $|y| < 2.8$[20] and $30\,\text{GeV} < p_{\text{T}} < 500\,\text{GeV}$.

3.8 Muons and the Muon Spectrometer

The conceptual layout of the muon spectrometer (MS) is shown in Fig. 3.17. This entire system [28, 94] is based on the magnetic deflection of muon tracks in the large superconducting toroid magnets. For $|\eta| < 1.4$, the magnetic bending is provided by the large barrel toroid; for $1.6 < |\eta| < 2.7$, the muon tracks are bent by two smaller end-cap magnets inserted into both ends of the barrel toroid; and in the transition region $1.4 < |\eta| < 1.6$ the bending is provided by both the end-cap and barrel fields. In the barrel (transition and end-cap) region, the muon tracks are measured in chambers arranged in three cylindrical (planes) layers parallel (perpendicular) to the beam axis. The coverage and number of channels are summarized in Table 3.3.

The Monitored Drift Tubes (MDT) [109] (1163 chambers) and Cathode Strip Chambers (CSC) [22] (32 chambers) provide the precision measurements for the system. Over most of the η range, this is largely done by MDT, while CSC takes over for large pseudorapidities. For $\eta < 2.4$, the trigger chambers have the unique role of providing bunch crossing identification, well-defined p_{T} trigger thresholds, and measure the muon coordinate in a direction orthogonal to the precision-tracking chambers. The trigger chambers are composed of Resistive Plate Chambers (RPC) [13] and Thin Gap Chambers (TGC) [206].

[20] Rapidity.

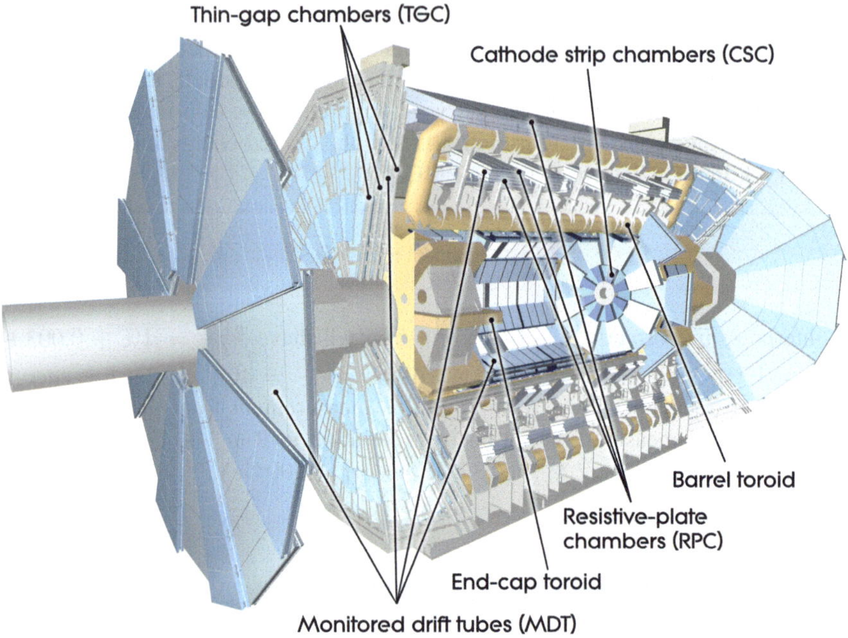

Fig. 3.17 A cut-away view of the ATLAS muon system [28]

Table 3.3 Summary of the components of the muon spectrometer, their coverage in η, and the number of readout channels

Detector	Coverage	Number of channels
MDT	$\|\eta\| < 2.0$	354,000
CSC	$2.0 < \|\eta\| < 2.7$	31,000
RPC	$\|\eta\| < 1.05$	373,000
TGC	$1.05 < \|\eta\| < 2.4$	318,000

Chapter 4
Trigger and Data Acquisition

This chapter provides an introduction to the Trigger and Data Acquisition (TDAQ) system in ATLAS. The trigger system is a crucial component of the experiment, responsible for selecting events of interest at a recording rate of approximately 1 kHz from up to 40 MHz of proton–proton collisions corresponding to 25 ns bunch spacing of the LHC.

I will describe an overview and motivation triggering in Sect. 4.1, the subsystems of the trigger system in Sect. 4.2 during Run 2, a brief description of a trigger menu Sect. 4.3, and discuss the data and simulated samples for the thesis analysis in Sect. 4.4. Finally, I close off this chapter by discussing the instrumentation upgrades in Sect. 4.2.1 for Run 3 and beyond.

4.1 Overview

During Run 1, the trigger system [31, 32, 58, 60, 73] of the ATLAS experiment operated at instantaneous luminosities of $8 \times 10^{33}\text{cm}^{-2}\,\text{s}$ with center-of-mass energies $\sqrt{s} = 7$–$8\,\text{TeV}$. Since 2015, the start of Run 2, the center-of-mass energy nearly doubled to 13 TeV. This increase coupled with higher luminosity and more proton–proton interactions per bunch crossing[1] requires an efficient trigger system to maintain rates low enough to record data while maintaining the physics impact. The rest of this chapter will only describe the system from 2015 and beyond.

The TDAQ system shown in Fig. 4.1 consists of a hardware-based first-level trigger (L1) and a software-based high-level trigger (HLT). The L1 trigger decision

[1]More pile-up.

© Springer Nature Switzerland AG 2020

G. Stark, *The Search for Supersymmetry in Hadronic Final States Using Boosted Object Reconstruction*, Springer Theses,
https://doi.org/10.1007/978-3-030-34548-8_4

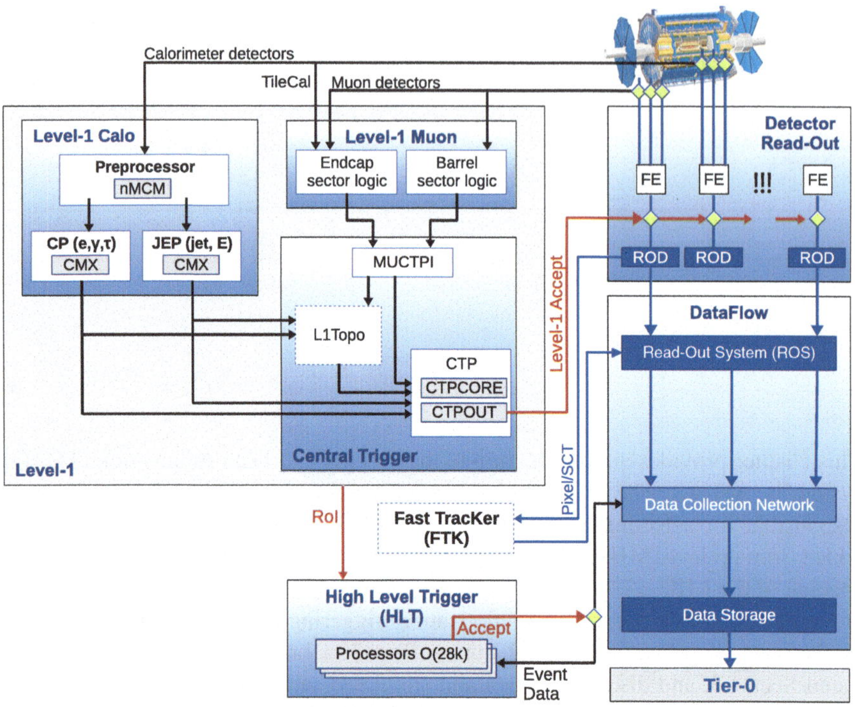

Fig. 4.1 A schematic overview of the ATLAS TDAQ system in Run 2 with emphasis on the components relevant for triggering [85]. The main subsystems shown are the Level-1 (L1) trigger, the High-Level Trigger (HLT), and the Fast TracKer (FTK)

is formed by the Central Trigger Processor (CTP)[2] which receives inputs primarily from L1 calorimeter trigger (L1Calo) and L1 muon trigger (L1Muon). The other role the CTP is preventative to protect front-end readout buffers from overflowing. This is known as dead-time and comes in two forms: simple and complex. Simple dead-time is the amount of time needed to allow the readout windows to process data.[3] Complex dead-time is the determined based on the rate in which the downstream front-end buffers can empty out.[4] The decision by the L1CTP is called the L1 accept (L1A) at a maximum rate of 100 kHz. Events are buffered in the read-out system to be processed by the HLT which receives region-of-interests (ROIs) from the L1 subsystems for locally based reconstruction. After the HLT acceptance at a maximum rate of 1 kHz,[5] the events are transferred to local storage to be

[2]The CTP has a configurable lookup table mapping combinations of input signals to an output decision.

[3]In the start of 2015, this was set to 100 ns, or 4 bunch crossings.

[4]This concept of rate-limiting is often otherwise called a "token bucket" or a "leaky bucket."

[5]This is a detector readout limitation.

exported out to the Tier-0 computing facility[6] at CERN for offline reconstruction, described in more detail in Chap. 5. Each event is 1–2 MB in size which means the readout system writes out 1–2 GB s^{-1} to disk. At the end of the day, the ATLAS detector can only save one event for every 40,000 produced at the LHC. The trigger system is crucial and optimized to increase the chance of selecting the interesting, rare events for offline physics analysis.

4.2 The TDAQ Subsystems

4.2.1 Level-1 Trigger

The Level-1 Trigger is composed of two main subsystems: L1Calo and L1Muon. The L1 trigger decision, L1A, is based on the outputs of the muon spectrometer Sect. 3.8 and the calorimeters Sect. 3.7. As the decisions need to be made quickly, all of the reconstruction algorithms are implemented in the hardware, and in some cases on FPGAs. The logic of L1Calo is much more complicated, compared to L1Muon, as it tries to identify electrons, photons, taus, jets, and calculate the missing transverse energy. As such, I expand more on the details of the L1Calo system as it will also provide the necessary context for discussing the instrumentation upgrades described in Sect. 4.2.1. In the rest of this section, I describe the different components that go into the L1 CTP to make a decision, accepting 1 out of every 400 events.

4.2.1.1 Level-1 Calorimeter Trigger

The L1Calo trigger receives inputs from the hadronic and electromagnetic calorimeters as described in Sect. 3.7. As seen in Fig. 4.1, the inputs from the calorimeter need to be preprocessed by the preprocessor system [92]. This preprocessor digitizes and calibrates the analog signals from the calorimeter detectors. In particular, a bunch-by-bunch pedestal subtraction scheme enables a significant rate reduction of the triggers used. The bunch-by-bunch correction accounts for the increased trigger rates at the beginning of a bunch train caused by the combination of in-time and out-of-time proton–proton collision events convolved with the electronic pulse signal from LAr [29, fig. 2].

The preprocessor outputs are used as inputs to the Cluster Processor (CP) and Jet/Energy Processor (JEP) subsystems in parallel. These outputs are approximately 7000 trigger towers, of granularity $\Delta\eta \times \Delta\phi = 0.1 \times 0.1$, as shown in Fig. 4.2. Figure 4.2 depicts the elements used for the electron/photon and tau/hadron

[6]These computers are running Scientific Linux CERN (SLC), a publicly available operating system for scientific computing.

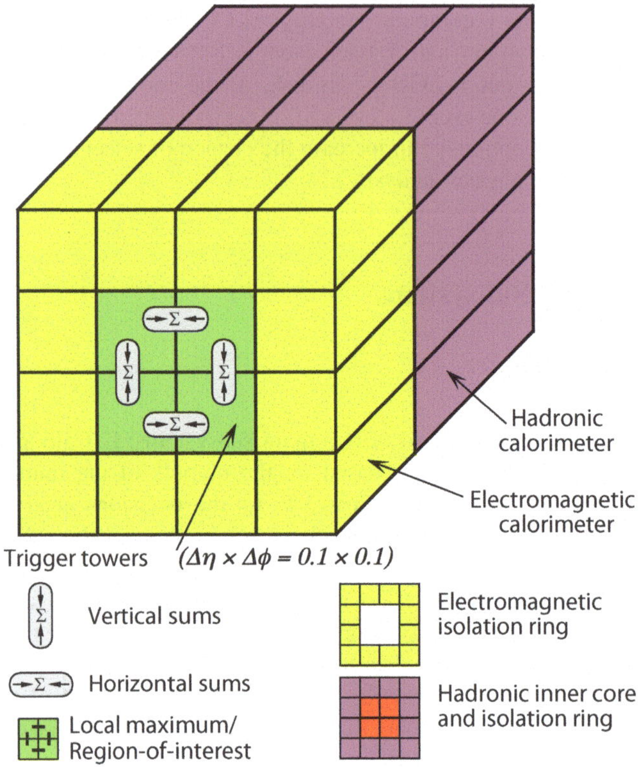

Fig. 4.2 A schematic view of the trigger towers used as input to the L1Calo trigger algorithms [85]

algorithms. They are based on a sliding window of 4×4 trigger towers in both the electromagnetic and hadronic calorimeters to form six basic calculables [161][7]:

1. four 2×1 trigger tower regions are formed to measure the transverse energy of the electromagnetic showers (the vertical and horizontal sums inside the green)
2. a hadronic core (in red) of the four hadronic towers, behind the electromagnetic towers, used for isolation criteria in the hadronic calorimeters
3. four hadronic clusters which are the sum of the previous two items (1) and (2) to measure the transverse energy of hadronic showers
4. an electromagnetic isolation ring (in yellow) which consists of the twelve electromagnetic towers surrounding the core (in green), used for isolation criteria in the electromagnetic calorimeters

[7] Since these firmware algorithms need to run in nanoseconds, the energy sums are scalar energy sums, rather than vector sums.

5. a hadronic isolation ring (in purple) which consists of the twelve hadronic towers surrounding the core (in red), used for isolation criteria in the hadronic calorimeters
6. a 2×2 ROI which is summed over both the electromagnetic and hadronic layer to identify candidate ROIs

These six calculables are used by the algorithms to identify triggerable objects. The transverse energy thresholds are configurable for different η regions[8] to account for the varying detector energy responses. There are two main modules here: Cluster Processor Module (CPM) and Jet/Energy Module (JEM).

CPM To identify a 2×2 ROI as an electromagnetic trigger candidate or a hadronic trigger candidate, the electron/photon (tau/hadron) algorithm searches for narrow, highly energetic showers in the electromagnetic calorimeters: applying a threshold on the electromagnetic tower region sums, isolation criteria by applying a maximum energy threshold, and making sure the showers do not (do) penetrate into the hadronic calorimeters, respectively. If these conditions are met, the window is said to contain the respective trigger candidate.

JEM To identify a 2×2 "jet" ROI, sliding trigger tower windows of sizes 4×4 and 8×8[9] look for ROIs where the summed electromagnetic and hadronic transverse energy exceed a predefined threshold surrounding the ROI which is a local maxima. If these criteria are met, then the window is said to contain a "jet" trigger candidate.

In the case where electronic saturation occurs, a trigger candidate is produced. These ROIs are sent to the Common Merger Extended Module (CMX) as Trigger OBjects (TOBs) which handles the logic of counting and identifying the position/energy to L1 topological processor (L1Topo). A TOB contains the transverse energy sum, η-ϕ coordinates, and isolation thresholds (where relevant), as well as identifying information to specify what kind of TOB it is.[10] Additionally, these ROIs are also sent to the HLT to seed the trigger algorithms there.

4.2.1.2 Level-1 Muon Trigger

In parallel to L1Calo in Sect. 4.2.1.1, the L1Muon system looks for coincidences in different layers of the muon chambers. Extra logic exists to reject muons that do not originate from the primary vertex. Events with muons with large transverse energy compared against a predefined threshold are selected and sent to L1Topo, through the Muon-to-CTP interface (MUCTPI), for a decision.

[8]Granularity here is $\Delta\eta = 0.1$.

[9]A 4×4 (8×8) window size is 0.4×0.4 (0.8×0.8) in $\Delta\eta \times \Delta\phi$.

[10]Identifying information such as a `JET` TOB or a `EM` TOB, etc.

4.2.2 HLT

After the L1A, the events are processed by the HLT using finer-granularity calorimeter information, measurements from the MS, and tracking information[11] from the ID. The HLT runs on a processing farm to perform a subset of offline event reconstruction. The lower input rate and software-based reconstruction allows for more complex and configurable trigger decisions. There are, however, some computational limitations such as track reconstruction that necessitates an upgrade, FTK, being commissioned[12] in 2017 to allow for full tracking information at the HLT. The FTK is briefly described in Sect. 4.2.2.1

4.2.2.1 FTK

A new FTK system [240] will provide global ID track reconstruction at the L1 trigger rate using lookup tables stored in custom associative memory chips for the pattern matching capabilities for every event that contains a L1A. The FPGA-based track fitter performs a fast linear fit and the tracks are made available to HLT, allowing the use of tracks at a rate much higher than capable with a CPU-based system. Since the FTK provides a hardware-based tracking solution that can handle the challenge of high luminosity, it allows ATLAS to maintain trigger thresholds such as the ability to reconstruct and identify secondary vertices.[13]

4.3 Trigger Menu

The trigger menu [85] defines a list of L1 and HLT triggers and consists of five different flavors of triggers:

1. primary triggers—used for physics analyses and are typically unprescaled[14]
2. support triggers—used for efficiency measurements, performance measurements, and monitoring

[11]Note that the tracking information from the ID is not available at L1.

[12]This is a rather generic term that refers to both the process of making the reconstruction algorithms work as intended (commissioning of reconstructed objects) and the process of understanding how what is reconstructed in the detector corresponds to what actually happened in the detector (detector commissioning).

[13]This is important in analyses that are sensitive to *b*-tagging, for example, but this is explained more in Chap. 5.

[14]If a trigger is prescaled, this means the trigger rate is purposefully decreased in order to keep the output rate manageable.

3. alternative triggers—used for experimental/new triggers that overlap significantly with primary triggers—but could be useful for a specific analysis or purpose
4. backup triggers—like primary triggers but with tighter selections and a lower expected rate
5. calibration triggers—used for calibration effort

When the LHC beams are colliding, multiple trigger menus are defined and available to be used. The rate and bandwidth constraints of the ATLAS detector, TDAQ, and offline computing are dependent on the luminosity and average number of proton–proton collisions. Therefore, the menu[15] is defined for a given range of luminosity that provides an expected output rate during detector operation. The two most relevant constraints are 100 kHz for L1 and 1 kHz for HLT, the former is limited by ATLAS readout capability and the latter is limited by offline computing power. Trigger names all have the same pattern, as also used throughout this thesis later on, which consist of:

1. Trigger level: `L1` or `HLT`
2. Multiplicity: `SINGLE`, `MULTI`, $n \in \mathbb{Z}$
3. Object type: `el` for electron, `mu` for muon, `j` for jet, `xe` for missing transverse energy, `te` for transverse energy
4. Threshold value in GeV
5. Seeded L1 trigger (if describing an HLT trigger)

So, for example:

- `HLT_MU20_L1MU15` describes an HLT trigger requiring a 20 GeV muon candidate which is seeded by an L1 trigger requiring a 15 GeV muon candidate
- `HLT_xe70` describes an HLT trigger requiring 20 GeV of missing transverse energy

4.4 Data and Simulated Event Samples

The data used in this analysis were collected by the ATLAS detector from *pp* collisions produced by the LHC at a center-of-mass energy of 13 TeV and 25 ns proton bunch spacing over the 2015 and 2016 data-taking periods. The full dataset corresponds to an integrated luminosity of 36.1 fb^{-1} after the application of beam, detector, and data-quality requirements. The uncertainty in the combined 2015 + 2016 integrated luminosity is 2.1%. It is derived, following a methodology similar to that detailed in [72], from a preliminary calibration of the luminosity scale using x–y beam-separation scans performed in August 2015 and May 2016. Events are required to pass a $E_{\mathrm{T}}^{\mathrm{miss}}$ trigger with thresholds of 70 GeV, 100 GeV, and 110 GeV at

[15] It can be, and is, also defined for different types of bunch grouping from the LHC.

the HLT level for the 2015, early 2016, and late 2016 datasets, respectively. These triggers are fully efficient for events passing the preselection defined in Sect. 7.3, which requires the offline reconstructed $E_{\mathrm{T}}^{\mathrm{miss}}$ to exceed 200 GeV. There are on average 24 inelastic pp collisions (see Sect. 3.3.1) in the dataset.

Samples of monte-carlo (MC) simulated events are used to model the signal and background processes in this analysis, except multijet processes, which are estimated from data. SUSY signal samples in which each gluino decays as $\tilde{g} \rightarrow t\bar{t}\tilde{\chi}_1^0$ were generated with up to two additional partons using MADGRAPH5_aMC@NLO [17] v2.2.2 at LO with the NNPDF 2.3 [102] PDF set. These samples were interfaced to PYTHIA v8.186 [242] for the modeling of the parton showering, hadronization, and underlying event.

The dominant background in the signal regions is the production of $t\bar{t}$ pairs with additional high p_{T} jets. For the generation of $t\bar{t}$ and single-top quarks in the Wt-channel and s-channel the POWHEG-BOX [14] v2 event generator with the CT10 [199] PDF set in the matrix element calculations was used. Electroweak t-channel single-top quark events were generated using the POWHEG-BOX v1 event generator. This event generator uses the four-flavor scheme for the NLO matrix elements calculations together with the fixed four-flavor PDF set CT10f4. For all processes involving top quarks, top quark spin correlations are preserved. In the t-channel, top quarks were decayed using MadSpin [24]. The parton shower, fragmentation, and the underlying event were simulated using PYTHIA v6.428 [241] with the CTEQ6L1 PDF set [224]. The h_{damp}[16] parameter in POWHEG, which controls the p_{T} of the first additional emission beyond the Born level and thus regulates the p_{T} of the recoil emission against the $t\bar{t}$ system, was set to the mass of the top quark (m_{top} = 172.5 GeV). All events with at least one leptonically decaying W boson are included. Single-top and $t\bar{t}$ events in which all top quarks decay hadronically do not contain sufficient $E_{\mathrm{T}}^{\mathrm{miss}}$ to contribute significantly to the background.

Smaller backgrounds in the signal region come from the production of $t\bar{t}$ pairs in association with $W/Z/h$ bosons and possibly additional jets, and production of $t\bar{t}t\bar{t}$, W/Z+jets, and $WW/WZ/ZZ$ (diboson) events. Other potential sources of background, such as the production of three top quarks or three gauge bosons, are expected to be negligible. The production of $t\bar{t}$ pairs in association with electroweak vector bosons W and Z was modeled by samples generated at LO using MADGRAPH5_aMC@NLO v2.2.2 and showered with PYTHIA v8.186, while samples to model $t\bar{t}H$ production were generated using MADGRAPH5_aMC@NLO v2.2.1 and showered with HERWIG++ [101] v2.7.1. These samples are described in detail in [74]. MADGRAPH5_aMC@NLO was also used to simulate the $t\bar{t}t\bar{t}$ production and the showering was performed with PYTHIA v8.186. The W/Z+jets processes were simulated using the SHERPA v2.2.0 [172] event generator, while SHERPA v2.1.1 was used to simulate diboson production processes. Matrix elements for the W/Z+jets and diboson processes were calculated using Comix [171] and

[16]This is a parameter that will be varied for theory systematics, as described in Sect. 7.7.

OpenLoops [128] and merged with the SHERPA parton shower [238] using the ME+PS@NLO prescription [182]. The SHERPA diboson sample cross-section was scaled down to account for its use of $\alpha_{\text{QED}} = 1/129$ rather than $\alpha_{\text{QED}} = 1/132$, corresponding to the use of current Particle Data Group [218] parameters, as input to the G_{μ} scheme [75]. Samples generated using MADGRAPH5_aMC@NLO v2.2.2 were produced with the NNPDF 2.3 PDF set and W/Z+jets samples were generated with the NNPDF 3.0 PDF set [103], while all other samples used CT10 PDFs.

All simulated event samples were passed through the full ATLAS detector simulation using GEANT4 [11]. The simulated events are reconstructed with the same algorithm as that used for data. For all samples, except the ones generated using SHERPA, the EVTGEN v1.2.0 program [200] was used to simulate the properties of the bottom- and charm-hadron decays. All PYTHIA v6.428 samples used the PERUGIA2012 [243] set of tuned parameters (tune) for the underlying event, while PYTHIA v8.186 and HERWIG++ showering were run with the A14 [47] and UEEE5 [169] underlying-event tunes, respectively. In-time and out-of-time pile-up interactions from the same or nearby bunch crossings were simulated by overlaying additional pp collisions generated by PYTHIA v8.186 using the A2 tune [33] and the MSTW2008LO parton distribution function set [208] on top of the hard scattering events. Details of the sample generation and normalization are summarized in Table 4.1. Additional samples with different event generators and settings are used to estimate systematic uncertainties in the backgrounds, as described in Sect. 7.7.

The signal samples are normalized using the best cross-section calculations at NLO in the strong coupling constant, adding the resummation of soft gluon emission at next-to-leading-logarithm (NLL) accuracy [111–113, 197, 198]. The nominal cross-section and the uncertainty are taken from an envelope of cross-section predictions using different PDF sets and factorization and renormalization scales, as described in [119]. The cross-section of gluino pair-production in these simplified models is $14 \pm 3\,\text{fb}^{-1}$ for a gluino mass of 1.5 TeV, falling to $1.0 \pm 0.3\,\text{fb}^{-1}$ for 2 TeV mass gluino. This is also summarized in Table 4.1.

Finally, contributions from multijet background are estimated from data using a procedure described in [43], which performs a smearing of the jet response in data events with well-measured $E_{\text{T}}^{\text{miss}}$ (the so-called seed events). The response function is derived in Monte Carlo dijet events and is different for b-tagged and non-b-tagged jets.

The specific list of samples used in the analysis are shown in Appendix F.

4.5 ATLAS Trigger System Phase-I Upgrade

The current L1Calo trigger system functions with relatively high-resolution identification of events with a wide variety of objects including electron, photons, tau-leptons, and jet objects, along with missing transverse energy. A detailed description of LAr Calorimeter Phase-I Upgrade design can be found in [12, 203]

Table 4.1 List of event generators used for the different processes

Process	Event Generator + fragmentation/hadronization	Tune	PDF set	Cross-section order
SUSY signal	MadGraph5_aMC@NLO v2.2.2 + Pythia v8.186	A14	NNPDF2.3	NLO+NLL [111–113, 119, 197, 198]
$t\bar{t}$	Powheg-Box v2 + Pythia v6.428	PERUGIA2012	CT10	NNLO+NNLL [153]
Single top	Powheg-Box v1 or v2 + Pythia v6.428	PERUGIA2012	CT10	NNLO+NNLL [191–193]
$t\bar{t}W/t\bar{t}Z$/4-tops	MadGraph5_aMC@NLO v2.2.2 + Pythia v8.186	A14	NNPDF2.3	NLO [155]
$t\bar{t}H$	MadGraph5_aMC@NLO v2.2.1 + Herwig++ v2.7.1	UEEE5	CT10	NLO [18]
Diboson WW, WZ, ZZ	Sherpa v2.1.1	Default	CT10	NLO [75]
W/Z+jets	Sherpa v2.2.0	Default	NNPDF3.0	NNLO [130]

Information is given about the underlying-event tunes, the PDF sets, and the pQCD highest-order accuracy used for the normalization of the different samples

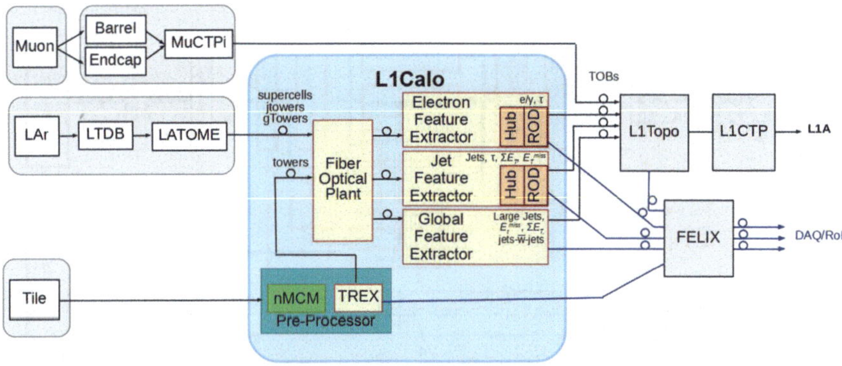

Fig. 4.3 The L1Calo system following the completion of the Phase-I upgrade at the start of Run 3 [114]. The new elements include the L1Calo PreProcessor Module (PPM), Optical Plant, Hub, ReadOut Driver (ROD), and the three FEXs: electron Feature EXtractor (eFEX), jet Feature EXtractor (jFEX), and gFEX

following the schedule of the LHC upgrade described in Sect. 3.2. In order to maintain a high trigger acceptance and trigger rate for these objects, LAr plans to provide finer granularity by means of super-cells which are up to $\Delta\eta \times \Delta\phi = 0.025 \times 0.1$ providing information for each calorimeter layer. The planned L1Calo Feature EXtractors (FEXs) will take advantage of this increased granularity.

Figure 4.3 shows the LAr Trigger Digitizer Boards (LTDBs) will digitize and transmit the input calorimeter signals to LATOME cards. The FPGA on each LATOME card will reconstruct the transverse energy of each super-cell and then transmits this information to each of the FEXs. Both jFEX and eFEX are meant to provide similar, yet improved functionality for the CPMs and JEMs while gFEX [114] is a completely new addition that will be added as part of the Phase-I upgrade. After a L1A, the LAr Digital Processing System (LDPS) and FEXs deliver output data to the TDAQ readout chain via Front-End Link EXchange (FELIX) [19], a multi-purpose routing device that interfaces the various ATLAS sub-detectors to the data acquisition system. In order to allow for appropriate commissioning of these new FEXs, the LTDB will also send the legacy trigger towers ($\Delta\eta \times \Delta\phi = 0.1 \times 0.1$) to the current L1Calo preprocessor and along the path described in Sect. 4.2.1.1.

In the following subsection, I describe how gFEX plans to fit in the system and Chicago's unique role in this forward-facing project for Phase-I.

4.5.1 The Global Feature Extractor Module

The gFEX concept was introduced in mid-2013 and I joined the team shortly. This is an on-going project and the physics motivations are presented in Chap. 9. A block diagram of the gFEX module is shown in Fig. 4.4 with a constructed board

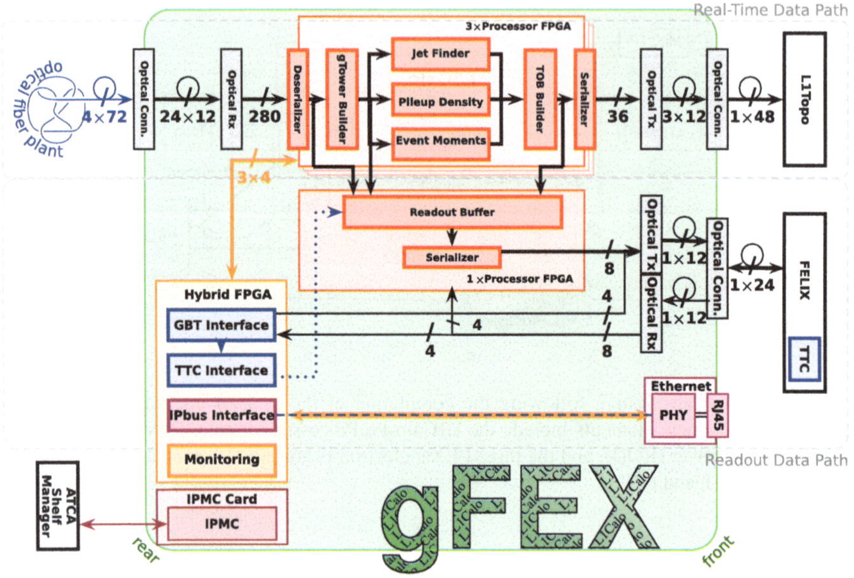

Fig. 4.4 A block diagram of the gFEX module [114]. Shown are the real-time (to L1Topo) and readout (to FELIX) data paths. The Zynq+® is also shown

in Fig. 4.5. A special feature of this subsystem is that it receives data from the entire calorimeter with a single electronics module. This maximizes trigger capability and flexibility for future trigger menus. The requirement of a single module imposes unique constraints on the design of this board. In order to be installed by the ATLAS detector and allow for reasonable temperature/power usage, up to four FPGAs can reasonably fit on the board.

There are three large Processor FPGAs (pFPGAs) for data processing and a Zynq+® from Xilinx that combines an FPGA and a CPU into a System-on-Chip (SoC). Chicago's role is to provide slow-control and monitoring of gFEX through the Zynq+®. The design is for each of the three pFPGAs to have 100 high-speed links[17] for input calorimeter data and output TOBs, ROIs, and calorimeter data. Unlike the design of other FEXs, gFEX allocates more fibers to carrying calorimeter output than for intra-FPGA communications. Each pFPGA has 2π azimuthal coverage for a given slice in η and executes all feature identification algorithms. Two pFPGAs handle the central $|\eta| < 2.5$ region receiving data from EMCal and hadronic calorimeter (HCal) with the third pFPGA handling the forward $2.5 < |\eta| < 4.9$ region receiving data from EMCal, HCal, and FCal.

The FPGA on the Zynq+®, known as the Zynq FPGA (zFPGA), is also employed in the calculation of global quantities. Towers from LATOME containing

[17] Also known as fibers.

Fig. 4.5 A picture of the gFEX v4 board at Brookhaven National Lab. The three pFPGAs have not been placed yet. The Zynq+® is placed with a heat-sink and fan on top in the lower-right. This board has successfully loaded the custom Linux kernel I built along with the "ironman" slow-control and monitoring software

Table 4.2 Latency profile of the real-time trigger path for the gFEX [114]

Latency (BC)	pFPGA	zFPGA
2.0	Receiver and deserialization	
1.0	Demultiplexing and synchronization	
5.0	Primitive processing (Algorithms)	
1.0	TOB selection for output	Global fragment transfers
1.0	TOB selection for output	Global TOB processing
1.0	Multiplexing	Multiplexing
2.0	Transmitter and serialization	Transmitter and serialization
2.0	Optical fiber to L1Topo (10 m)	Optical fiber to L1Topo (10 m)
15.0	Total latency for gFEX	

The maximum latency envelope is 15 BCs including transmission to the L1Topo

calorimeter data are $\Delta\eta \times \Delta\phi = 0.2 \times 0.2$, known as gCaloTowers, are summed across layers to form gTowers which are used as inputs to algorithms on the pFPGAs that need to run in 5 Bunch Crossings (BCs) latency.

Physics objects and observables are reconstructed by very fast fixed-latency algorithms in firmware running on the pFPGAs and zFPGA. These identified features are used in the L1 trigger decision. These algorithms have 5 BCs to run, out of a tentative total latency of 15 BCs given to gFEX for Run 3, as described in Table 4.2. These algorithms currently include tower building to form gTowers,[18] pile-up suppression, and calculations for: jet multiplicity, jet substructure, total and missing transverse energy. An illustration of gTower segmentation and gBlock formation is shown in Fig. 4.6.

More algorithms are being designed and tested through the entire gFEX development and commissioning process. More details on some of the physics studies being performed are shown in Chap. 9. After these algorithms finish processing, the zFPGA steps in the role of forming TOBs which contain global quantities, such as pile-up and missing transverse energy, and quantities related to jet candidates, such as jet multiplicity, found with the algorithms. These TOBs are sent to L1Topo for a decision. A trigger menu will be designed based on the rates of certain objects. For example, Fig. 4.7 shows a study of rates that I did for gFEX where there was an average of 80 and 200 proton–proton collisions per event, high luminosity and high pile-up.

For example, in order to obtain a 10 kHz rate with a single gTower trigger, a 80 GeV threshold needs to be set for events with an average of 80 proton–proton collisions, while a 95 GeV threshold needs to be set for events with an average of 200 proton–proton collisions.

[18]While gTowers are useful from a physics perspective, latency requirements forces us to be a little bit clever about the design of the algorithms on firmware. A contiguous group of gTowers are formed into gBlocks, as long as the scalar sum of gTower energy is greater than a threshold, which are used as common inputs to the algorithm firmware.

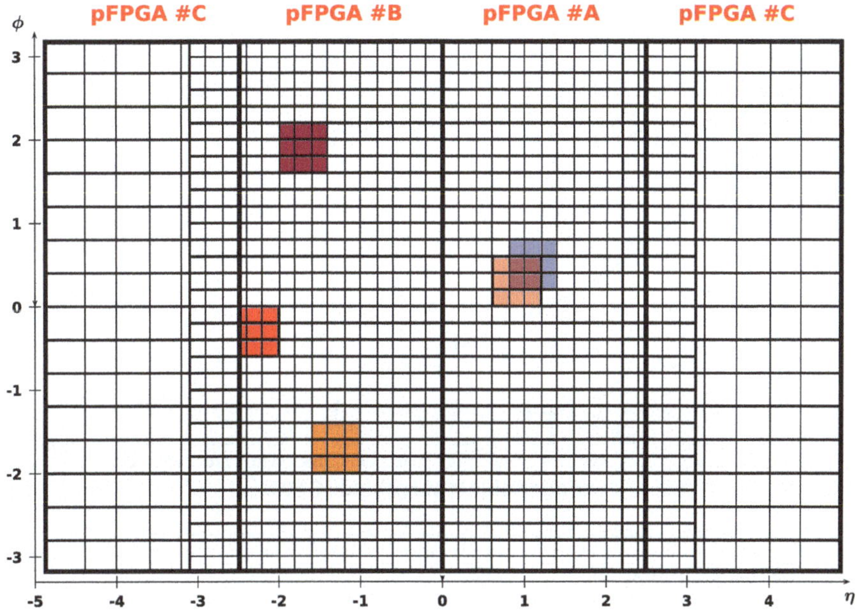

Fig. 4.6 gTower segmentation in the calorimeter [114]. Note the special gTower coverage at $2.4 < |\eta| < 2.5$ and $3.1 < |\eta| < 3.2$. The central regions covered by pFPGA 1 and pFPGA 2 contain primarily gTowers of size up to $\Delta\eta \times \Delta\phi = 0.2 \times 0.2$ and the forward regions have up to $\Delta\eta \times \Delta\phi = 0.4 \times 0.4$. gBlock formation is also illustrated in colors here as contiguous blocks of gTowers, often 3×3. Note that gBlocks in an event are allowed to overlap. Also note that not all gBlocks have the same size

If one were to design a trigger menu based on these results from gFEX for $\langle\mu\rangle = 200$, `L1_GFEX_1T95`, `L1_GFEX_2T55`, and `L1_GFEX_4B35` are possible items providing a 10 kHz rate. This study is preliminary and further work needs to be done to understand the rates of the TOBs provided by gFEX for commissioning and trigger menu definition. Some studies on the efficiency of the trigger are provided in Chap. 9.

4.5.2 Slow Control and Monitoring of gFEX

In order to ensure that the gFEX module functions properly through the entirety of an experiment, it is important to be able to flag problematic data and to detect when the board is under excessive load before it affects other subsystems and the experiment as a whole. The gFEX has a complete snapshot of the calorimeter information for each event and a Zynq+® capable of processing this data. Calorimeter information can be sent[19] to the Zynq+® for further processing based upon:

[19] At a rate less than the bunch crossing rate of 40 MHz.

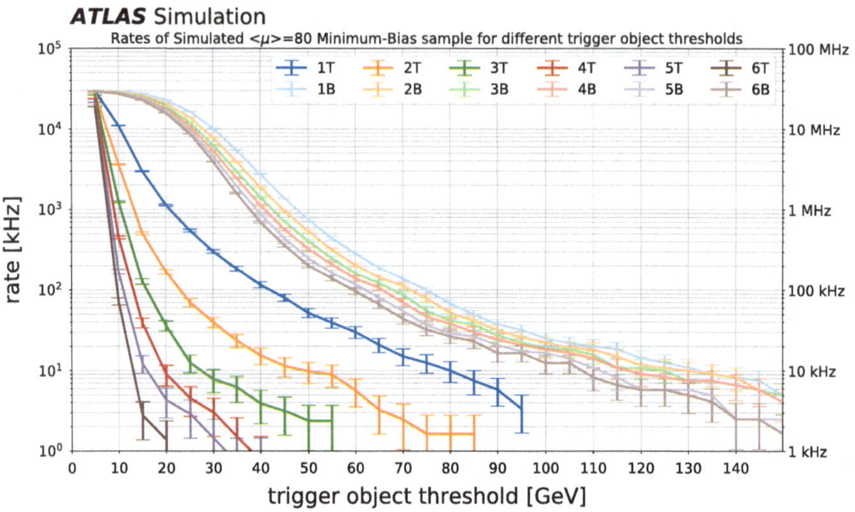

(a)

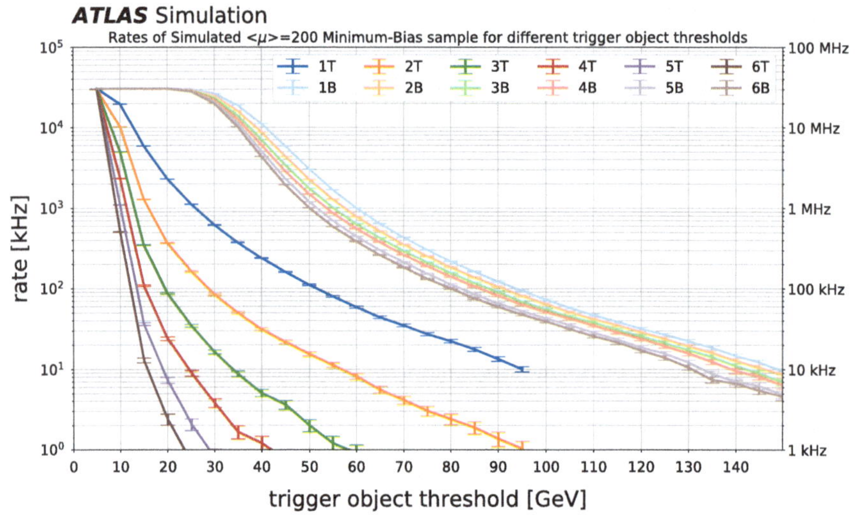

(b)

Fig. 4.7 gFEX rates shown for a simulated sample with (**a**) 80 and (**b**) 200 proton–proton interactions, representing potential data events for HL-LHC. Each curve corresponds to a different trigger item requiring one, two, or more gFEX trigger objects (gTowers [dark curves], gBlocks [light curves]) with the x-axis representing the energy threshold required for the trigger object and the y-axis showing the potential rate for the given selection. A lower threshold can provide a highly efficient trigger while maintaining rates based on the readout capabilities of the ATLAS detector

- an error flag from the calorimeter data (such as a failed checksum)
- a signal or flag from the FEX algorithms
- a period clock
- an external signal

This last bullet point is of note as this external signal is a command sent from the ATLAS control room over a networking interface. The monitoring framework will be capable of sampling the data to provide reports of the health of gFEX at various levels. For example, the health of the board can be reported through histograms of calorimeter channels with errors; errors in an event can be detected by emulating the FEX algorithms on the CPU of the Zynq+®; or time-dependent errors such as flagging errors correlated amongst many BCs.

Having a SoC provides many benefits by allowing an operating system to be loaded directly on gFEX and programmed using modern programming languages, such as Python, to provide a maintainable and flexible interface. Unlike the firmware written for FPGA which requires updates and changes when using a different chip, software remains largely the same as the kernel loads and unloads the necessary drivers to interface with the different hardware. It is important to gFEX to switch hardware as the full module and final design is realized as a series of incrementally improved boards for testing different components. Since joining gFEX, I have developed "meta-l1calo"—a full suite of tools for compiling a Linux kernel from scratch incorporated with specific tools such as Python and I2C (I^2C) drivers; and ironman—an open-sourced, single-threaded monitoring framework written in Python to be used on SoCs in L1Calo to connect control and monitoring requests with hardware in a transport-neutral way. The technical details of "meta-l1calo" and ironman are described in Appendix C. Because of this flexibility, you can customize the gFEX to present itself as a representational state transfer application programming interface (RESTful API), borrowing from modern techniques used to power the internet today, such as querying for monitoring data to be populated by navigating with your browser as demonstrated in Fig. 4.8. This software has been successfully deployed on multiple iterations of gFEX boards and can reduce the learning curve for slow-control and monitoring on custom SoCs.

As ironman is single-threaded, one can take advantage of Multi-Processor SoCs (MPSoCs) like Zynq+® to run different, multiple instances of ironman can run in parallel with different functionalities. An instance of ironman can run slow-control for the I^2C and clock configuration, separately from an instance running monitoring of the on-board temperature and power usage. This separation of concerns and modularity allows the board to function for the next 20 years with minimal expert intervention.

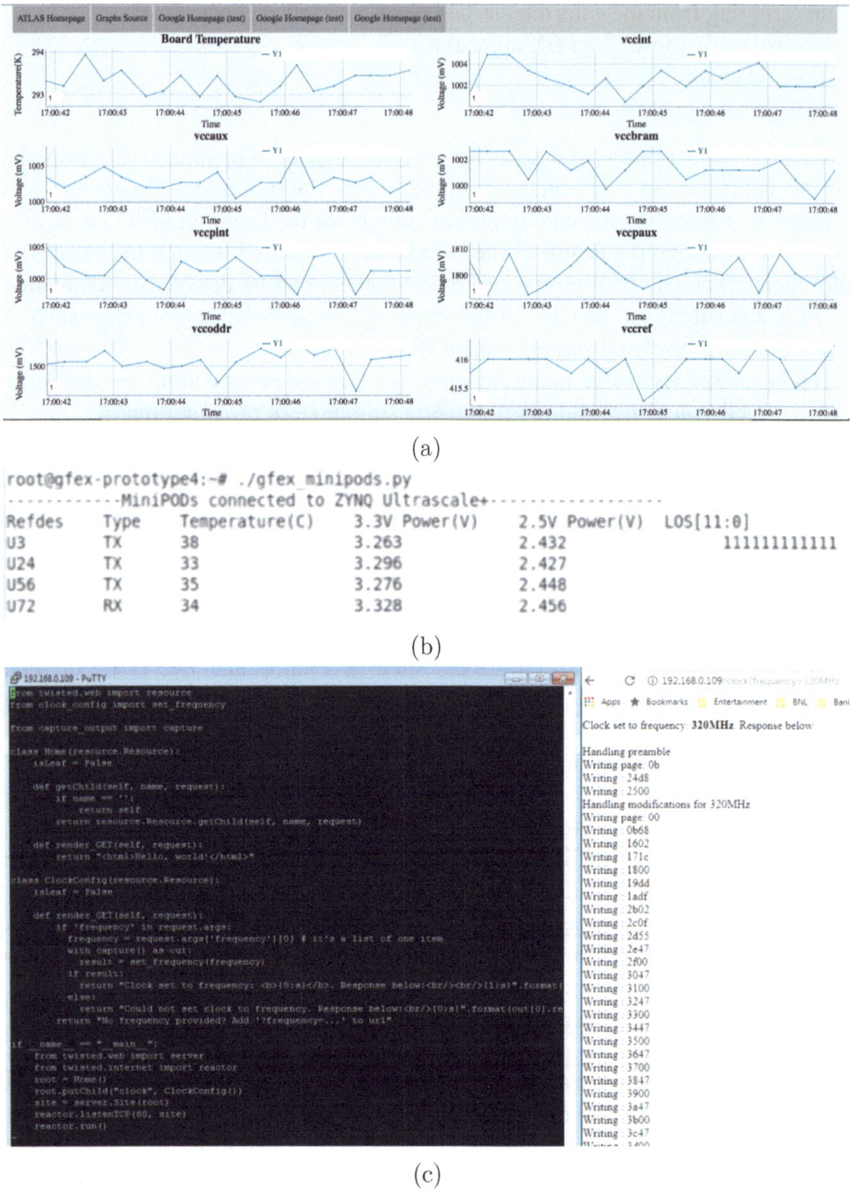

Fig. 4.8 Demonstration of `ironman`'s flexibility and adaptability to various programming languages and data transfer models. The flexibility is demonstrated by (**a**) a website with JavaScript polling the board, (**b**) direct access over SSH with Python, and (**c**) a RESTful server running on the board

4.5.3 *Trigger-Aware Analysis Software*

An integral part of introducing instrumentation to the ATLAS detector is having the ability to monitor the gFEX performance in order to spot trouble before it affects the experiment as a whole. The gFEX can also be used to monitor actual physics, for example, we can keep track of when the missing transverse energy is higher than normal for several sequential bunch crossings. Monitoring is controlled by the Zynq+® which receives and process the calorimeter data and interacts with the other components of the L1Calo system. The Zynq+® will also interact with non-L1Calo systems like external servers for logging.

Trigger-aware analysis has not yet been implemented. The readout size for every gTower is approximately 16.4 kB/event (15 bit $\times$ 1120 gTowers). Only reading out the TOBs would be less than 1.75 kB/event. This information could be read out directly to a dedicated stream for analysis of the gFEX TOBs offline.

Chapter 5
Event Reconstruction

The goal of particle physics experiments is to reconstruct and measure the outgoing particles produced in proton–proton collisions to describe the hard scatter process. After an event is accepted by the ATLAS trigger systems to be recorded to disk, the objects of interest such as electrons, muons, and jets must be reconstructed from the low-level detector signals. These complex objects, meant to be representative of the true SM particle, are built from some of the low-level detector signals, such as muon spectrometer tracks or energy depositions in the electromagnetic or hadronic calorimeters. As the LHC is a hadron collider, the LHC tends to produce colored final states through the collisions of gluons. Many BSM physics models contain these hadronic objects which are crucial to reconstruct accurately, amidst the initial and final state radiation and multiple simultaneous proton–proton collisions. Once reconstructed, the measured properties of these objects may be calibrated to a particular energy scale.

This chapter describes in detail the methods of standard ATLAS event reconstruction used for jets (Sect. 5.1), *b*-jets (Sect. 5.2), muons (Sect. 5.3), electrons and photons (Sect. 5.4), taus (Sect. 5.5), and missing transverse momentum (Sect. 5.6).

5.1 Jets

The first question with a particle physics detector is how to measure the hadronic final state. The difficult in reconstructing quark and gluons is because, due to the nature of QCD described in Sect. 2.1.3, colored particles cannot be observed directly in the detector. Quarks and gluons get "smeared" by the showering process and "obscured" through the hadronization process. The solution is to build objects called **jets**, the name for collimated sprays of particles produced by quarks and gluons as they shower and hadronize. The final step to connect theory in Sect. 2.1.3 to

© Springer Nature Switzerland AG 2020

G. Stark, *The Search for Supersymmetry in Hadronic Final States Using Boosted Object Reconstruction*, Springer Theses,
https://doi.org/10.1007/978-3-030-34548-8_5

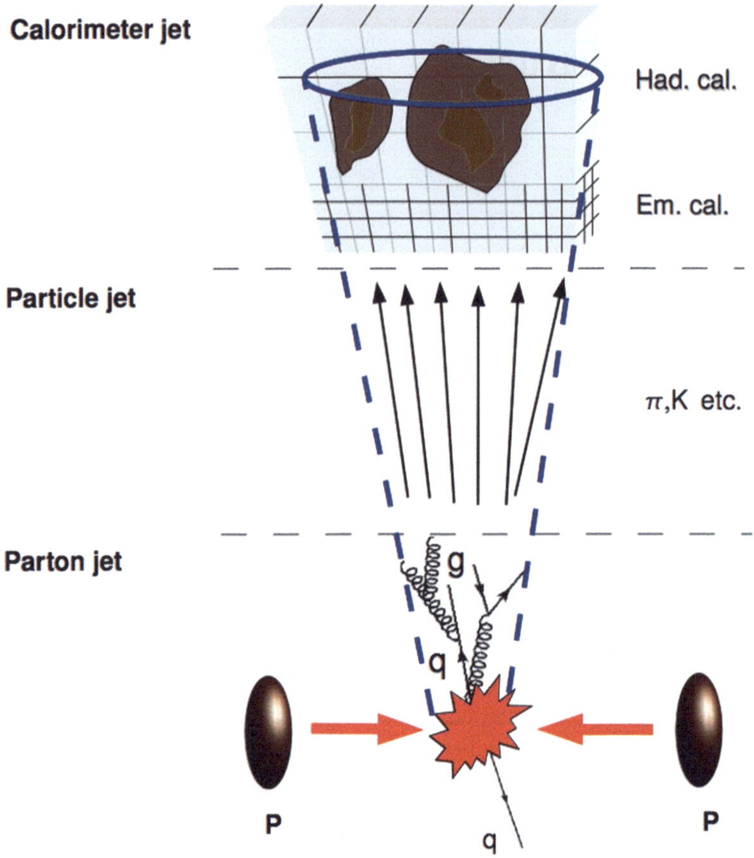

Fig. 5.1 The evolution of the partonic system which demonstrates how jets are formed and measured by the calorimeters in ATLAS [255]

the calorimeters in Sect. 3.7 in order to measure jets are to cluster inputs from the calorimeter as shown in Fig. 5.1. Proton–proton collisions produce partons which shower (parton jet) and then hadronize (particle jet) into colorless objects which deposit their energies in the HCal and EMCal.

From these energy depositions, one can group them up to form jets, as shown in the candidate event from my SUSY search (Fig. 5.2). In the event display, the six jets are represented by cones around the groups calorimeter energy deposits, but this is where things start getting tricky. How do we deal with the underlying event that consists of initial and final state radiation, of pile-up, of the multiple, simultaneous proton–proton collisions that can obscure the physics of the hard scatter? How can we start to identify more than two or three jets cleanly even as we move to a higher luminosity with more proton–proton interactions per BC?

The rest of this section is dedicated to describing the goal of jet clustering algorithms (Sect. 5.1.1), calibrating the energy of jets (Sect. 5.1.2), characterizing

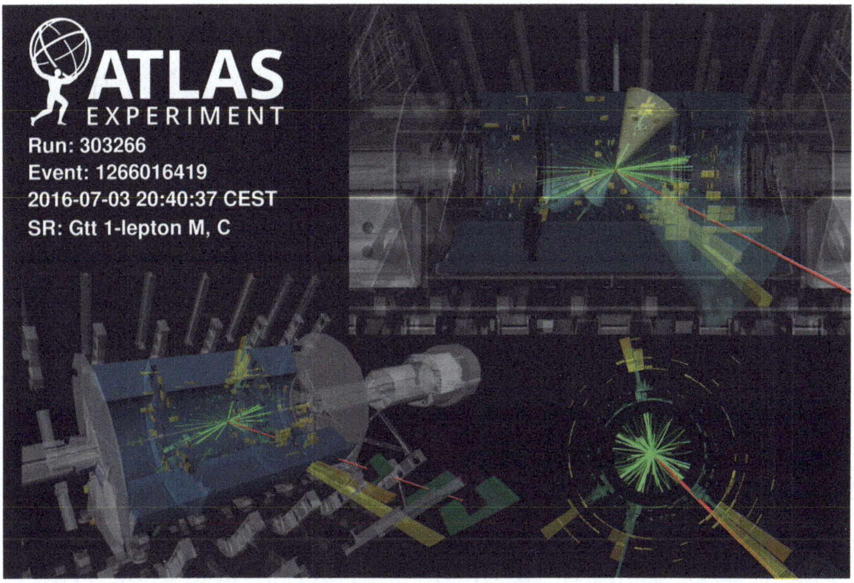

Fig. 5.2 A candidate supersymmetry event with 6 jets (shown as cones) and a muon track which is colored red [82]. The b-quark tagged jets are colored blue, while the non-b-quark tagged jets are colored yellow. The size of the jet cones are proportional to the measured and calibrated p_T of the jet. ID tracks with $p_T > 1$ GeV are colored green and their brightness is proportional to their p_T. The LAr and Tile are colored teal and yellow, respectively, and their length is proportional to the measured transverse energy deposit

the uncertainty of the jet calibrations (Sect. 5.1.4), and some kinematic properties of jets at $\sqrt{s} = 13$ TeV (Sect. 5.1.5).

5.1.1 Jet Algorithms

The shower of particles in the calorimeter originating from the fragmentation and hadronization of quarks and gluons produces objects known as jets. However, jets are not unique, and are instead defined based on the clustering algorithm, its parameters, and its inputs. In the Snowmass Accords of 1990 [185], there was a first attempt to define a set of requirements jet algorithms needed to fulfill including:

- simple to implement in an experimental analysis;
- simple to implement in the theoretical calculation;
- defined at any order of perturbation theory;
- yields finite cross-section at any order of perturbation theory;
- yield s cross-section that is insensitive to hadronization.

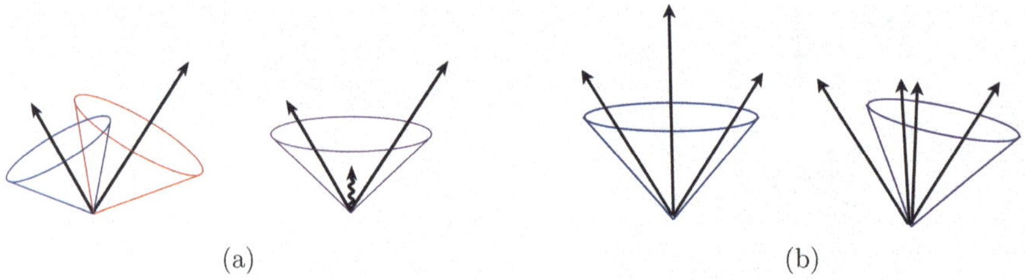

Fig. 5.3 Illustration of the (a) infrared sensitivity and (b) collinear sensitivity of a cursory designed jet algorithm [188]

Over the past 25 years, many algorithms have come and gone. The so-called cone algorithms [236] have become the most popular option by experimental physicists [239]. Typically, these algorithms collect all particles within some stable, seeded cone defined by a size parameter R, the cross-sectional area of the cone in the η-ϕ plane. These "simple" cone algorithms were used with mixed success as they did manage to satisfy many of the requirements for jet algorithms; however, they were found to not be **infrared-safe** (IR-safe) and **collinear-safe** as shown in Fig. 5.3. IR-safety is a theoretical guideline that adding or removing soft particles should not change the result of jet clustering. In practice, the underlying event and pile-up activity should not affect the jet final state in hard scattering. Collinear safety states that the splitting of a large p_{T} particle into two collinear particles should not affect the jet finding or the number of jets identified.

For example, a seeded cone algorithm fails to be collinear-safe, as the choice of seed is often the one with the most energy, and accidental collinear fragmentation can suppress the seed. Similarly, cone algorithms that varied the center of the jet as it clustered more inputs together would not be IR-safe as the addition of soft gluon radiation could shift the center enough to overlap with an existing jet. Both IR-safety and collinear-safety have dictated the guidelines for the next generation of cone algorithms. By changing focus from seeds and the cones around seeds, to relationships between close-by inputs, one can satisfy both of the safety requirements. Instead of identifying inputs based on their distance to a seed, identify pairs of inputs based on their distance parameter. If two inputs have a distance parameter below some threshold, they are merged and the process continues until no merges are possible [236]. This was able to solve the safety requirements by merging together both collinear particles as well as soft particles. This next generation, known as sequential recombination algorithms, was defined in six steps depending on two parameters, R (size parameter) and P (algorithm choice, explained further below):

1. For each protojet i, define the **jet-beam distance measure**,

$$d_i = p_{\mathrm{T},i}^{2P}. \tag{5.1}$$

2. For each pair of protojets $i, j \forall i \neq j$, define the **jet-jet distance**,

$$d_{i,j} = \min(d_i, d_j) \left(\frac{\Delta R_{i,j}^2}{R^2} \right). \tag{5.2}$$

3. Minimize the set of d_i and $d_{i,j}$ so far and call this $d_{\min}$.
4. If $d_{\min}$ belongs to a protojet i (the set of d_i), then the protojet is not mergeable, removed as an input, and is defined as a jet.
5. If $d_{\min}$ belongs to the distance between protojets i, j (the set of $d_{i,j}$), then the two protojets are removed, merged into a single protojet k, and added as an input.
6. This repeats until there are no more remaining protojets i.

The choice of the parameter P corresponds with the choice of particular algorithm which is applied.

$P = 1$ This defines the k_t algorithm [129]. Softest protojets are clusters first into harder ones. As soft radiation is prioritized, this algorithm can be susceptible to energy fluctuations from pile-up and detector noise. This typically forms irregularly shaped jets.

$P = 0$ This defines the Cambridge-Aachen (C/A) algorithm [158]. This omits all trace of p_T-dependence in the clustering and depends only on angular properties. This was still susceptible to some of the same problems at the k_t algorithm, being sensitive to soft gluon radiation. Also like k_t algorithm, this typically forms irregularly shaped jets as well.

$P = 1$ This defines the anti-k_t algorithm [125, 126]. This algorithm prefers the hardest protojets and clusters them first. This is similar to the cone algorithms because it prefers hardest protojets which are seed-like.

The `FastJet` software package [126] contains the technical execution of the jet clustering algorithms and was able to reduce the complexity of jet finding to $\mathcal{O}(n \ln n)$ for n inputs [123]. Figure 5.4 depicts the three different algorithm choices for $P = 0, \pm 1$. For this thesis, the anti-k_t algorithm is used with an R parameter $R = 0.4$ which is a typical parameter for small-radius jets in ATLAS. Larger jets are typically $R = 1.0$ or $R = 1.2$, are more sensitive to pile-up contributions, but are not used in this analysis.

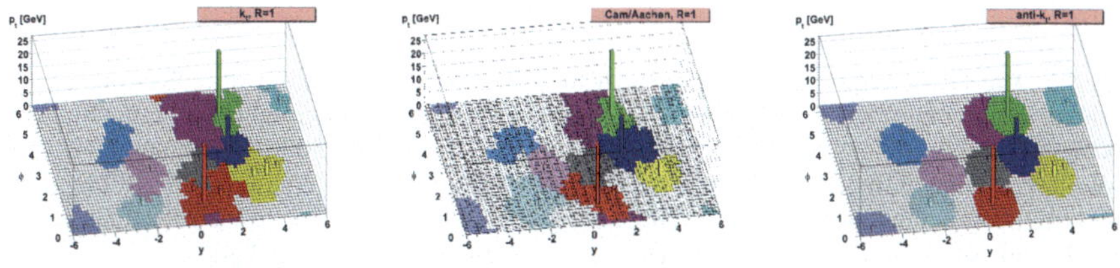

(a) (b) (c)

Fig. 5.4 A sample parton-level Monte Carlo simulated event illustrating the shapes and areas of the resulting hard jets for $R = 1.0$. (**a**) $P = 1$, k_t, (**b**) $P = 0$, C/A, and (**c**) $P = -1$, anti-k_t [125]

The last consideration is about the inputs to these jet algorithms which form different kinds of jet collections for a given algorithm. For example, to produce **truth jets**, inputs are simulated particles by a Monte Carlo generator. These are generally used to study the performance of algorithms without detector effects and to calibrate/study the resolution of other reconstructed jets. Jets can be formed from tracks, **track jets**, inputs are the hits in the ID which correspond to trajectories of charged particles. Another set of jets are **calorimeter jets**,[1] inputs are energy deposits left in the calorimeter for both neutral and charged particles.

Given the computing budget for ATLAS and the ability to do offline reconstruction quickly for analyses, using the energy measurements in the calorimeter at the cell level, readouts, is computationally intensive. Calorimeter cells are also very sensitive to pile-up and a single quark could shower over many cells. In pre-2011 ATLAS operations, the cell noise was almost entirely electronic noise. Now that pile-up is significant, the noise contribution from pile-up is also dominating as shown in Fig. 5.5 and summarized in Eq. (5.3) [204].

$$\sigma_{\text{noise}} = \begin{cases} \sigma_{\text{noise}}^{\text{electronic}} & \text{(2010 and prior)} \\ \sigma_{\text{noise}}^{\text{electronic}} \oplus \sigma_{\text{noise}}^{\text{pile-up}} & \text{(2011+)} \end{cases} \tag{5.3}$$

In Eq. (5.3), $\sigma_{\text{noise}}^{\text{electronic}}$ is the electronic noise and $\sigma_{\text{noise}}^{\text{pile-up}}$ is the noise from pile-up determined with Monte Carlo simulations for $\langle \mu \rangle \geq 8$.

Given this noise contribution, calorimeter cells were clustered together to form **Three-dimensional topological clusters** (or topoclusters, for short) using an algorithm designed to maintain a certain amount of cell signal to the average expected noise in the electromagnetic calorimeters [86]. These topoclusters are seeded by

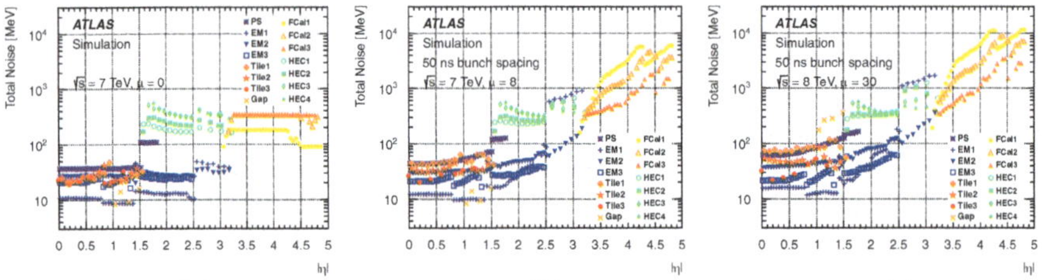

(a) (b) (c)

Fig. 5.5 The energy-equivalent cell noise in the ATLAS calorimeters on the electromagnetic scale as a function of $|\eta|$ in the detector for the (**a**) 2010 configuration with $\langle \mu \rangle = 0$, (**b**) 2011 configuration with $\langle \mu \rangle = 8$, and (**c**) 2012 configuration with $\langle \mu \rangle = 30$ [86]. The various colors indicate the noise in the first layer of the FCal and the three layers of the LAr EMCal, the three layers of the Tile, the four layers of the HEC, and the three modules of the FCal. The higher levels in 2011 and 2012 compared to 2010 indicate the contribution from increased pile-up

[1]Colloquially known as "reco jets," short for reconstructed jets.

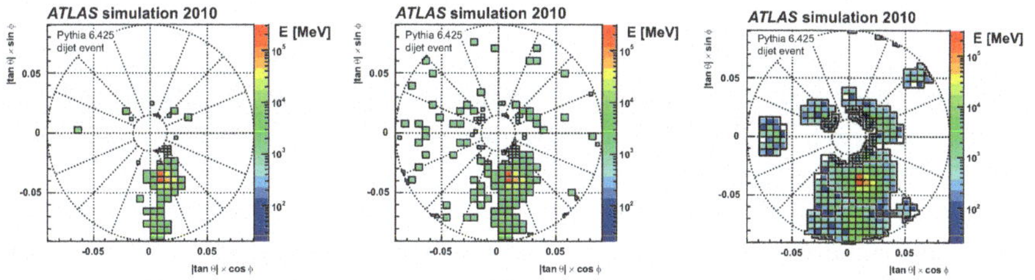

(a) (b) (c)

Fig. 5.6 Stages of topocluster formation in the first module of the FCal for a simulated dijet event with at least one jet entering the calorimeter is shown for $\sqrt{s} = 8\,\text{TeV}$ [86]. Shown are (**a**) seed cells for topoclustering, $S > 4$, (**b**) growth cells, $N > 2$, and (**c**) boundary cells, $P > 0$. Cells which are not colored, but contained inside a topocluster boundary are cells with negative energy

cells with a large signal-to-noise ratio,[2] S, and then these seeds are summed with the scalar energy measured in adjacent cells, N, up to a boundary condition, P. This algorithm is shown in Eq. (5.4) and the three particular parameters $S > 4$, $N > 2$, $P > 0$ were chosen and optimized using studies with ATLAS test-beam data [204].

$$\left|E_{\text{cell}}^{\text{EM}}\right| > S\sigma_{\text{noise, cell}}^{\text{EM}} \quad \text{(primary seed threshold)} \tag{5.4a}$$

$$\left|E_{\text{cell}}^{\text{EM}}\right| > N\sigma_{\text{noise, cell}}^{\text{EM}} \quad \text{(threshold for growth control)} \tag{5.4b}$$

$$\left|E_{\text{cell}}^{\text{EM}}\right| > P\sigma_{\text{noise, cell}}^{\text{EM}} \quad \text{(boundary cell filter)} \tag{5.4c}$$

Figure 5.6 shows the three stages of topoclustering described by Eq. (5.4). Since negative energies are allowed to join topoclusters, primarily caused by the pulse shaping of the LAr calorimeter and caused by pile-up, are expected to partially cancel the positive fluctuations by pile-up. Finally, topoclustering also improves the performance of the calorimeter by suppressing isolated fluctuations due to pile-up and electronic noise.

5.1.2 Jet Calibrations

The jets formed from topoclusters are reconstructed at the electromagnetic scale as described in Sect. 5.1.2.1 are still not ready for use by analyses. A series of further corrections are derived from both Monte Carlo simulation and data[3] to account for

[2]Here, the noise is the expected noise σ_{noise} in Eq. (5.3).

[3]Data-derived corrections are called **in-situ** corrections.

the non-compensating[4] nature of the ATLAS calorimeters [61, 62]. Calibrations derived using both Monte Carlo and data are applied in sequence as described in Sect. 5.1.3 in order to get the energy **scale** of the jet as close to the scale of the particle. Section 5.1.4 describes the systematics and uncertainties associated with these corrections and calibrations that need to be considered by analyses using the calibrated jets. Note that MC simulation which is built on test beam data, prior collision data, and theoretical considerations is not a perfect description of the detector response; hence, the need for in-situ calibrations after all other considerations.

5.1.2.1 Topocluster Calibration

As topoclusters are used as inputs for the anti-k_t clustering algorithm in this analysis, the first step is to calibrate the topoclusters. The intention of the calibration scheme for topoclusters is to provide a calorimeter signal for physics object reconstruction that is agnostic to the kind of object being reconstructed. In addition, the signal from hadronically interacting particles will always be smaller than the signal from electromagnetically interacting particles depositing the same amount of energy due to the non-compensating nature of the hadronic calorimeters. Finally, one needs to account for energy losses due to dead material and inefficiencies of the clustering procedure itself. Therefore, a weight is assigned to each calorimeter cell based on the probability $P_{\mathrm{clus}}^{\mathrm{EM}}$ of the topocluster to have been generated by an electromagnetic shower, using the kinematics of the topocluster:

$$w_{\mathrm{cell}}^{\mathrm{cal}} = P_{\mathrm{clus}}^{\mathrm{EM}} \cdot \underbrace{1}_{w_{\mathrm{cell}}^{\mathrm{EM}}=1} + \left(1 - P_{\mathrm{clus}}^{\mathrm{EM}}\right) \cdot \underbrace{\frac{E_{\mathrm{cell}}^{\mathrm{dep}}}{E_{\mathrm{cell}}^{\mathrm{EM}}}}_{w_{\mathrm{cell}}^{\mathrm{Had}}} . \tag{5.5}$$

By default, all topoclusters are reconstructed at the electromagnetic scale,[5] so $w_{\mathrm{cell}}^{\mathrm{EM}} = 1$. $w_{\mathrm{cell}}^{\mathrm{Had}}$ is the ratio of the energy deposited per cell to the same energy reconstructed at the electromagnetic scale. Neutral pion showers, $\pi^0 \to \gamma\gamma$, are used to calibrate the electromagnetic likelihood; positively charged pions, $\pi^+ \to \pi^0 + e^+ + \nu_e$, are used to train the hadronic likelihood.

For Run I in ATLAS, $0 < P_{\mathrm{clus}}^{\mathrm{EM}} < 1$ and jets built from topoclusters were known as LCW jets (local cell weighting).[6] In Run 2, the default jets were EM jets[7] built from topoclusters at the electromagnetic scale with $P_{\mathrm{clus}}^{\mathrm{EM}} = 1$.

[4]The energy measured by the detectors is not the full energy of the particle that is being detected/measured.

[5]What this means is that the measured signal from the electromagnetic cells and hadronic cells are used with no other cell-level corrections.

[6]Colloquially called "LCTopo."

[7]Colloquially called "EMTopo."

5.1.3 Jet Energy Calibration

The jet calibration procedure summarized in Fig. 5.7 is intended to correct for several detector effects that affect the jet energy measurement:

- dead material: energy lost in the dead material of the detector, e.g. inactive absorbers and instrumentation
- non-compensation: difference in detector response between hadrons, leptons, and photons; specifically the response is lower for hadrons
- punch-through: energy leakage where the hadron showers deposit energy outside of the calorimetry system
- pile-up: energy originating from additional proton–proton collisions in the detector (part of the underlying event which includes initial and final state radiation)
- minimum threshold: hardware limits on energy deposits
- out-of-cone: inefficiencies in reconstruction due to not capturing the full particle shower in the jet

The following subsections briefly describe each stage, more detailed information is found in [87].

5.1.3.1 Jet Origin Correction

In this stage, the jet direction is corrected for the difference between the default ATLAS detector origin, the geometric center of the detector, and the actual position of the primary proton–proton interaction. In reality, particles point back to vertices at the main interaction point. Therefore the primary vertex of the event, especially in the case of multiple proton–proton interactions, is identified by the vertex consisting of the largest $\sum \vec{p}_{\mathrm{T}}^{\,2}$ of tracks and the origin of the jet is corrected to point back towards this primary vertex. This correction improves the angular resolution of jets

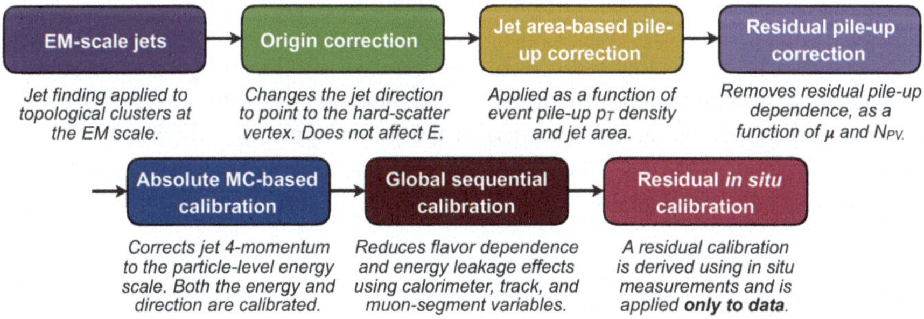

Fig. 5.7 Calibration stages for EM-scale jets [87]. Other than the origin correction, each stage of the calibration is applied to the four-momentum of the jet

with a small effect on jet p_T. After this correction, the jets are said to be at the **origin corrected scale**.

5.1.3.2 Pile-Up Correction

In this stage, the average additional energy due to multiple proton–proton interactions is subtracted from the jet using an area-based subtraction scheme [124]. The average energy density is calculated using k_t $R = 0.4$ jets described in Eq. (5.6) as the median energy density calculated from the area A^i and transverse momentum p_T^i of each jet i.

$$\rho = \text{median} \left(\frac{p_T^i}{A^i} \right) \Bigg|_{k_t \text{ jets}} . \tag{5.6}$$

ρ represents the pile-up energy density of the calorimeter. Note an interesting feature is that while topoclustering mitigates pile-up correction partially by allowing for negative-energy cells to cancel out positive-energy cells, the k_t clustering only permits positive-energy topoclusters in the calculation of pile-up energy density. The inclusion of negative energies in the calculation of pile-up is an interesting topic that will be explored more in Chap. 9.

The p_T of each jet is corrected by a variety of factors shown in Eq. (5.7). One factor is to estimate the amount of pile-up in a jet of area A using $\rho \times A$; one factor is a simulation-based residual correction based on the position of the jet in pseudorapidity $\alpha(\eta)$ and number of primary vertices N_{PV}; and one factor is based on the average number of interactions per bunch crossing μ for the position of the jet in pseudorapidity $\beta(\eta)$. The different functions $\alpha(\eta)$ and $\beta(\eta)$ are determined from simulation and allows for both in-time and out-of-time pile-up effects to be accounted for as shown in Fig. 5.8. The residual correction factors α and β are taken from a fit to the purple bands.

$$p_T^{\text{corr}} = p_T^{\text{EM}} - \rho \times A - \alpha(\eta) \times (N_{PV} - 1) - \beta(\eta) \times \mu \tag{5.7}$$

Notice that there are four sources of uncertainty introduced by this calculation by N_{PV}, μ, ρ, and an uncertainty due to the p_T-dependence of the correction itself. These uncertainties are included as part of the jet energy scale (JES) uncertainties discussed in Sect. 5.1.4. After the correction is applied to jets, the jets are said to be at the **pile-up corrected scale**.

5.1.3.3 MC-Based Correction

This correction is also known as the JES which is meant to correct the response of the jet energy and jet η in the detector back to the truth level. A correction is

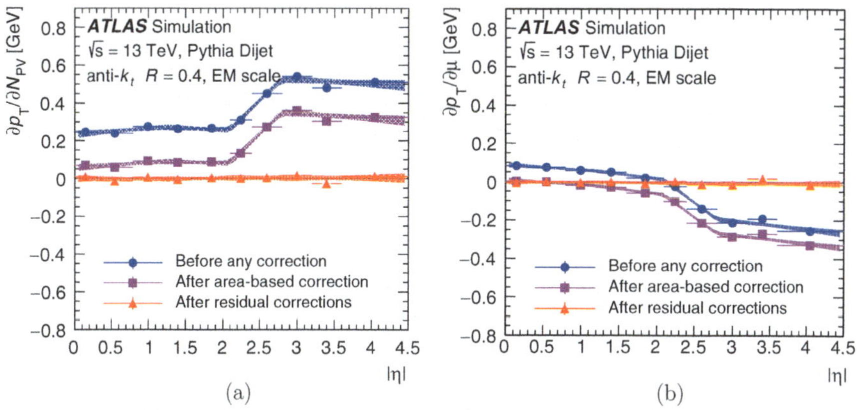

Fig. 5.8 Dependence of the EM-scale anti-k_t jet p_{T} on (**a**) in-time pile-up and (**b**) out-of-time pile-up as a function of $|\eta|$ for $p_{\mathrm{T}}^{\mathrm{truth}} = 25\,\mathrm{GeV}$ [87]. The dependence is shown before pile-up corrections (blue), after area-based pile-up correction (purple), and after residual corrections (red) using N_{PV} and μ

derived from truth information in Monte Carlo simulations in both p_{T} and η, due to both the non-compensating nature of the ATLAS calorimeters and the changing geometry as a function of η. Reconstructed jets are first required to be isolated which requires that the minimum ΔR between pairs of jets is no less than $2.5 \times R$. In events with isolated jets, the reconstructed jets are matched to the corresponding truth jet using a ΔR association requiring that the truth jet and reconstructed jet are less than $0.75 \times R$. If a reconstructed jet has no matching truth jet, it is discarded.

$$\mathcal{R}^{\mathrm{jet}} = \frac{E_{\mathrm{reco}}^{\mathrm{jet}}}{E_{\mathrm{truth}}^{\mathrm{jet}}} \tag{5.8}$$

The jet energy response (see Eq. (5.8)) of isolated, reconstructed jets in QCD multijet simulation is binned in energy of the matched truth jet $E_{\mathrm{truth}}^{\mathrm{jet}}$ [8] (see Fig. 5.9) and pseudorapidity η_{det} in the detector. Each bin produces a Gaussian distribution which is fit and the mean value is extracted, $\langle \mathcal{R}^{\mathrm{jet}} \rangle$. This peak value is used to transform $\langle E_{\mathrm{truth}}^{\mathrm{jet}} \rangle \mapsto \langle E_{\mathrm{reco}}^{\mathrm{jet}} \rangle$ for each η_{det} bin, known as the "numerical inversion" procedure to derive corrections in reconstructed jets from truth jets.

$$E_{\mathrm{EM+JES}}^{\mathrm{jet}} = \frac{1}{\mathcal{F}\left(E_{\mathrm{reco}}^{\mathrm{jet}}\right)} E_{\mathrm{EM}}^{\mathrm{jet}} \tag{5.9}$$

[8] We bin in truth jet energy, rather than reco-jet energy to remove a dependence of the calibration on the reco-jet p_{T} spectrum which includes detector-level effects that almost certainly introduce a bias.

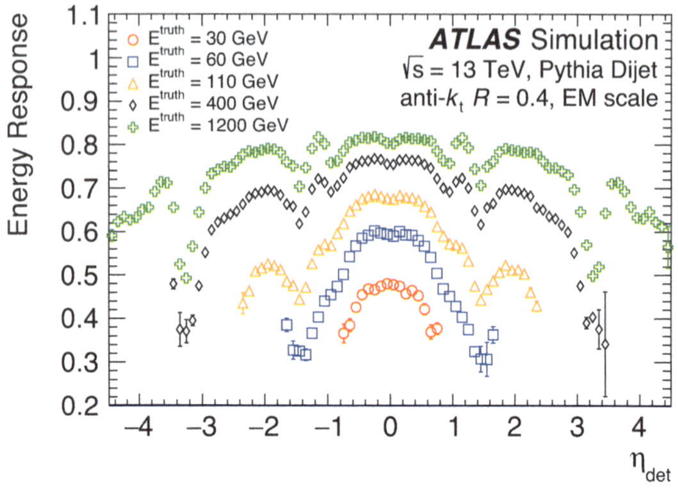

Fig. 5.9 The average jet energy response as a function of η_{det} for jets of a truth energy of 30, 60, 110, 400, and 1200 GeV[87]. The energy response is shown after origin and pile-up corrections are applied. This shows the size of the calibration constants for jets built from topoclusters at the EM-scale

Finally, each entire η bin is fit to $\langle \mathcal{R}^{\mathrm{jet}} \rangle$, $\langle E^{\mathrm{jet}}_{\mathrm{reco}} \rangle$ to produce the jet calibration function $\mathcal{F}_{\mathrm{calib}}(E^{\mathrm{jet}}_{\mathrm{reco}})$ which is inverted to obtain the correction (see Eq. (5.9)) and bring the jets to the **EM+JES scale**. Figure 5.10 shows the corrections done on the jet η derived in a similar fashion, but with the response defined as $\mathcal{R}^{\mathrm{jet}} = \eta^{\mathrm{jet}}_{\mathrm{reco}} - \eta^{\mathrm{jet}}_{\mathrm{truth}}$.

5.1.3.4 Global Sequential Calibration

Following the previous calibrations and corrections on the origin, pile-up, and jet energy scale; there is still an observed dependence on the longitudinal and transverse features of the jet. In particular, quarks and gluons shower and hadronize differently which means that quark and gluon jets will have a different response in the detector. As gluons split into pairs of quarks, gluon-initiated jets tend to have a high multiplicity of soft signals. Alternatively, quark-initiated jets will often include hadrons with a higher fraction of the jet p_{T} that penetrate deeper into the calorimeter. As the goal of these jet calibrations is to be independent of the "type" of jet, removing these dependencies is important. This particular procedure, known as Global Sequential Calibration (GSC), was explored during the Run I calibration effort [41] which identified five variables that are used to improve the resolution of the JES. Each of these variables exists in a subset of the detector and corrections are applied sequentially:

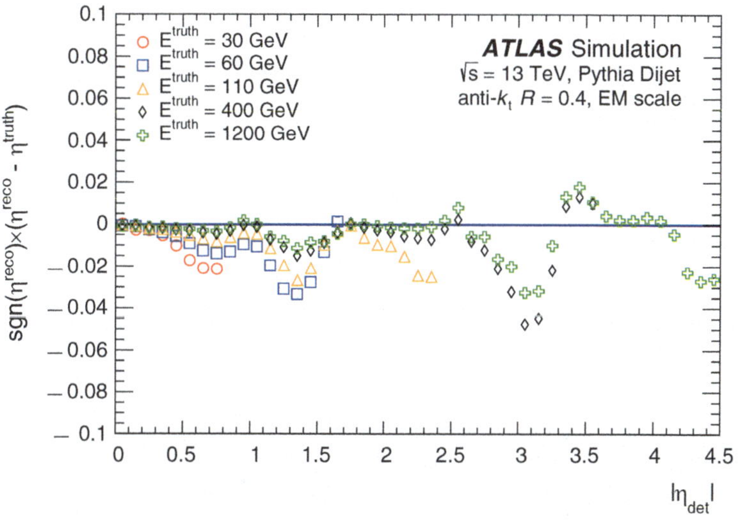

Fig. 5.10 The signed difference between truth jet η^{truth} and the reconstructed jet η^{reco} due to biases in the jet reconstruction [87]. This bias is addressed with an η correction applied as a function of $|\eta_{\text{det}}|$. The effect of changing geometries, such as barrel end-cap transition region around $|\eta_{\text{det}} \sim 1.4|$ and endcap-forward transition region around $|\eta_{\text{det}}| \sim 3.1$ can clearly be seen

1. f_{Tile0}, $|\eta_{\text{det}}| < 1.7$: the fraction of jet energy in the first layer of Tile
2. f_{LAr3}, $|\eta_{\text{det}}| < 3.5$: the fraction of jet energy in the third layer of the EMCal
3. n_{trk}, $|\eta_{\text{det}}| < 2.5$: the number of tracks with $p_{\text{T}} > 1$ GeV associated to the jet
4. $\mathcal{W}_{\text{trk}}$, $|\eta_{\text{det}}| < 2.5$: the width of the tracks associated to the jet, weighted by their p_{T}
5. n_{segments}, $|\eta_{\text{det}}| < 2.7$: the number of muon tracks associated to the jet

Figure 5.11 shows the distribution of the five GSC variables. The variables used are properties that characterize the longitudinal and transverse topology of the energy deposited by the jet [41]. For example, a large energy deposit in calorimeter layers closest to the interaction point indicates a shower that developed early, leading to a lower detector response in the calorimeters as a fraction of that jet energy would not have reached the calorimeters.

5.1.3.5 In-Situ Calibration

Following the JES and GSC calibration steps, a data-driven approach, in-situ calibration, is applied to account for differences in jet responses between data and Monte Carlo simulation. The corrections are designed to correct for the energy scale differences between data and Monte Carlo, as Monte Carlo is already calibrated at this stage to have the correct energy scale with respect to truth jets. The differences can arise from imperfections in the simulation of the hard scatter event, pile-up, jet

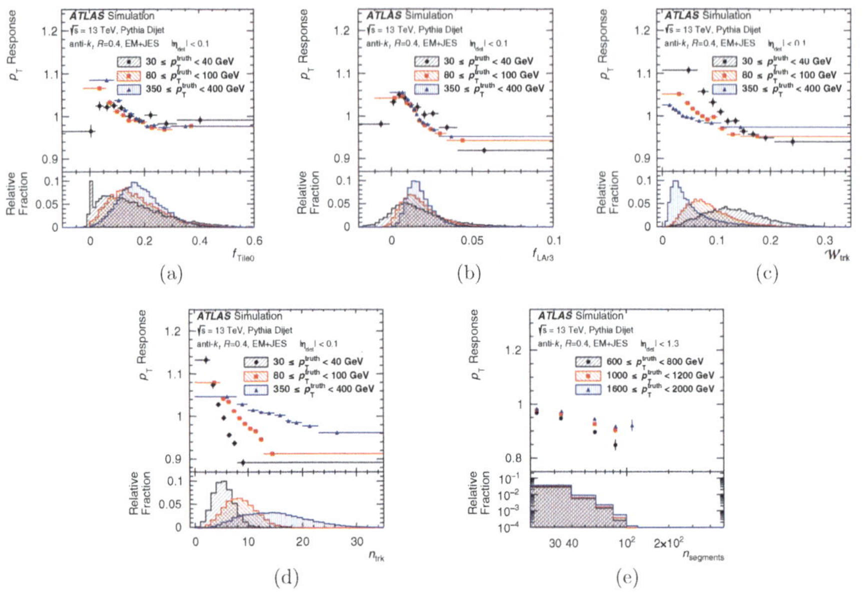

Fig. 5.11 The average jet response in Monte Carlo simulation as a function of the GSC variables for three different ranges of $p_{\mathrm{T}}^{\mathrm{truth}}$ [87]. (**a**) and (**b**) are shown with no GSC corrections applied. (**c**) and (**d**) are shown with the first two corrections applied. (**e**) is shown with the previous four corrections applied. (**a**)–(**d**) are shown for $|\eta_{\mathrm{det}}| < 0.1$ while (**e**) is shown for $|\eta_{\mathrm{det}}| < 1.3$

formation, and so on. These differences are quantified by a jet balancing approach, where the p_{T} of a jet is balanced against other well-measured reference objects, $\mathcal{R} = p_{\mathrm{T}}^{\mathrm{jet}}/p_{\mathrm{T}}^{\mathrm{object}}$. There are four in-situ calibrations performed:

1. η-intercalibration: correct the response of jets in the forward region $0.8 < |\eta| < 4.9$ using well-measured jets in the central region $|\eta| < 0.8$
2. jet-balance using Z+jet ($Z^0 \to e^+e^-$, $Z^0 \to \mu^+\mu^-$) with a well-reconstructed Z^0 boson
3. jet-balance using γ+jet with a well-reconstructed photon
4. jet-balance using multijets with a high p_{T} jet balanced against a system of low p_{T} jets (multijet balance)

The jet p_{T} response of the three jet-balance in-situ calibrations are shown in Fig. 5.12. Each of these techniques is statistically combined, in overlapping regions of jet p_{T}, into a total calibration as shown in Fig. 5.13. Each method is weighted by p_{T} bin based on the statistical power, such that a method's weight is increased in p_{T} regions of smaller relative uncertainty and smaller bin size, in order to maximize the precision in each p_{T} region. The inverse of this ratio is taken as the **in-situ correction**.

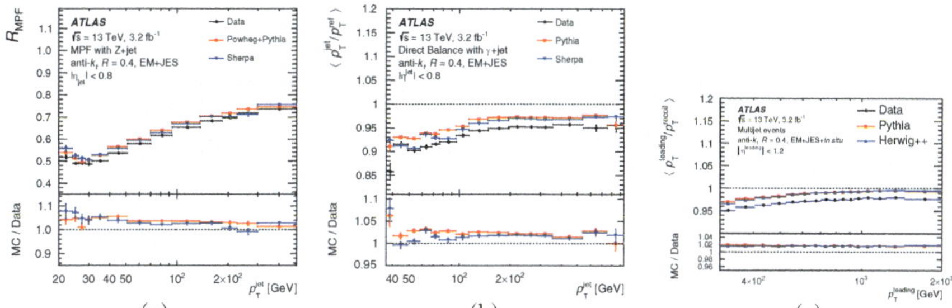

(a) (b) (c)

Fig. 5.12 The average jet p_T response of EM+JES jets calibrated up to the η-intercalibration as a function of jet p_T for (**a**) Z+jet events, (**b**) γ+jet events, and (**c**) multijet balance [87]

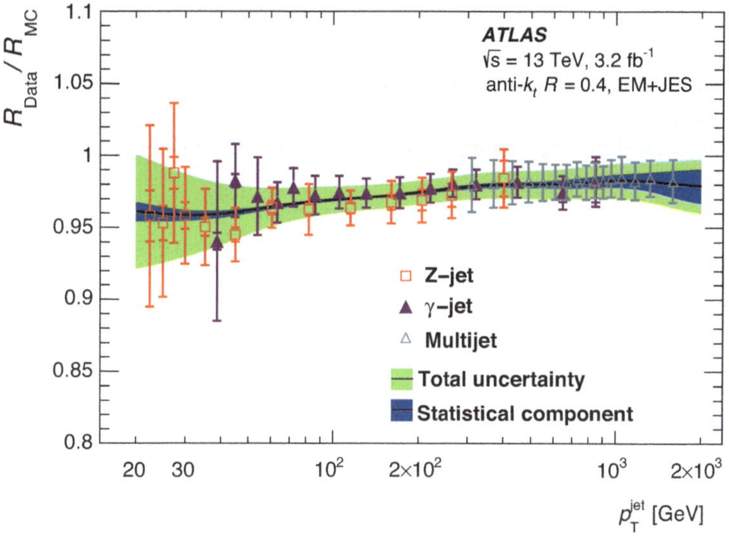

Fig. 5.13 The in-situ calibrations for Z+jet, γ+jet, and multijet balance are statistically combined to create a total calibration and uncertainty [87]. The final derived correction (black line) and its statistical uncertainty band (dark blue) and total uncertainty band (light green) are shown

5.1.4 *Uncertainties*

At the end of the day, there are 79[9] JES systematic uncertainties propagated from the individual calibrations and studies described in Sect. 5.1.2 [87].

[9]There are 80 total, but one of them is for a type of simulation not used in this thesis analysis and does not apply.

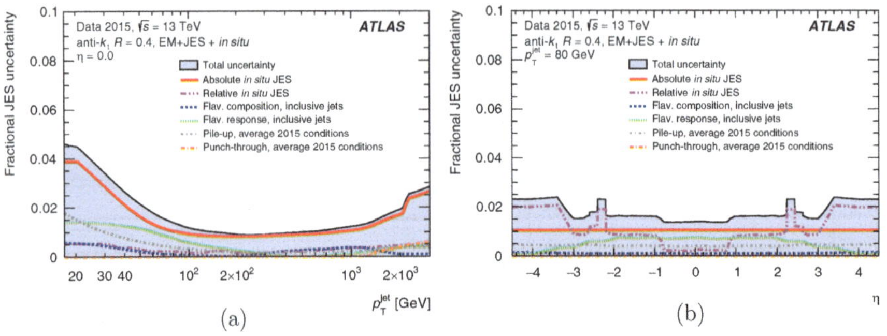

Fig. 5.14 Combined uncertainty of JES calibration as a function of (**a**) jet p_T at $\eta = 0$ and (**b**) η at $p_T = 80\,\text{GeV}$ [87]

- 67 come from the jet-balance in-situ calibration procedures in Sect. 5.1.3.5;
- 3 from the η-intercalibration;
- 4 come from pile-up uncertainties in Sect. 5.1.3.2;
- 3 come from differences in jet response and compositions of gluons, light-quarks, and b-quark-initiated jets
- 1 comes from the GSC punch-through correction;
- and 1 comes from uncertainty on high jet $p_T > 2\,\text{TeV}$ beyond the reach of in-situ methods.

The full combination of uncertainties is shown in Fig. 5.14 and is largest at low p_T starting at 4.5% decreasing to 1% at 200 GeV. It rises after due to the statistical uncertainties of the in-situ calibrations which end at 2 TeV, hence the sharp increase. In terms of η, the uncertainty is fairly constant across the detector and reaches a maximum of 2.5%[10] for the forward jets.

While the 79 uncertainties provide the most accurate understanding of the JES uncertainty, implementing and evaluating them all is computationally intensive. A reduced set of Nuisance Parameters (NPs) is made available through an eigen-decomposition in a way to preserve the correlations observed with all 79 [87]. Four reduced sets of NPs are formed after a global-reduction procedure and grouping in a way to preserve correlations in low-p_T (`JES2`), medium-p_T (`JES3`), and high-p_T (`JES4`) kinematic regimes, as well as one that provides general representation in all kinematic regimes (`JES1`).

5.1.5 *Jet Kinematics*

This section is meant to provide a brief summary of some of the kinematic properties of $R = 0.4$ anti-k_t jets formed from EM-scale topoclusters with both

[10]There is a feature around $2.0 < |\eta| < 2.6$ due to the non-closure uncertainty of the η-intercalibration.

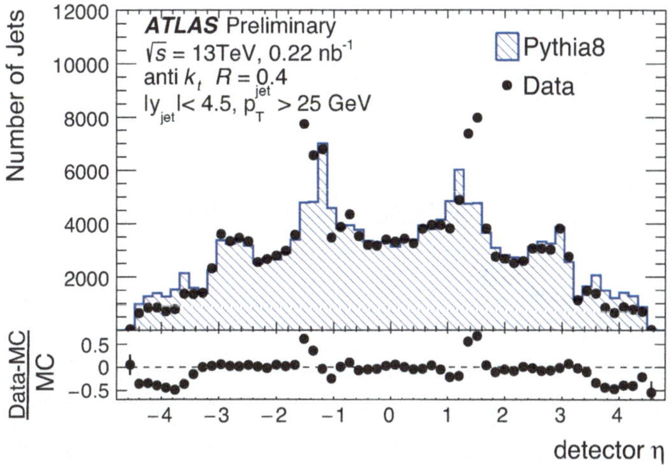

Fig. 5.15 Jet kinematic distributions for η_{det} for jets above $p_{\mathrm{T}} > 25$ GeV [63]. The Monte Carlo prediction is normalized to the data and the error s reflect the statistical uncertainty

origin correction (Sect. 5.1.3.1) and pile-up correction (Sect. 5.1.3.2) applied to the reconstructed jets. A few representative kinematic distributions of the topoclusters used as jet inputs and the kinematics of the jets themselves are shown, with more in [63]. Figure 5.15 shows the distribution of the number of jets as a function of η_{det} with disagreements observed in the gap region between the barrel and extended barrels of the Tile ($|\eta_{\mathrm{det}}| \sim 1.4$) and the forward region ($3.1 < |\eta| < 4.9$). For example, in the gap region, a larger number of high p_{T} jet events are observed in data than what is predicted by Monte Carlo simulations. Figure 5.16 shows a distribution of the mean number of constituents for each jet binned in jet η. This plot is meant to emphasize the lower granularity in the forward regions and larger topocluster sizes. The primary differences between data and simulation are due to the modeling of the soft-energy components of the jets. And finally, Fig. 5.17 shows the very nice modeling of the minimum ΔR between jets for which the bulk of the distribution agrees to better than 10%. This observable is used to define the isolation criteria for the MC-based calibration.

5.2 Flavor Tagging of Jets

In general, the jet reconstruction algorithms do not identify the type of parton that initiated a given jet. However, there are a few exceptions that rely on the tracking information from the ID. For example, one could roughly identify jets formed from hadronic τ decays since τ is at the mass of hadrons and can decay with one track (e.g. $\tau^- \rightarrow W^-\nu_\tau \rightarrow \pi^-\pi^0\nu_\tau$) or three tracks (e.g. $\tau^- \rightarrow \pi^+\pi^-\pi^-\nu_\tau$) [218]. The

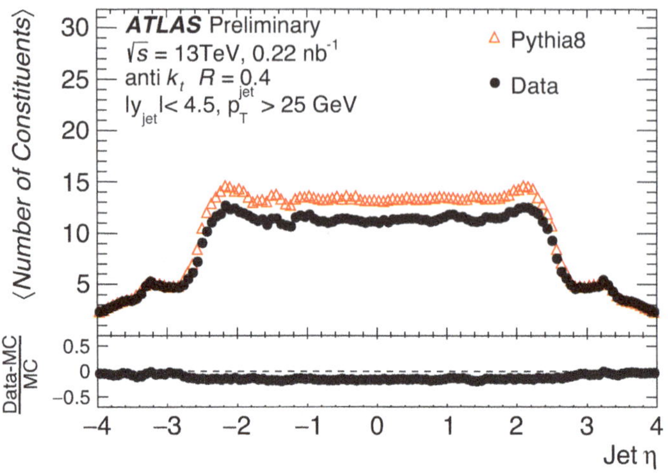

Fig. 5.16 Topocluster multiplicity for jets with $p_T > 25$ GeV is shown as a function of jet η [63]. The error bars reflect the statistical uncertainty

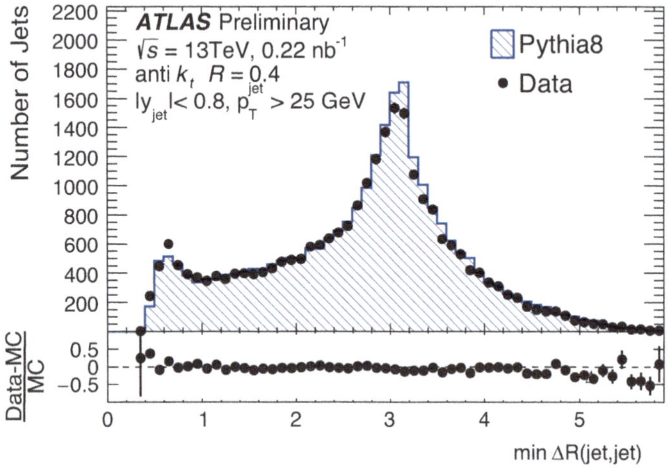

Fig. 5.17 The minimum distance in ΔR to a jet with $p_T > 25$ GeV and $|y| < 0.8$ [63]. The Monte Carlo prediction is normalized to the data. Requiring $R = 0.4$ jets to be isolated requires the minimum $\Delta R > 1.0 = 2.5 \times R$

particular identification we are concerned with though are jets associated with B-hadrons. In particular, these are hadrons that are relatively long-lived[11] and decay primarily via weak interactions. This identification is known as b-tagging and is part of the flavor tagging efforts of the ATLAS collaboration. The most vital input

[11] B-hadrons have a lifetime $\sim$1.5 ps ($c\tau \sim 450\,\mu$m) compared to top quarks with a mean lifetime $\sim 10^{-25}$ s.

required for b-tagging are the charged particle tracks reconstructed in the ID which has an acceptance $|\eta| < 2.5$. These B-hadrons will typically decay a few mm away from the primary vertex, inside the ID which has a radius of about 1 m. This secondary decay creates a **secondary vertex** and the b-tagging algorithms take advantage of this to identify the displaced tracks of B-hadrons [26, 64, 88, 89]. There are three basic algorithms whose outputs are used as inputs to the standard multivariate discriminant algorithm (`MV2`) that is used in ATLAS analyses for Run II:

- an impact parameter-based algorithm (`IP2D`, `IP3D`),
- a secondary vertex reconstruction algorithm (`SV`),
- and a decay chain multi-vertex reconstruction algorithm, JetFitter (`JF`).

5.2.1 Impact Parameter Tagging Algorithms

The typical B-hadron usually has at least one vertex displaced from the primary vertex. To parameterize the minimum distance between the displaced track and the primary vertex, the transverse impact parameter d_0 is defined in the $r - \phi$ plane, while the longitudinal impact parameter $z_0 \sin\theta$ is defined in the longitudinal plane. B-hadrons will typically have large impact parameters due to their relatively long lifetimes. If a secondary vertex is identified as "behind" the primary vertex, the sign of the impact parameter is negative and is normally due to background and to calibrate the light quark tag rate [81]. Two impact parameter significances can be defined for transverse d_0/σ_{d_0} and longitudinal $z_0 \sin\theta/\sigma_{z_0 \sin\theta}$. Figure 5.18 shows distributions of the signed significances for each impact parameter, with well-measured tracks being weighted up and poorly measured tracks being weighted down. A log-likelihood ratio discriminant shown in Fig. 5.19 is computed using the number of tracks of a given jet along with the probability density functions (b-flavor, light-flavor) derived from distributions of the impact parameters. This log-likelihood ratio discriminant is used as input to the multivariate algorithm described later in this section.

5.2.2 Secondary Vertex Finding Algorithm

The secondary vertex finding algorithm [30] (`SV`) explicitly reconstructs a displaced secondary vertex within the jet. From all the vertices associated with a jet, these are filtered to only look at those with two tracks. Vertices with a pair of two tracks are rejected if they likely originate from the decay of some non-B-hadron or hadronic interactions with the detector material. From the remaining tracks, all possible two-track vertices are formed and required to be significantly displaced from the primary interaction vertex by requiring the sum of the impact parameter significances of the tracks in the two-track vertex to be higher than 2 [30]. Altogether, eight kinematic

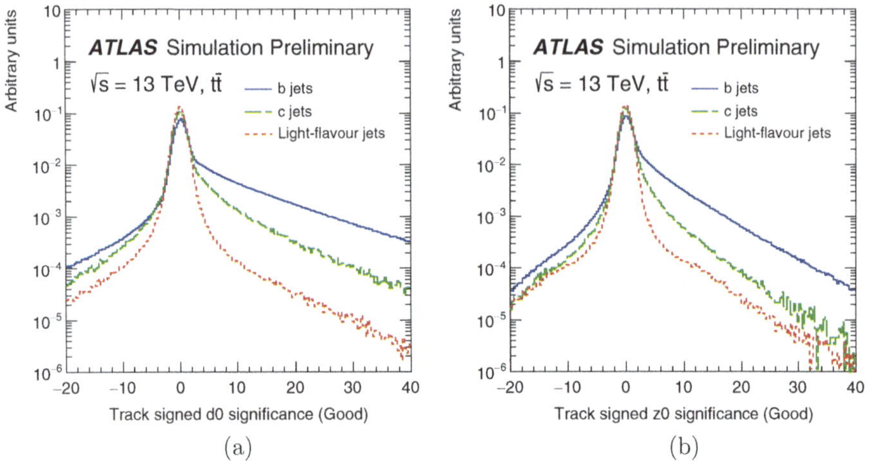

Fig. 5.18 The (**a**) transverse and (**b**) longitudinal signed impact parameter significance of tracks in $t\bar{t}$ events for b-flavor (blue), c-flavor (green), and light flavor (red) jets [26]. The tail shown in the b jets line (blue) is due to the long lifetime of B-hadrons

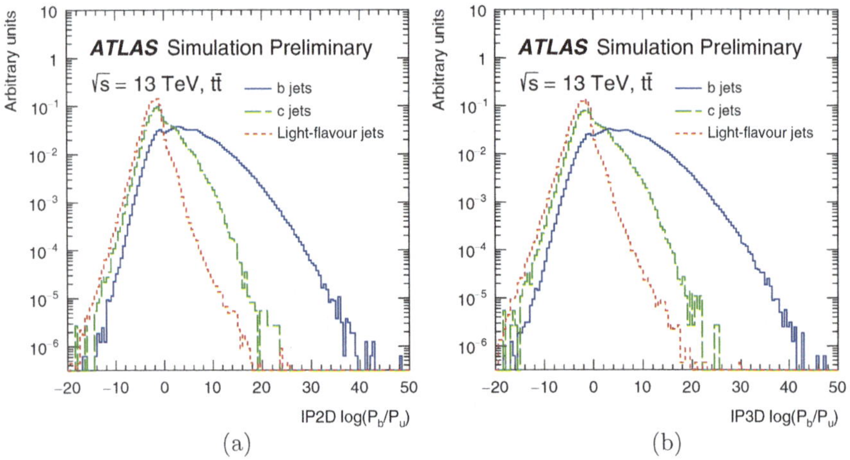

Fig. 5.19 The log-likelihood ratio for the (**a**) IP2D and (**b**) IP3D b-tagging algorithms in $t\bar{t}$ events for b-flavor (blue), c-flavor (green), and light flavor (red) jets [26]. The log-likelihood is calculated as the ratio of b-flavor and light-flavor probability density functions. Jets with no tracks are assigned a large negative value in an underflow bin which is not shown on this plot

properties of the two-track vertices and the tracks at the reconstructed secondary vertex are used as inputs to the multivariate algorithm. Distributions of two of these inputs are shown in Fig. 5.20 for the number of two-track vertices identified and the number of tracks at the reconstructed secondary vertex.

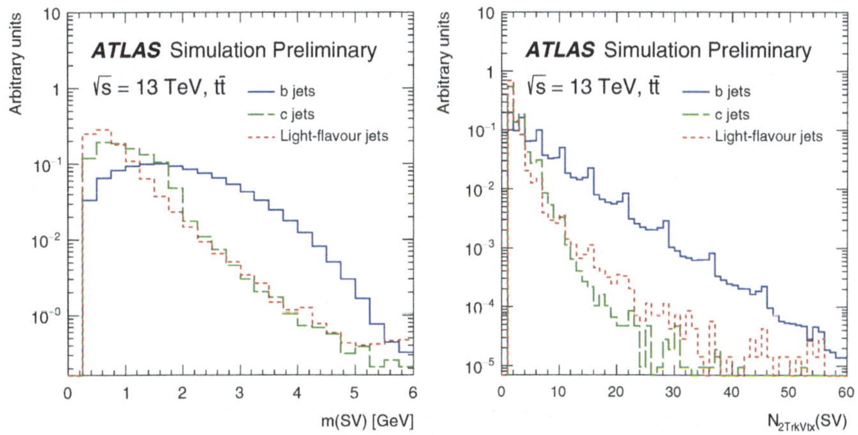

(a) (b)

Fig. 5.20 Properties of the secondary vertices reconstructed by the SV algorithm for *b*-flavor (blue), *c*-flavor (green), and light flavor (red) jets [26]. Shown are (**a**) the invariant mass of the tracks associated with the reconstructed vertex and (**b**) the number of two-track vertices reconstructed within the jet

5.2.3 Decay Chain Multi-Vertex Algorithm

The decay chain multi-vertex reconstruction algorithm, JetFitter [220], uses the topologies of B-hadron and C-hadron decay vertices inside the jet to reconstruct the full B-hadron decay chain. A Kalman filter is used to identify a line along which the B-hadron and C-hadron decay vertices lie along to approximate the B-hadron flight path and identify potential secondary vertices. This algorithm seeks to identify the two tracks of the B-hadron and C-hadron decays. The eight kinematic properties of the reconstructed two-track vertices and tracks at the reconstructed secondary vertex are used as inputs to the multivariate algorithm. Distributions of two of these inputs are shown in Fig. 5.21 for the number of two-track vertices identified and the number of tracks at the reconstructed secondary vertex.

5.2.4 Multivariate Algorithm

The p_{T} and η of the jet, along with the three outputs from IP2D and IP3D, eight outputs from SV and from JF, make up the 24 input variables that go into a multivariate classifier [183], using a Boosted Decision Tree (BDT) algorithm. This achieves a better discrimination than any of the basic algorithms described previously. The tagger, called `MV2`, is trained on jets from $t\bar{t}$ Monte Carlo simulation. Three such taggers were developed for ATLAS in Run II called `MV2c00`, `MV2c10`, `MV2c20` with the names indicating the c-jet fraction in the background. `MV2c10`

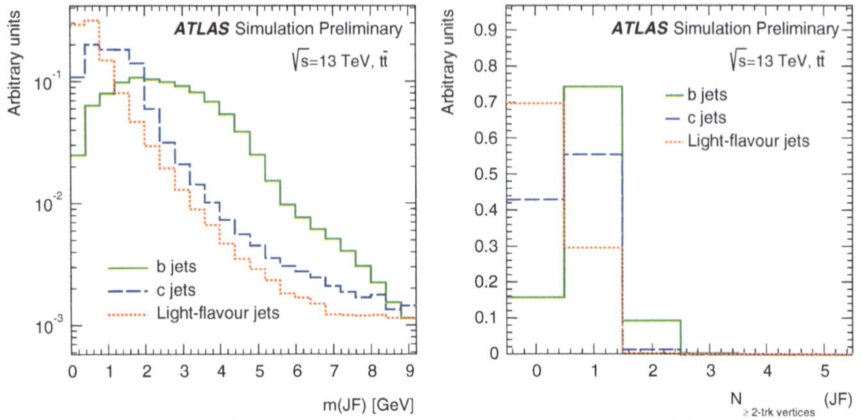

(a) (b)

Fig. 5.21 Properties of the secondary vertices reconstructed by the JF algorithm for b-flavor (blue), c-flavor (green), and light flavor (red) jets [26]. Shown are (**a**) the invariant mass of the tracks associated with the reconstructed vertex and (**b**) the number of two-track vertices reconstructed within the jet

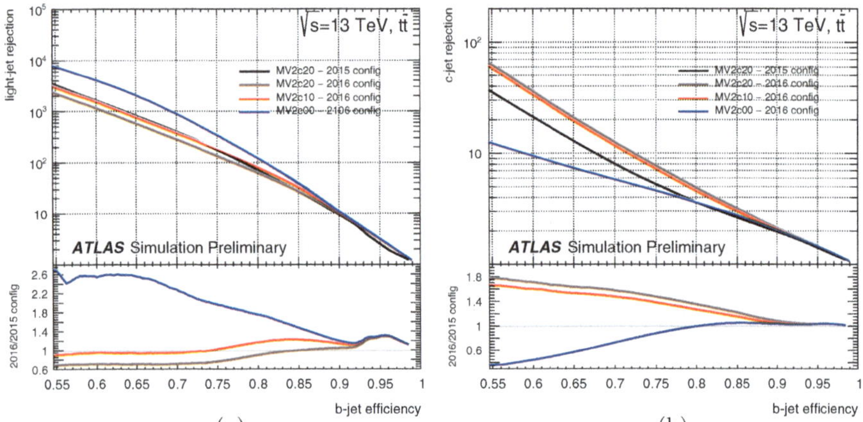

(a) (b)

Fig. 5.22 (**a**) Light-flavor jet and (**b**) c-flavor jet rejection versus b-flavor jet efficiency for the 2015 and 2016 configurations of the MV2 b-tagging algorithm is shown evaluated for $t\bar{t}$ events [26]

is the tagger used in this thesis analysis and indicates that the tagger was trained on a sample whose background composition is 10% c-flavor jets and 90% light-flavor jets. Figure 5.22 shows the performance of the optimized MV2 algorithms in rejecting light-flavor jets and c-flavor jets as a function of the b-jet efficiency. It should be noted that while the MV2c20 algorithm does provide better c-flavor jet rejection, it does so at the expense of a reduced light-flavor jet rejection. Because of

Table 5.1 Operating points for the MV2c10 b-tagging algorithm, including benchmark numbers for the efficiency and rejection rates [26]

	Efficiency [%]	Rejection [%]		
Cut value	b-jet	c-jet	Light-jet	τ
0.9349	60	34	1538	184
0.8244	70	12	381	55
0.6459	77	6	134	22
0.1758	85	3.1	33	8.2

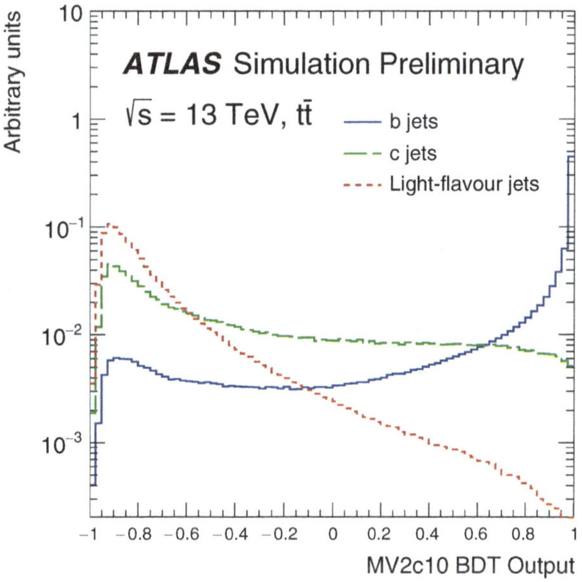

Fig. 5.23 The MV2c10 output for b-flavor (blue), c-flavor (green), and light flavor (red) jets evaluated with $t\bar{t}$ events [26]

this consideration and that this thesis analysis is sensitive to light jets, the MV2c10 tagger was chosen as the standard b-tagging discriminant.

From this, one can define four sets of standard working points shown in Table 5.1 for b-tagging identified by picking a b-jet efficiency. Figure 5.23 shows the output of the MV2c10. Operating points are defined by a single cut value on the discriminant output distribution and are chosen to provide a specific b-jet efficiency on an inclusive $t\bar{t}$ sample. For example, the 77% working point has a rejection factor of 6 and of 134 on charm and light jets, respectively.

Finally, in order for this tagger to be useful for physics analyses, scale factors need to be derived using data to account for differences between simulation and data. For Run I, this was done in [34, 35, 48, 49]. Correction factors are applied to the simulated event samples to compensate for differences between data and simulation in the b-tagging efficiency for b, c and light jets. The correction for b-jets is derived from $t\bar{t}$ events with final states containing two leptons, and the corrections are consistent with unity with uncertainties at the level of a few percent

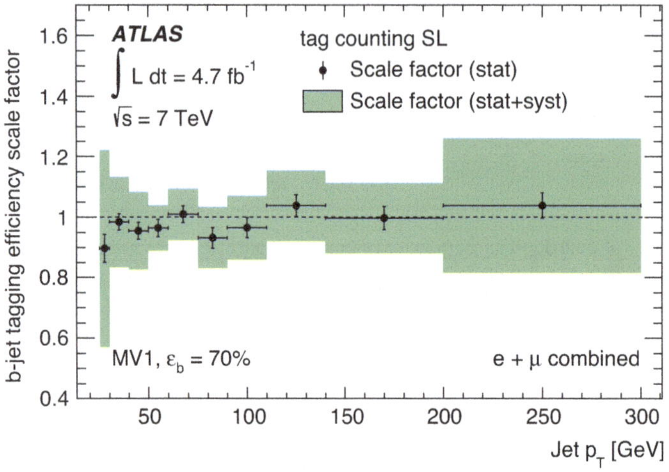

Fig. 5.24 The data/simulation scale factors for the `MV1` algorithm at 70% b-jet tagging efficiency [76]. The error bars show the statistical uncertainties while the green band indicates the total uncertainty. Please note that the algorithm here is different from the `MV2` algorithm used in this thesis analysis. The performance studies for the current algorithm are still being done

over most of the jet p_{T} range. An example of these correction factors for the Run-I algorithm[12] are shown in Fig. 5.24 along with a total systematic uncertainty that needs to be considered in this thesis analysis which is very sensitive to the b-jet tagging algorithm due to the large number of expected b-jets in the final state.

5.3 Muons

Muons are one of the simplest particles to identify in the ATLAS detector. As muons traverse the entire detector, reconstructed tracks from both the ID and the MS are used. Four different muon types are defined depending on which subdetectors are used in the reconstruction in order of decreasing priority:

1. Combined (CB) muon: tracks are reconstructed independently in the ID and MS, and a combined track fit is performed by adding or removing tracks from the MS to improve the fit quality.
2. Segment-tagged (ST) muons: a track in the ID is classified as a muon if it associated with a track segment in the MDT or the CSC. This is primarily used for low p_{T} muons that do not traverse the entire MS.

[12] The Run-2 algorithm performance is still undergoing study and will not be public in time for this thesis.

3. Calorimeter-tagged (CT) muons: a track in the ID is classified as a muon if it is associated with an energy deposit in the calorimeter compatible with a low-ionizing particle.
4. Extrapolated (ME) muons: the muon trajectory is reconstructed based only on the MS track in at least two layers and ensuring that it originates from the interaction point. This is mainly used to extend the acceptance for muon reconstruction in the region outside the ID from $2.5 < |\eta| < 2.7$.

From the muon-classified tracks, **muon quality** requirements are placed on tracks from each portion of the subdetector which amount to requiring a specific number of hits in each subcomponent [77]. Four muon quality identificates are used:

medium Default selection for muons in ATLAS. This is the quality criteria used in the thesis analysis for identifying muons. Only CB and ME tracks are used with at least 3 CB track hits and at least 3 ME layers.

loose Designed to maximize reconstruction efficiency, primarily for reconstructing Higgs boson candidates in the four-lepton final state [65]. All muon types are used.

tight Designed to maximize the purity of muons. Only CB muons with hits in at least two layers of the MS and satisfying the "medium" selection are used.

high-p_T Designed to maximize the momentum resolution for tracks with $p_T >$ 100 GeV, primarily for high-mass W' and Z' resonances [50, 51]. CB muons passing the "medium" selection and having at least three layers of the MS are selected.

Muons are further calibrated to data using the well-studied resonances $J/\Psi \rightarrow \mu\mu$ and $Z \rightarrow \mu\mu$. Figure 5.25 shows the combined uncertainty in quadrature of this calibration effort as a function of the p_T of the reconstructed muons. $J/\Psi \rightarrow \mu\mu$ targets the low-p_T region with $Z \rightarrow \mu\mu$ targeting the high-p_T region. The total systematic uncertainty in muon reconstruction is less than 2% across the board.

As muons from the decay of heavy particles such as the W, Z, h are often produced isolated from other particles, a set of **muon isolation** requirements are also placed around each muon candidate. Two primary isolation-based variables are used, one is track-based ($p_T^{\text{varcone30}}$) and one is calorimeter-based ($E_T^{\text{topocone20}}$). Each isolation criteria looks at the scalar sum of transverse momentum in a cone of $\Delta R < 0.2$ or $\Delta R < 0.3$[13] and compares it to the transverse momentum of the muon candidate. The isolation is effective at suppressing muons produced from processes such as meson decay in flight and heavy-flavor decay. There are seven isolation working points that are defined for use by analyzers which differ primarily on a cut of the ratio between the energy of the muon candidate and the surrounding "background" energy. This isolation requirement is defined to ensure a flat efficiency of around 99% across the whole electron transverse energy and muon transverse momentum ranges. This is described in more detail in Sect. 6.2.3.

[13]Look at the name of the isolation variable to know the cone size.

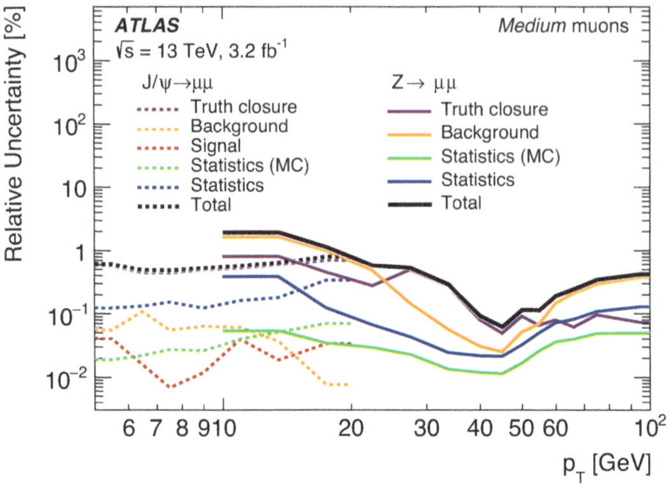

Fig. 5.25 Total uncertainty in the efficiency scale factor for "medium" muons as a function of p_{T} as obtained from $Z \to \mu\mu$ (solid lines) and $J/\Psi \to \mu\mu$ (dashed lines) decays [77]. The combined uncertainty is the sum in quadrature of the individual contributions

5.4 Electrons and Photons

Both electrons and photons are reconstructed based on the electromagnetic shower in the LAr electromagnetic calorimeter described in Sect. 3.7. The electromagnetic calorimeter is divided into a grid of 3×5 towers of size $\Delta\eta \times \Delta\phi = 0.025 \times 0.025$ to scan for areas of local maxima which are used to seed clusters. The clusters are then matched to a well-reconstructed ID track. The existence of this matched track and its properties are used to subsequently identify the cluster as being consistent with a prompt electron, a photon conversion, or an unconverted photon [53, 71]. A photon carries no electric charge and will not produce a track in the ID: label as an unconverted photon. A converted photon is one which has a secondary vertex because of the decay into an electron-positron pair: label the cluster as a converted photon if the matched track is extrapolated from a secondary vertex. Otherwise, an electron would have a matched track that is extrapolated from the primary vertex and so is labeled as a prompt electron.

Similar to muons, electrons have **electron quality** and **electron isolation** identifications. Three levels of identification working points are provided for **electron quality** called `Loose`, `Medium`, `Tight` which are selections on the discriminant of a multivariate analysis to provide electron identification using a likelihood-based method. In addition to the identification criteria, isolation criteria are defined using two primary isolation-based variables: one is track-based ($p_{\mathrm{T}}^{\mathrm{varcone20}}$) and one is calorimeter-based ($E_{\mathrm{T}}^{\mathrm{topocone20}}$). Each isolation criteria looks at the scalar sum of transverse momentum in a cone of $\Delta R < 0.2$ or $\Delta R < 0.3$[14] and compares it to the transverse momentum of the electron candidate. The isolation is effective at disentangle prompt electron candidates[15] from other non-isolated electron candidates.[16]

Electrons and photons are calibrated in a similar procedure to muons (Sect. 5.3) to derive data-driven scale factors using $J/\Psi \rightarrow ee$, $Z \rightarrow ee$, and $Z \rightarrow \ell\ell\gamma$ processes. These corrections ensure uniformity in the electromagnetic response across separate regions of the detector and will introduce systematic uncertainties including the mismodeling between simulation and data.

5.5 Taus

While tau leptons are not used directly in this thesis or the analysis in this thesis, I would like to devote a short section to explaining their relevance. Tau leptons are charged leptons, but they are very different from electrons and muons. From an experimental point of view, tau leptons decay into other types of particles before entering the detector. They can decay hadronically around 60% of the time into hadrons plus neutrinos and 40% of the time leptonically to electrons or muons plus neutrinos. The leptonic decays are not distinguishable from electrons and muons described in Sects. 5.3 and 5.4; the hadronic decays present as multiple hadronic showers matched to tracks in the ID. As taus can present with a secondary vertex, they can fake b-tagged jets and this is described more previously in Sect. 5.2.

5.6 Missing Transverse Momentum

The last crucial object is the missing transverse momentum which represents the overall transverse momentum imbalance in the event, commonly written out as $E_{\mathrm{T}}^{\mathrm{miss}}$ or colloquially "MET."[17] Conservation of momentum in the plane transverse to the beam axis implies that the vector transverse momenta of the collision products should sum to zero. An imbalance implies the existence of weakly interacting stable particles, such as neutrinos in the SM or many supersymmetric particles such as neutralinos in the BSM. The measurement of $E_{\mathrm{T}}^{\mathrm{miss}}$ is also affected by poorly reconstructed objects, visible particles that escape the detector unseen, or particles that otherwise fail to be reconstructed. While the $E_{\mathrm{T}}^{\mathrm{miss}}$ is reconstructed offline, it

---

[14]Look at the name of the isolation variable to know the cone size.

[15]Prompt electron candidates come from heavy-resonance decays such as $W \rightarrow e\nu_e$, $Z \rightarrow ee$.

[16]Non-isolated candidates include electrons from photon conversion, from heavy-flavor hadron decays, and light hadrons mis-identified as electrons.

[17]Not to be confused with the New York Mets.

can and often is reconstructed again at an analysis-level with extra refinement such as the precise removal of objects that overlap or to specify the inputs of all visible particles identified in an analysis. For this thesis analysis, the $E_{\mathrm{T}}^{\mathrm{miss}}$ is reconstructed again and the procedure is described in more detail in Sect. 6.2.6. In general, $E_{\mathrm{T}}^{\mathrm{miss}}$ is defined in Eq. (5.10)

$$-E_{\mathrm{T}}^{\mathrm{miss}} = \sum_{e} p_{\mathrm{T}} + \sum_{\gamma} p_{\mathrm{T}} + \sum_{\tau} p_{\mathrm{T}} + \sum_{\mu} p_{\mathrm{T}} + \sum_{j} p_{\mathrm{T}} + \sum_{\mathrm{soft}} p_{\mathrm{T}}. \tag{5.10}$$

As described in [66, 68], a baseline set of selection criteria are applied to all the visible, reconstructed objects that enter the $E_{\mathrm{T}}^{\mathrm{miss}}$ calculation, such as quality criteria on the leptons and p_{T} cuts on the objects. However, one special term is the "soft" term which comes in two different forms: track and cluster. Track soft-terms (TST) are ID tracks, extrapolated from the primary vertex, which are not within $\Delta R < 0.05$ of an electron or photon, $\Delta R < 0.2$ of a tau, or matched with a combined muon (Sect. 5.3) or jet. Similarly, Cluster soft-terms (CST) are built from energy depositions in the calorimeter not associated with reconstructed physics objects. By construction,[18] this is sensitive to pile-up which makes it a less suitable choice for high-luminosity environments. For the thesis analysis, as described in Sect. 6.2.6, the $E_{\mathrm{T}}^{\mathrm{miss}}$ is reconstructed using track soft-terms.

The performance of $E_{\mathrm{T}}^{\mathrm{miss}}$ shown in Fig. 5.26 is studied by comparing to $Z \rightarrow \mu\mu$ and $W \rightarrow e\nu_e$ processes [68]. $Z \rightarrow \mu\mu$ has no real $E_{\mathrm{T}}^{\mathrm{miss}}$ and is a good choice to study the performance of $E_{\mathrm{T}}^{\mathrm{miss}}$ due to the precise measurements of the kinematics of the Z boson. Likewise, $W \rightarrow e\nu_e$ is provides genuine $E_{\mathrm{T}}^{\mathrm{miss}}$

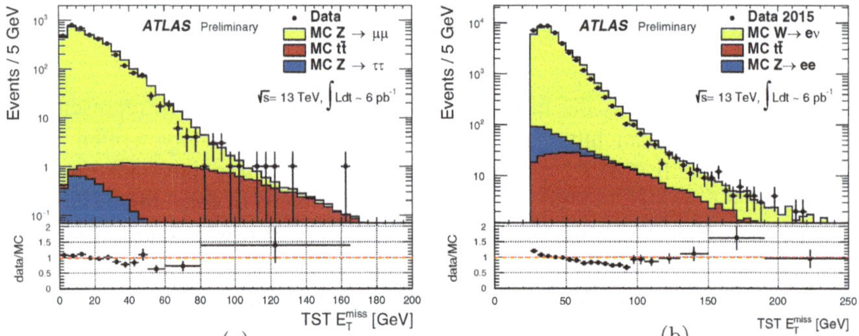

(a) (b)

Fig. 5.26 The distribution of reconstructed track soft-term $E_{\mathrm{T}}^{\mathrm{miss}}$ is shown for (**a**) $Z \rightarrow \mu\mu$ topologies and (**b**) $W \rightarrow e\nu_e$ topologies [68]. The agreement between data and simulation for $W \rightarrow e\nu_e$ is notably worse with respect to $Z \rightarrow \mu\mu$ in the low $E_{\mathrm{T}}^{\mathrm{miss}}$ region likely due to the missing QCD multijet background not included in the studies

[18] A pun!.

through the neutrinos from the hard scatter interaction and helps validate the scale and direction of reconstructed $E_{\mathrm{T}}^{\mathrm{miss}}$ [68]. Systematics from all the included input objects is propagated through the $E_{\mathrm{T}}^{\mathrm{miss}}$ calculation. The soft-term has systematics associated with the data/simulation scale factor that is derived using comparisons with $Z \rightarrow \ell\ell$.

Chapter 6
Boosted Object Reconstruction

After a detailed exposition of the various reconstruction techniques in Chap. 5, further selections and quality criteria are applied to define a set of baseline and candidate reconstructed objects that will be used in the analysis and the kinematic variables to discriminate between signal and background and maximize the sensitivity for a discovery of new physics. An introduction to estimating the size parameter of large radius jets is described in Sect. 6.1. The rest of this brief chapter describes in detail the further selections on the reconstructed objects for: jets (Sect. 6.2.1), b-tagged jets (Sect. 6.2.2), leptons[1] (Sect. 6.2.3), the procedure for removing energy overlaps in the reconstructed objects (Sect. 6.2.4), large radius re-clustered jets (Sect. 6.2.5), and the reconstructed missing transverse momentum (Sect. 6.2.6).

6.1 Size of Boosted Jets

Consider a two-body relativistic decay as shown in Fig. 6.1 with $m_{\pi^0} \gg m_\gamma = 0$. The question one might pose is: *What is the opening angle between the two photons?* Since a (real) photon has no mass, its energy and momentum are the same $E_\gamma = |\vec{p}_\gamma|$. To transform between the lab frame (LAB) and the rest frame (RF) of the π^0, the boost is described along z by $\gamma = \frac{E_{\pi^0}}{m_{\pi^0}}$ and $\beta = v_{\pi^0} = \frac{|\vec{p}_{\pi^0}|}{E_{\pi^0}}$. Applying this boost to each photon (denoted by $\pm$) to transform from the rest frame to the lab frame, we find in each component

[1]The term "lepton" exclusively refers to electron or muon in this thesis.

© Springer Nature Switzerland AG 2020

G. Stark, *The Search for Supersymmetry in Hadronic Final States Using Boosted Object Reconstruction*, Springer Theses,
https://doi.org/10.1007/978-3-030-34548-8_6

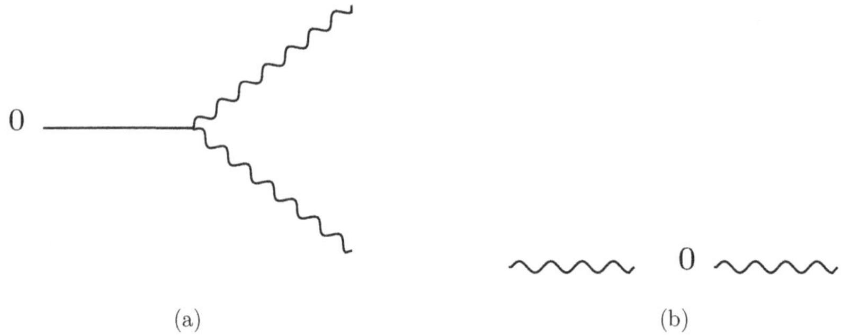

Fig. 6.1 A Feynman diagram of the two-body decay of a neutral pion π^0 to two photons γ in the (**a**) lab frame and (**b**) rest frame

$$E_{\pm,\parallel}^{\text{LAB}} = \gamma\left(E_{\parallel}^{\text{RF}} \pm \beta E^{\text{RF}}\right) = \frac{1}{2}\gamma m_{\pi^0} \tag{6.1}$$

$$E_{\perp}^{\text{LAB}} = E_{\perp}^{\text{RF}} = \frac{1}{2} m_{\pi^0} \sin\theta^{\text{RF}}, \tag{6.2}$$

where θ^{RF} is the angle that the photon makes with the z-axis and $E_{\parallel}$, $E_{\perp}$ refer to a longitudinal and perpendicular component of the γ energy with respect to the momentum of the neutral pion. Therefore, the total energy of the photons in the lab frame is sum in quadrature

$$E_{\pm}^{\text{LAB}} = \frac{1}{2} m_{\pi^0} \sqrt{\gamma^2\left(\cos\theta^{\text{RF}} \pm \beta\right)^2 + \sin^2\theta^{\text{RF}}} = \frac{1}{2} m_{\pi^0}\left(1 \pm \beta\cos\theta^{\text{RF}}\right). \tag{6.3}$$

Unfortunately, the algebra starts to get a little bit hairy, so it is surprisingly useful to define

$$g = \left|\frac{E_+ - E_-}{E_+ + E_-}\right| = \beta\cos\theta^{\text{RF}} \tag{6.4}$$

which simplifies Eq. (6.3) so that we get $E_{\pm}^{\text{LAB}} = \frac{1}{2} m_{\pi^0}(1 \pm g)$. So to determine the angle θ between the photons, we calculate the 4-vector of the total system

$$m_{\pi^0}^2 = (\gamma_+ + \gamma_-)^2 = 2E_+E_- - 2E_+E_-\cos\theta \quad \Rightarrow \quad \cos\theta = \frac{2E_+E_- - m_{\pi^0}^2}{2E_+E_-}. \tag{6.5}$$

This equation can then simplify down to

$$\cos\theta = \frac{\gamma^2(1-g^2)-2}{\gamma^2(1-g^2)} = \frac{\gamma^2(\beta^2-g^2)-1}{\gamma^2(1-g^2)} = \frac{E_{\pi^0}^2(1-g^2)-2m_{\pi^0}^2}{E_{\pi^0}^2(1-g^2)}$$

$$= 1 - \frac{2m_{\pi^0}^2}{E_{\pi^0}^2(1-g^2)} \tag{6.6}$$

where the identity $\gamma^2 - \beta^2\gamma^2 = 1$ is used. Now, a "small-angle approximation" is used in the case of a boosted regime where the photons are collimated in the lab frame. This allows us to further simplify Eq. (6.6). In small $g \to 0, \theta$, we can expand θ out

$$\cos\theta \approx 1 - \frac{1}{2}\theta^2 = 1 - \frac{2}{\gamma^2} \tag{6.7}$$

and this gives us $\theta = \frac{2}{\gamma}$, or as more commonly known among experimental physicists in ATLAS

$$\Delta R \sim \frac{2m}{p_{\mathrm{T}}}. \tag{6.8}$$

What this equation tells us is that the higher the p_{T} of a parent particle (such as a neutral pion or a top quark), the smaller the opening angle, the smaller the ΔR between decay particles, and the more collimated the particles are. It turns out that this works very well in practice, as shown in Fig. 6.2 for $Z' \to t\bar{t}$ decays where the angular separation is shown as a function of the parent particle in a 2-body decay for the top quark decay and the hadronic W boson decay. It is here that one can see for $p_{\mathrm{T}}^{\mathrm{top}} > 350\,\mathrm{GeV}$, the decay products of the top quark tend to have a separation of $\Delta R < 1.0$. As will be described in the next section, this technique is effective for capturing the full decay of boosted objects with significant substructure inside a single large radius jet.

6.2 Objects

This section will detail the definitions of objects used in the analysis from the perspective of the physics of the detector. Corrections derived from data control samples are applied to simulated events to account for differences between data and simulation in the reconstruction efficiencies, momentum scale, and resolution of leptons,[2] in the efficiency and fake rate for identifying b-jets, and in the efficiency for rejecting jets originating from pile-up interactions.

[2]The term "lepton" exclusively refers to electron or muon in this thesis.

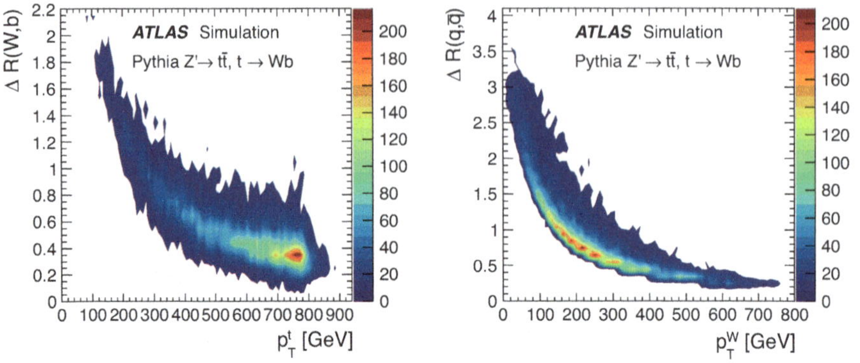

(a) $\Delta R(W, b)$ (b) $\Delta R(q, \bar{q})$

Fig. 6.2 The angular separation between the (**a**) W boson and b-quark in top decays and (**b**) light quarks in W boson decays as a function of the p_{T} of the parent particle [44]. Both distributions are at the generator level and do not include effects due to initial and final state radiation, or the underlying event

Interaction vertices from the proton–proton collisions are reconstructed from at least two tracks with $p_{\mathrm{T}}^{\mathrm{track}} > 0.4\,\mathrm{GeV}$, and are required to be consistent with the beamspot envelope. The primary vertex is identified as the one with the largest sum of squares of the transverse momenta from associated tracks [69]

$$\sum p_{\mathrm{T}}^{\mathrm{track}\,2}. \tag{6.9}$$

Identifying the primary vertex is also important to mitigate pile-up contamination in the calculation of $E_{\mathrm{T}}^{\mathrm{miss}}$.

A series of basic selection criteria are applied to electrons, muons, and jets to define the baseline candidates for an event. From these objects, an overlap removal procedure is applied to prevent double-counting. More stringent requirements are then applied to the overlap-removed, baseline objects to select the final candidates that will be used for the calculation of the kinematic observables in Sect. 7.2.1. The details of the object selections and overlap removal procedure are given below in more detail.

6.2.1 *Small-Radius Jets*

Small-radius (small-R) jets[3] are objects that have been reconstructed from three-dimensional topo-clusters [86] in the hadronic and electromagnetic calorimeters

[3]Unless otherwise specified, "jets" will always refer to the candidate, overlap-removed (see Sect. 6.2.4), small-R jets.

using the anti-k_t jet algorithm [125, 126] with a radius parameter of $R = 0.4$. Before we can make use of these jets, they need to be properly calibrated to compensate and account for reconstruction-based limitations such as:

- dead material: energy lost in the dead material of the detector, e.g., inactive absorbers and instrumentation
- non-compensation: difference in detector response between hadrons, leptons, and photons; specifically the response is lower for hadrons
- punch-through: energy leakage where the hadron showers deposit energy outside of the calorimetry system
- pile-up: energy originating from additional proton–proton collisions in the detector (part of the underlying event which includes initial and final state radiation)
- minimum threshold: hardware limits on energy deposits
- out-of-cone: inefficiencies in reconstruction due to not capturing the full particle shower in the jet

Each topo-cluster is calibrated to the electromagnetic scale response prior to jet reconstruction. The reconstructed jets are then calibrated to the particle level by the application of a jet energy scale (JES) derived from $\sqrt{s} = 13\,\text{TeV}$ data and simulations [90]. Quality criteria are imposed to reject events that contain at least one jet arising from non-collision sources or detector noise [67]. Further selections are applied to reject jets that originate from pile-up interactions by means of a multivariate algorithm, jet-vertex-tagger (JVT) [52], using information about the tracks matched to each jet [78]. For jets with $p_{\text{T},i}^{\text{jet}} < 60\,\text{GeV}$ and located centrally in the detector, where pile-up jets are prominent, we require JVT > 0.59. The baseline jets are required to have $p_{\text{T},i}^{\text{jet}} > 20\,\text{GeV}$ and $|\eta| < 2.8$. After resolving overlaps with baseline leptons (electrons and muons), candidate jets are a subset of the baseline jets that have no overlaps and a stricter cut on transverse momentum, $p_{\text{T},i}^{\text{jet}} > 30\,\text{GeV}$.

6.2.2 *b*-Tagged Jets

From the candidate jets described in Sect. 6.2.1, we apply a multivariate *b*-tagging algorithm, `MV2c10`, using information about the [26, 76]

- impact parameters of inner detector tracks matched to the jet
- presence of displaced secondary vertices
- reconstructed flight paths of *b*-hadrons and *c*-hadrons inside the jet

Of the candidate jets, we require a tighter cut on $p_{\text{T},i}^{\text{jet}} > 30\,\text{GeV}$ with $|\eta| < 2.5$ as this is the region of the calorimeter where the ID provides tracking information. The `MV2c10` algorithm provides a discriminant which a selection is applied on to provide a specific efficiency for the *b*-tagging algorithm. In this analysis, a working point was chosen corresponding to an efficiency of 77%, determined from a sample of simulated $t\bar{t}$ events.[4] This working point was found to be optimal for the statistical significance of the search for a majority of the phase space, compared to the 60, 70, and 85%[5] working points.

6.2.3 Leptons

All baseline leptons are required to have $p_{\mathrm{T}} > 20\,\mathrm{GeV}$ along with the specified identification and quality criteria.[6] Baseline electrons are reconstructed from energy clusters in the EMCal and ID tracks. These electrons are required to have $|\eta| < 2.47$ and pass a set of "loose" quality and identification criteria [53, 71]. Baseline muons are reconstructed from matching tracks in the ID and muon spectrometer. These muons are required to have $|\eta| < 2.5$ and pass a set of "medium" quality and identification criteria [77].

Leptons are selected from the candidates that survive the overlap removal procedure if they fulfill a requirement on the $\sum p_{\mathrm{T}}^{\mathrm{track}}$ of additional ID tracks in an "isolation" cone around the lepton track. This isolation requirement is defined to ensure a flat efficiency of around 99% across the whole electron transverse energy and muon transverse momentum ranges.

The angular separation between the lepton and the b-jet ensuing from a semi-leptonic top quark decay narrows as the p_{T} of the top quark increases. This increased collimation is taken into account by continuously changing the radius of the isolation cone on an event-by-event basis as

$$\min_{\text{leptons}}\left(0.2, \frac{10\,\mathrm{GeV}}{p_{\mathrm{T}}^{\mathrm{lep}}}\right), \tag{6.10}$$

where $p_{\mathrm{T}}^{\mathrm{lep}}$ is the lepton p_{T} expressed in GeV.

Selected electrons are further required to meet the "tight" quality criteria [53, 71]. Selected muons do not have tightened quality criteria. Leptons are matched to the primary vertex by requiring the transverse impact parameter d_0 of the associated ID track to satisfy

[4] At the 77% working point, the corresponding rejection factors against jets originating from c-quarks, τ-leptons, and light quarks & gluons are 6, 22, and 134, respectively [82].

[5] The optimization did sometimes favor the 85% working point, but 77% was not significantly worse. On top of this, there are some benefits using a lower efficiency working point for background estimation due to the enhanced purity of the flavor composition.

[6] The muon and electron definition choices were optimized in the previous version of the analysis [79].

$$\left|\frac{d_0}{\sigma_{d_0}}\right| < 5 \qquad \text{for electrons and} \tag{6.11a}$$

$$\left|\frac{d_0}{\sigma_{d_0}}\right| < 3 \qquad \text{for muons,} \tag{6.11b}$$

where σ_{d_0} is the measured uncertainty of the transverse impact parameter. The longitudinal impact parameter z_0 is required to satisfy

$$|z_0 \sin\theta| < 0.5\,\text{mm}. \tag{6.12}$$

Both the transverse d_0 and longitudinal z_0 impact parameters are defined with respect to the selected primary vertex.

6.2.4 Overlap Removal

In order to make sure the event is reconstructed accurately, one needs to account for double-counting of energy in the detector. For example, if a high-p_T muon radiates some energy in the hadronic calorimeter and then is subsequently captured by the muon spectrometer. I could reconstruct a jet with very few ID tracks and detect a muon, but this jet is formed from final state radiation rather than as part of the hadronization of the parton shower so I should remove this "jet" if it "overlaps" with the muon. Steps are taken to remove "fake" objects as well as energy overlap in the ATLAS detector.

Objects are considered to overlap if they lie less than a distance ΔR from each other

$$\Delta R = \sqrt{(\Delta\eta)^2 + (\Delta\phi)^2} \tag{6.13}$$

where ΔR defines the distance in rapidity y and azimuthal angle ϕ. Overlaps between candidate objects are removed sequentially. The overall procedure is as follows:

1. Electron-Muon: Overlaps between electron and muon candidates are removed. Only electrons get removed at this stage.
2. Electron-Jet: Overlaps between electron and jet candidates are removed. Both electrons and jets can get removed at this stage.
3. Muon-Jet: Overlaps between remaining jets and muon candidates are removed. Both muons and jets can get removed at this stage.

Electron candidates that lie a distance $\Delta R < 0.01$ from muon candidates are removed to suppress contributions from muon bremsstrahlung.

Overlap removal between electron and jet candidates aims to resolve two sources of ambiguity:

1. remove jets that are formed primarily from the showering of a prompt electron; and
2. remove electrons that are produced in the decay chains of hadrons.

Consequently, any non-b-tagged jet whose axis lies $\Delta R < 0.2$ from an electron is discarded. Electrons with $E_{\mathrm{T}} < 50\,\mathrm{GeV}$ are discarded if they lie $\Delta R < 0.4$ from the axis of any remaining jet and the corresponding jet is kept. For higher E_{T} electrons, the latter removal is performed using a variable threshold

$$\Delta R = \min\left(0.4, 0.04 + \frac{10\,\mathrm{GeV}}{E_{\mathrm{T}}^{\mathrm{electron}}}\right) \tag{6.14}$$

to increase the acceptance for events with collimated top quark decays.[7]

Overlap removal between muon and jet candidates aims to resolve two sources of ambiguity:

1. remove jets[8] that appear in close proximity to muons due to high-p_{T} muon bremsstrahlung while retaining the overlapping muon; and
2. remove muons that are likely to have originated from the decay of hadrons while retaining the overlapping jet.

If the angular distance between a muon and a jet is small, $\Delta R < 0.2$, the jet is removed if it is not b-tagged and has fewer than three matching ID tracks. Muons overlapping with remaining jets are removed if

$$p_{\mathrm{T}}^{\mathrm{muon}} < 50\,\mathrm{GeV} \quad \text{and} \quad \Delta R < 0.4 \quad \text{or} \tag{6.15a}$$

$$p_{\mathrm{T}}^{\mathrm{muon}} > 50\,\mathrm{GeV} \quad \text{and} \quad \Delta R < \min\left(0.4, 0.04 + \frac{10\,\mathrm{GeV}}{p_{\mathrm{T}}^{\mathrm{muon}}}\right). \tag{6.15b}$$

6.2.5 Large-Radius Jets

The overlap-removed, candidate jets described in Sect. 6.2.1 are re-clustered [84, 213] into large-radius (large-R) jets[9] using the anti-k_t jet algorithm [125, 126] with a radius parameter of $R = 0.8$. Unlike regular large-R jets formed from topo-

[7] If you have a very boosted top quark, you often have a real electron close to a real b-jet. This is why both the electron and b-tagged jet would be kept.

[8] These jets usually have very few matching ID tracks.

[9] Unless otherwise specified, "large-R" jets will always refer to the candidate, re-clustered, trimmed jets.

clusters and require a separate set of calibrations, the small-R jet calibrations[10] directly propagate through to the re-clustered large-R jets, eliminating the need to prepare specific calibration for each large-R jet collection considered for use in an analysis. These re-clustered jets are further groomed [44, 83, 196, 213] using a trimming[11] algorithm with $f_{\text{cut}} = 0.1$. Candidate large-R jets are required to have $p_{\text{T}}^{\text{jet}} > 100\,\text{GeV}$ and $|\eta| < 2.0$. The radius R and trimming parameter f_{cut} were optimized in the previous version of the analysis [79]. Scanning over $R = 0.8, 1.0, 1.2$ with $f_{\text{cut}} = 0.05, 0.10, 0.20$, the expected sensitivity of the analysis was maximized with $R = 0.8$, $f_{\text{cut}} = 0.1$. Therefore, any variable that can be constructed from these large-R jets to take advantage of the high likelihood for a contained top quark in each re-clustered jet will become a powerful variable to discriminate signal (four tops) over background (zero, one, and two tops).

6.2.6 Missing Transverse Momentum

The $E_{\text{T}}^{\text{miss}}$ in the event is defined as the magnitude of the negative vector sum ($\vec{p}_{\text{T}}{}^{\text{miss}}$) of the transverse momenta of all selected and calibrated baseline muons, electrons, and small-R jets in the event, with an extra "soft" term added to account for energy deposits that are not associated with any of these selected objects.

$$E_{\text{T}}^{\text{miss}} = \underbrace{-\sum \vec{p}_{\text{T}}}_{\text{baseline objects}} + \text{track soft term}. \tag{6.16}$$

This "soft" term is calculated from ID tracks matched to the primary vertex to make it more resilient to contamination from pile-up interactions [66, 68].

[10]The JES uncertainties are used to describe the mass uncertainty on the re-clustered jets. In the signal regions, less than 2% of these re-clustered jets were formed from a single small-R jet, so the mass of the re-clustered jet originates from the $p_{\text{T},i}^{\text{jet}}$ and separation between small-R jets.

[11]Trimming for re-clustered jets means to remove subjets where $p_{\text{T}}^{\text{subjet}} < f_{\text{cut}} p_{\text{T},i}^{\text{jet}}$. For this analysis, subjets with $p_{\text{T}} < 10\%$ of the re-clustered $p_{\text{T}}^{\text{jet}}$ were removed.

Chapter 7
Search for Massive Supersymmetry at 13 TeV

This chapter presents a search for supersymmetry involving pair-produced gluinos ($\tilde{g}$) decaying via off-shell top squarks ($\tilde{t}$) into the lightest neutralino ($\tilde{\chi}_1^0$) and third-generation SM particles: $\tilde{g} \rightarrow \tilde{t}_1\bar{t} \rightarrow t\bar{t}\tilde{\chi}_1^0$. This analysis was performed using $36.1\,\text{fb}^{-1}$ of data collected by the ATLAS experiment in 2015 and 2016. In the Run-I iteration of this analysis in 2014 [46], no significant excess of events over the expected SM background was observed and a lower limit on the $\tilde{g}$ mass was set to be at 1.4 TeV at the 95% confidence level for this signal model. I discuss the work done during the Run-2 iteration of this analysis, also observing no significant excess, and setting a much stronger lower limit on the $\tilde{g}$ mass to 2.1 TeV at the 95% confidence level.

My work described in the rest of the chapter is organized as follows: Sect. 7.1 describes the signal models, the datasets, and the MC samples used to model the signal and backgrounds; Sect. 7.2 describes the kinematic objects and event selections used in the various channels; Sect. 7.3 presents the preselection, background estimation, and reweighting techniques; Sect. 7.4 describes the details of the cut-and-count analysis performed, optimizing for signal discovery; Sect. 7.5 describes the regions defined as a result of the previous section; Sect. 7.6 describes the semi-data-driven normalization of MC to data; and Sect. 7.7 describes the main sources of systematic uncertainty. The next chapter presents the results of this search.

7.1 Searching for New Physics: A Counting Experiment

7.1.1 Signal Models

I performed a search for pair-produced gluinos decaying via top squarks in events with multiple jets originating from the hadronization of b-quarks, high $E_{\text{T}}^{\text{miss}}$, and

© Springer Nature Switzerland AG 2020

G. Stark, *The Search for Supersymmetry in Hadronic Final States Using Boosted Object Reconstruction*, Springer Theses,
https://doi.org/10.1007/978-3-030-34548-8_7

potentially additional light-quark jets and/or an isolated charged lepton. The results I present in Chap. 8 supersede the ones obtained previously using 3.2 fb^{-1} of data collected in 2015 using the same strategy [79].

Various simplified SUSY models [15, 16] are employed to optimize the event selection and/or interpret the results of the search. In terms of experimental signature, they all contain at least four b-jets originating from either gluino or top quark decays, and two $\tilde{\chi}_1^0$, which escape the detector unseen, resulting in high $E_{\mathrm{T}}^{\mathrm{miss}}$. Each top quark decays to a W boson and a b-quark as described in Sect. 2.1.5 and a final state with four tops will have at least four b-jets.

Gluinos are assumed to be pair-produced and to decay as $\tilde{g} \rightarrow \tilde{t}_1 \bar{t} \rightarrow t\bar{t}\tilde{\chi}_1^0$.[1] In all cases, the stop squarks are assumed to be off-shell in order to have simplified models with only two parameters: the gluino and $\tilde{\chi}_1^0$ masses.[2] All other sparticles are decoupled.

A simplified model is used to optimize the event selection and to interpret the results. In the Gtt model, illustrated in Fig. 7.1, each gluino undergoes an effective three-body decay $\tilde{g} \rightarrow t\bar{t}\tilde{\chi}_1^0$) via off-shell top squarks, with a branching ratio of 100%. The Gtt model has four b-jets and two $\tilde{\chi}_1^0$, and four W bosons originating from the top quark decays, $t \rightarrow Wb$, in its final state.

The technical implementation of the simulated samples produced from these models is described in Sect. 4.4.

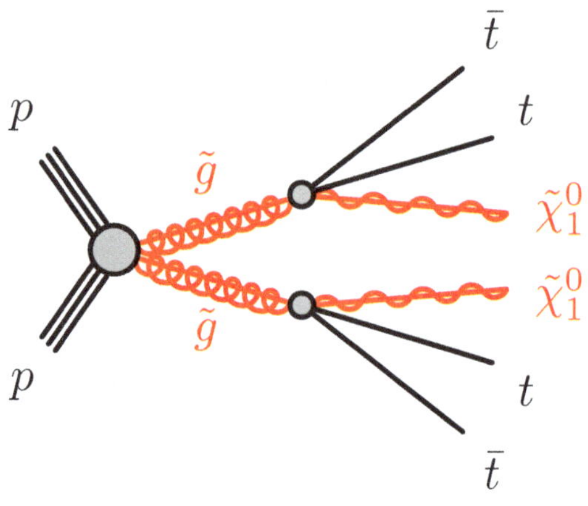

Fig. 7.1 The decay topology in the off-shell Gtt simplified model

[1] The charge conjugate process is implied.

[2] The analysis sensitivity is found to be mostly independent of the top squark mass, except when the top squark is very light [46].

7.2 Kinematic Variables and Event Selection

In ATLAS, many of the kinematics and selections are standardized through a central analysis framework. This section will detail the definitions used in the analysis from the perspective of the physics of the detector. Corrections derived from data control samples are applied to simulated events to account for differences between data and simulation in the reconstruction efficiencies, momentum scale and resolution of leptons (electrons and muons), in the efficiency and fake rate for identifying b-jets, and in the efficiency for rejecting jets originating from pile-up interactions.

There are three primary classes of event selection observables [143] that are sensitive to distinct features of SUSY processes:

- **Missing energy-type**: sensitive to the properties of the invisible states, e.g., how many neutralinos in the event, what is their mass, etc.;
- **Energy scale-type**: sensitive to the overall energy scale of the event, e.g., the mass of the gluino $m_{\tilde{g}}$;
- **Energy structure-type**: sensitive to the structure of the visible energy, e.g., how many partons are generated in the decay, how that energy is partitioned across the final state visible and invisible objects.

Observables that fall into each of these classes are used in the analysis search I performed and are discussed in the following sections.

7.2.1 Kinematic Variables

As described in Sect. 7.1.1, the final state of the signal is very busy. There can be 3 or 4 b-jets and, depending on the mass splitting between the $\tilde{g}$ and $\tilde{\chi}_1^0$, lots of missing transverse momentum with many highly energetic jets. It is useful to define some variables that discriminate the supersymmetry signal models against the Standard Model background. This section describes the definition of those variables that will be used to scan the phase space of the analysis in Sect. 7.4 to form search channels in Sect. 7.5 where the signal is expected to dominate over background.

7.2.1.1 Object Multiplicity

The number of jets, N_{jet}, and number of b-jets, $N_{b\text{-jets}}$, are counting variables which count the number of selected jets and number of selected b-jets in the event.

7.2.1.2 Effective Mass

The effective mass variable (m_{eff}) is defined as:

$$m_{\text{eff}} = \sum_{\text{jets}}^{N_{\text{jet}}} p_{\text{T}}^{\text{jet}} + \sum_{\text{leptons}}^{N_{\text{lepton}}} p_{\text{T}}^{\ell} + E_{\text{T}}^{\text{miss}}, \tag{7.1}$$

which is a scalar sum of the p_{T} of the selected objects in the event. It typically has a much higher value in pair-produced gluino ($\tilde{g}$) events than in background events.

7.2.1.3 Transverse Mass

In regions with at least one selected lepton, the transverse mass m_{T} composed of the p_{T} of the leading selected lepton (ℓ) and $E_{\text{T}}^{\text{miss}}$ is defined as:

$$m_{\text{T}} = \sqrt{2 p_{\text{T}}^{\ell} E_{\text{T}}^{\text{miss}} \{1 - \cos\left[\Delta\phi(\vec{p}_{\text{T}}^{\text{ miss}}, \vec{p}_{\text{T}}^{\ell})\right]\}}. \tag{7.2}$$

It is used to reduce the $t\bar{t}$ and W + jets background events in which a W boson decays leptonically. Neglecting resolution effects, the m_{T} distribution for these backgrounds has an expected upper bound corresponding to the W boson mass and typically has higher values for Gtt events. Another useful transverse mass variable is $m_{\text{T, min}}^{b\text{-jets}}$, the minimum transverse mass formed by $E_{\text{T}}^{\text{miss}}$ and any of the three highest-p_{T} b-tagged jets in the event:

$$m_{\text{T, min}}^{b\text{-jets}} = \min_{i \le 3}\left(\sqrt{2 p_{\text{T}}^{b\text{-jet}_i} E_{\text{T}}^{\text{miss}} \{1 - \cos\left[\Delta\phi(\vec{p}_{\text{T}}^{\text{ miss}}, \vec{p}_{\text{T}}^{\,b\text{-jet}_i})\right]\}}\right). \tag{7.3}$$

The $m_{\text{T, min}}^{b\text{-jets}}$ distribution has an expected upper bound corresponding to the top quark mass for $t\bar{t}$ events with a semi-leptonic top quark decay. Since the $\tilde{\chi}_1^0$, which produces $E_{\text{T}}^{\text{miss}}$ in SUSY events, is largely independent of the b-jet kinematics, the value of $m_{\text{T, min}}^{b\text{-jets}}$ can peak at larger values for signal processes.

7.2.1.4 Total Jet Mass

Another powerful variable is the total jet mass variable, M_J^{Σ} [38, 59], defined as:

$$M_J^{\Sigma} = \sum_{i \le 4} m_{J,i}, \tag{7.4}$$

where $m_{J,i}$ is the mass of the large-radius re-clustered jet i in the event. The decay products of a hadronically decaying boosted top quark can be reconstructed in a single large-radius re-clustered jet, resulting in a jet with a high mass. This variable typically has larger values for Gtt events than for background events. This is because

Gtt events contain as many as four[3] hadronically decaying top quarks while the background is dominated by $t\bar{t}$ events with one or two semi-leptonic top quark decays. The presence of numerous boosted and semi-boosted top quarks in the signal events leads to the formation of high-p_T, massive jets at the scale of $R \approx 0.8$. This variable is sensitive to the large degree of structure in signal events compared to background processes which are dominated by semi-leptonic and dileptonic $t\bar{t}$ events (Figs. 7.5 and 7.6).

7.2.1.5 Multijet Suppression

The requirement of a selected lepton, with the additional requirements on jets, E_T^{miss} and event variables described above, will make the multijet background negligible for the $\geq$1-lepton signal regions. For the 0-lepton signal regions, the minimum azimuthal angle $\Delta\phi_{min}^{4j}$ between $\vec{p}_T^{miss}$ and the p_T of the four leading small-R jets in the event, defined as:

$$\Delta\phi_{min}^{4j} = \min_{i\leq 4}\left(\left|\phi_{jet_i} - \phi_{\vec{p}_T^{miss}}\right|\right), \quad (7.5)$$

is required to be greater than 0.4. This requirement suppresses the multijet background, which can produce events with large E_T^{miss} if containing poorly measured jets or neutrinos emitted close to the axis of a jet. In particular, anything that is all-hadronic is typically mis-modeled[4] with Monte Carlo simulations so contributions from a multijet background are estimated using a data-driven approach "jet smearing" (described in [43]). This method performs a smearing of the jet response in data events with well-measured E_T^{miss} ("seed events") to estimate the tail of E_T^{miss} where the modeling is not as great. Figure 7.2 shows the impact of jet smearing on the modeling of multijet as compared to 2015 + 2016 data. This is even more important for b-jets because in the high tails of effective mass, most of the mis-measurement in the hadronic decays of multijets comes in the form of heavy-flavor decays. As Fig. 7.2 also shows, a selection of $\Delta\phi_{min}^{4j} > 0.4$ effectively suppresses a majority of the multijet background in the search, so the multijet background is typically called "reducible."

7.2.2 Event Selection

Two different types of selections are applied in succession for the analysis, event selection and then preselection. Two sets of preselection criteria targeting the 0-

[3]In the case of events with less than four re-clustered jets, all of them are used.

[4]The generators have different ways of calculating the non-perturbative QCD showers.

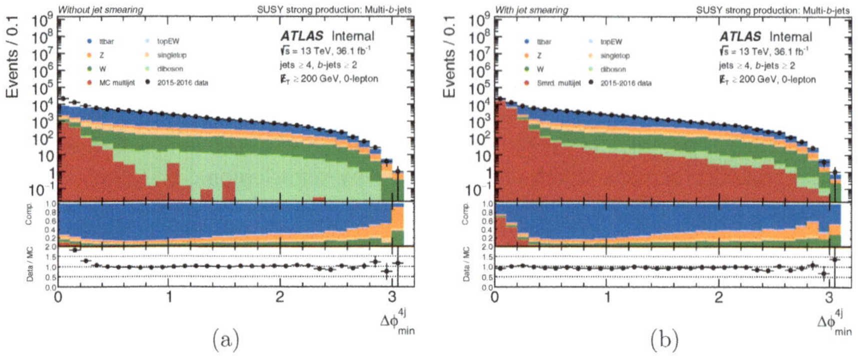

Fig. 7.2 Distributions of the $\Delta\phi^{4\text{j}}_{\text{min}}$ variable before and after jet smearing is applied for 2015–2016 data using a 0-lepton preselection, without the $\Delta\phi^{4\text{j}}_{\text{min}}$ selection applied. The left figure shows the distribution in log-scale of the number of events in each bin of $\Delta\phi^{4\text{j}}_{\text{min}}$. The right figure shows the background composition with each background contribution some fraction of the total background. The third panel shows the data/MC ratio where data is compared to the total estimated, pre-fit background. Looking at the background composition, it is clear that the multijet background (red) is reducible, as is evidenced by a $\Delta\phi^{4\text{j}}_{\text{min}} > 0.4$ selection that can be applied. (**a**) Without jet smearing. (**b**) With jet smearing

lepton and the 1-lepton channels are presented in Sect. 7.3. The modeling of the data in these regions is also discussed in that section. The event selection criteria, defined based on kinematic requirements for the objects defined in Sect. 6.2 and discriminating event-based variables described in Sect. 7.2.1, are used to further reject the background. There are a number of more common general cuts which are applied to all events described in the next few sections. These cuts are

- Good Runs List (GRL): applied to data only
- Tile, LAr, SCT: applied to data only
- Trigger: applied to both data and MC
- Jet Cleaning: applied to both data and MC
- Muon Cleaning: applied to both data and MC

7.2.2.1 Good Runs

ATLAS provides an XML file containing a list of events in data that have passed data quality criteria. This removes luminosity blocks with poor detector quality. In particular, this analysis is sensitive to b-tagging and runs with the Insertable B-Layer (IBL) turned off are not included. Two different GRLs are used: one for 2015 data and one for 2016 data.

7.2.2.2 Tile, LAr, and SCT

Following ATLAS recommendations, corrupted events due to errors in the Tile, LAr, or SCT are vetoed.

7.2.2.3 Trigger

The unprescaled[5] $E_{\mathrm{T}}^{\mathrm{miss}}$ triggers used for both the 0-lepton and 1-lepton analyses are `HLT_xe70` for 2015 data and `HLT_xe100_mht_L1XE50`/`HLT_xe110_mht_L1XE50` for early/late 2016 data, respectively. One important feature of the $E_{\mathrm{T}}^{\mathrm{miss}}$ triggers is to make sure that the selections in our analysis lie above the turn-on curve. This means that the trigger does not affect the selection efficiency when searching for signal over background.

The lowest unprescaled trigger for the 2015 dataset is 70 GeV and for 2016 dataset is 100 and 110 GeV. The efficiencies of these triggers are shown in Fig. 7.3 for the 2015 dataset and Fig. 7.4 for the 2016 dataset. These figures show the efficiency of the $E_{\mathrm{T}}^{\mathrm{miss}}$ trigger in both $t\bar{t}$ MC and 25 ns data for different requirements on the leading jet p_{T}. The trigger is plotted as a function of a modified, reconstructed $E_{\mathrm{T}}^{\mathrm{miss}}$. Typically, when $E_{\mathrm{T}}^{\mathrm{miss}}$ is reconstructed offline, it incorporates information from the muon spectrometer, but this information is not available in L1. To properly study the L1 $E_{\mathrm{T}}^{\mathrm{miss}}$ trigger, muons are subtracted from the reconstructed

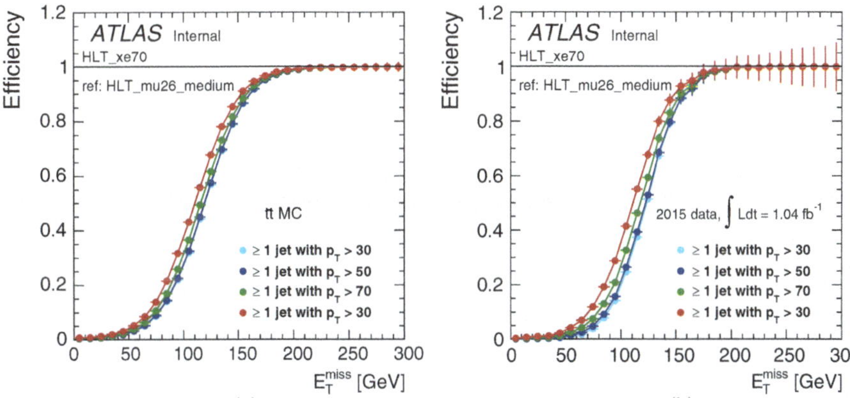

(a) (b)

Fig. 7.3 Efficiency of the 2015 $E_{\mathrm{T}}^{\mathrm{miss}}$ trigger for different requirements on the leading jet p_{T} [21]. The muons are subtracted from $E_{\mathrm{T}}^{\mathrm{miss}}$ to reproduce the L1 $E_{\mathrm{T}}^{\mathrm{miss}}$. The reference trigger used to select events is `HLT_mu26_medium`. (**a**) $t\bar{t}$ MC. (**b**) 25 ns data

[5] If a trigger is prescaled, this means the trigger rate is purposefully decreased in order to keep the output rate manageable.

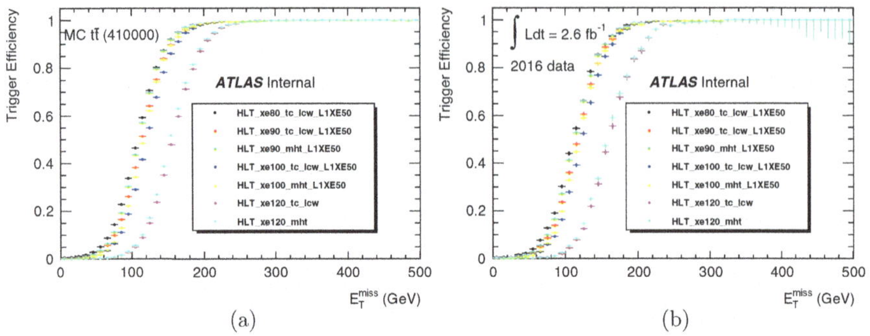

Fig. 7.4 Efficiency of the various 2016 $E_{\mathrm{T}}^{\mathrm{miss}}$ triggers as a function of the offline $E_{\mathrm{T}}^{\mathrm{miss}}$ [21]. The muons are subtracted from $E_{\mathrm{T}}^{\mathrm{miss}}$ to reproduce the L1 $E_{\mathrm{T}}^{\mathrm{miss}}$. The reference trigger used to select events is `HLT_mu26_ivarmedium`. One of the triggers used in the analysis for early 2016 data is shown here: `HLT_xe100_mht_L1XE50`. (**a**) $t\bar{t}$ MC. (**b**) 25 ns data

$E_{\mathrm{T}}^{\mathrm{miss}}$ to reproduce the L1 $E_{\mathrm{T}}^{\mathrm{miss}}$. Since our preselection, described in Sect. 7.3, requires $E_{\mathrm{T}}^{\mathrm{miss}} > 200$ GeV, you can see that both $E_{\mathrm{T}}^{\mathrm{miss}}$ triggers are fully efficient.

7.2.2.4 Jet Cleaning

Fake jets can arise from non-collision background or cosmic muons with a catastrophic energy loss in the calorimeters or from fake signals in the calorimeter, arising either from noise bursts or the presence of coherent noise. A set of cuts having a high rejection against fake jets while preserving an efficiency close to 100% for selected jets has been designed. The selection criteria are based on

- the timing of the calorimeter signal with respect to that of the BC,
- the quality of the fit on the calorimeter pulse shape,
- the fraction of jet energy belonging to specific calorimeter samples, and
- the charged momentum fraction[6] of the jet.

The selections are summarized at [67]. Events are vetoed if any of the baseline jets are determined to be "bad" based on the above criteria.

7.2.2.5 Muon Cleaning

Another set of event cleaning cuts come from identifying two types of muons, "fake" muons and cosmic-ray muons as described below. If either of these muons are identified in an event, the event is excluded from the analysis search.

[6]As measured by the ID relative to the calorimeter.

1. There are fake muons that can be reconstructed from high hit multiplicities in the muon spectrometer due to very energetic punch-through jets or from badly measured inner detector tracks in jets wrongly matched to muon spectrometer segments. Such fake muons contribute to fake $E_{\mathrm{T}}^{\mathrm{miss}}$. Events containing one or more fake muon candidates are rejected.
2. Events with muons arising from cosmic rays are also rejected to avoid the effects on the tails of the $E_{\mathrm{T}}^{\mathrm{miss}}$. Potential cosmic muons are identified after the muon-jet overlap removal (see Sect. 6.2.4) as muons with large longitudinal and transverse impact parameters. Events containing one or more muon candidates with

$$|d_0| > 0.2\,\mathrm{mm} \qquad \text{and} \tag{7.6a}$$

$$|z_0| > 1\,\mathrm{mm} \tag{7.6b}$$

are rejected to suppress cosmic rays.

7.3 Preselection Comparisons of Data/MC

This section contains distributions of the analysis observables described in Sect. 7.2.1, showing the pre-fit modeling of these observables in the following loose selection regions. All are triggered with the $E_{\mathrm{T}}^{\mathrm{miss}}$ triggers which are fully efficient in our offline $E_{\mathrm{T}}^{\mathrm{miss}}$ cut.

Preselection criteria,[7] as summarized in Table 7.1, in the 0-lepton and 1-lepton channels require $E_{\mathrm{T}}^{\mathrm{miss}} > 200\,\mathrm{GeV}$, in addition to the $E_{\mathrm{T}}^{\mathrm{miss}}$ trigger requirement, and at least four jets of which at least two must be b-tagged. The 0-lepton channel requires no selected leptons. The 1-lepton channel requires the event to contain at least one selected lepton. Because the 0-lepton channel requires no selected leptons, a $\Delta\phi_{\mathrm{min}}^{4\mathrm{j}}$ selection is additionally required to reduce the multijet background as described in Sect. 7.2.1.5.

Table 7.1 Definitions of the 0-lepton and 1-lepton preselection of the cut-and-count analysis. m_{eff}, $m_{\mathrm{T,\,min}}^{b\text{-jets}}$, and m_{T} have no preselection cuts for both channels

	0-lepton	1-lepton
N_{lepton}	=0	≥1
Trigger (Sect. 7.2.2.3)	70 GeV unprescaled	100 GeV unprescaled
$E_{\mathrm{T}}^{\mathrm{miss}}$	> 200 GeV	
N_{jet}	>4	
$N_{b\text{-jets}}$	>2	
$\Delta\phi_{\mathrm{min}}^{4\mathrm{j}}$	>0.4	–

[7]Preselection is designed to be 100% efficient for the signal processes under study.

In this analysis, correction factors need to be extracted to account for shape discrepancies in the m_{eff} spectrum between the data and the expected background for the 1-lepton preselection sample.[8] These factors are defined as the ratio of the number of observed events to the predicted number of background events in a given m_{eff} bin, in a signal-depleted region. This region is defined by applying the 1-lepton preselection criteria and requiring exactly two b-tagged jets and $m_{\mathrm{T,\,min}}^{b\text{-jets}} < 140\,\mathrm{GeV}$. This kinematic reweighting leads to correction factors ranging from 0.7 to 1.1. They are applied to the background prediction and the full size of the correction is taken as an uncertainty for both the background and signal events.

Figures 7.5 and 7.6 show the multiplicity of selected jets and b-tagged jets, the distributions of $E_{\mathrm{T}}^{\mathrm{miss}}$, m_{eff}, and M_J^{Σ} for events passing the 0-lepton or the 1-lepton preselection, respectively. Figures 7.5 and 7.6 also display the distribution of transverse mass: $m_{\mathrm{T,\,min}}^{b\text{-jets}}$ for 0-lepton channel and m_{T} for 1-lepton channel. The m_{eff} correction described above is applied in the 1-lepton channel.

The uncertainty bands include the statistical and experimental systematic uncertainties, as described in Sect. 7.7, but not the theoretical uncertainties in the background modeling.

The data and the predicted background are found to agree reasonably well at the preselection level after the kinematic reweighting described above. A discrepancy between data and prediction is observed for the number of b-tagged jets, but it has a negligible impact on the background estimate after the renormalization of the simulation in dedicated control regions with the same b-tagged jets requirements as the signal regions, as described in Sect. 7.5. Example signal models with enhanced cross-sections are overlaid for comparison.

7.4 Optimizations

The main analysis strategy documented in this thesis is a **cut-and-count** analysis, using partially overlapping single-bin signal regions (SRs), optimized to maximize the expected discovery power for benchmark signal models, and allowing for reinterpretation of the results. This is described in more detail in Sect. 7.4.1. In order to define these regions, it helps to scan the phase space of the kinematic variables to determine the strongest selections to maximize signal over background and this strategy is described in more detail in Sect. 7.4.2. After performing the optimization, the regions for the cut-and-count analysis are defined in Sect. 7.5.

[8]This effective mass reweighting is described in more detail in [82]. This was a shape discrepancy that was not observed in 0-lepton. The underlying cause is still not understood and the current, 2017 iteration of this analysis continues to study this more.

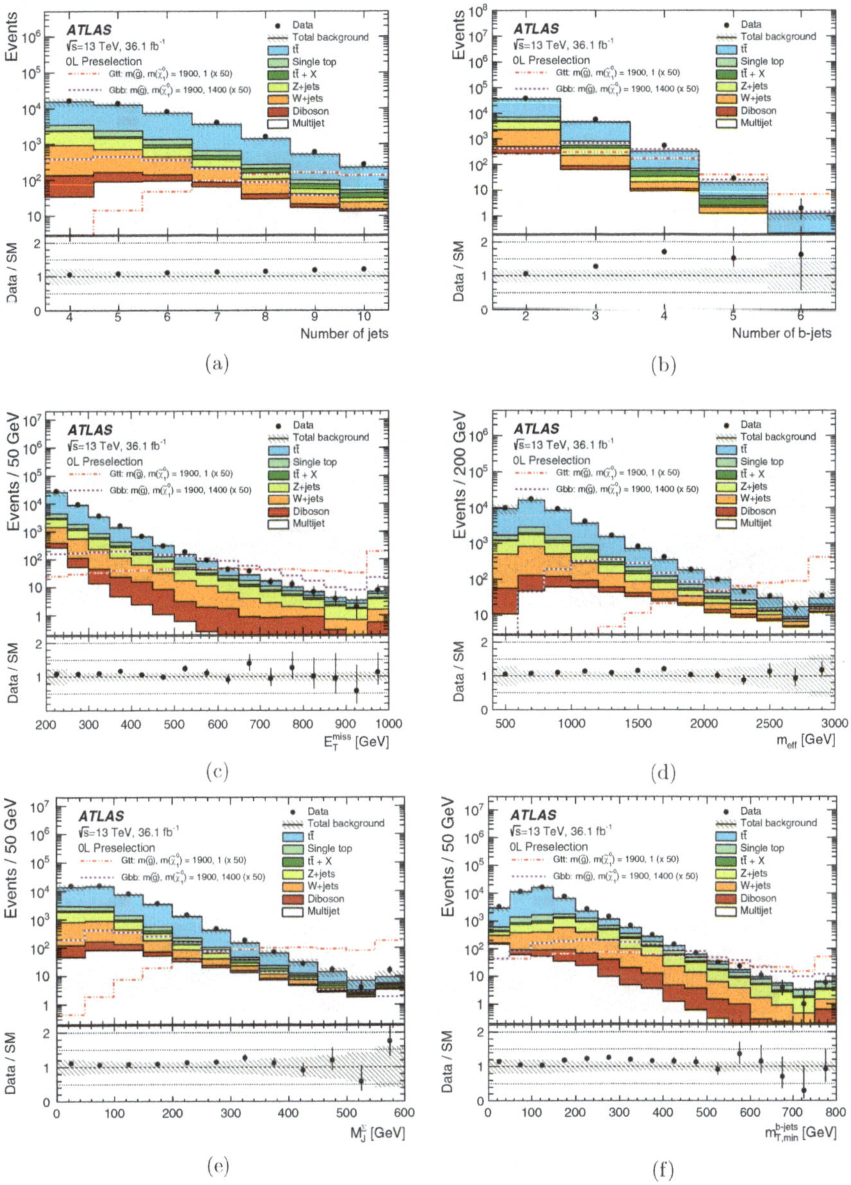

Fig. 7.5 Distributions of discriminating variables for events passing the 0-lepton preselection criteria. The statistical and experimental systematic uncertainties (as defined in Sect. 7.7) are included in the uncertainty band. The last bin includes overflow events. The lower part of each figure shows the ratio of data to the background prediction. All backgrounds (including $t\bar{t}$) are normalized using the best available theoretical calculation described in Sect. 4.4. The background category $t\bar{t}+X$ includes $t\bar{t}W/Z$, $t\bar{t}h$, and $t\bar{t}t\bar{t}$ events. Example signal models with cross-sections enhanced by a factor of 50 are overlaid for comparison. (**a**) N_{jet}. (**b**) $N_{b\text{-jets}}$. (**c**) $E_{\text{T}}^{\text{miss}}$. (**d**) m_{eff}. (**e**) M_J^{Σ}. (**f**) $m_{\text{T, min}}^{b\text{-jets}}$

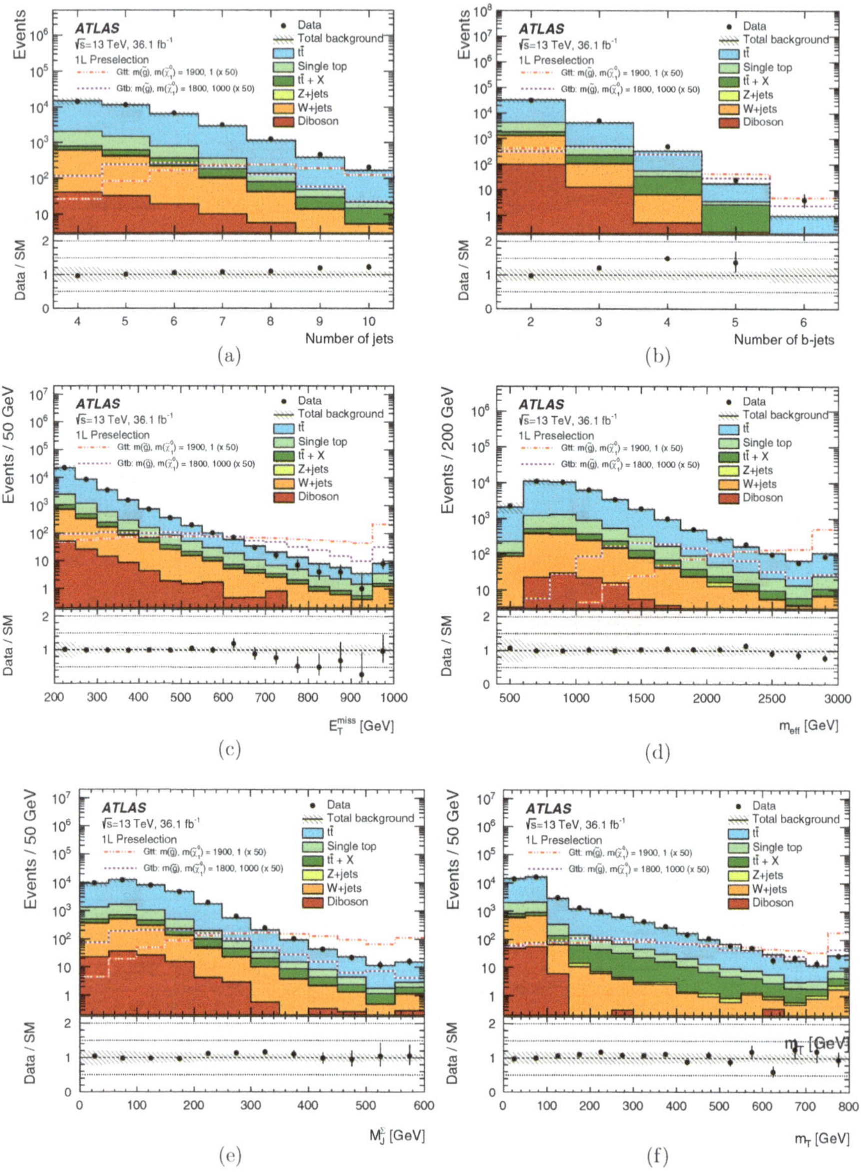

Fig. 7.6 Distributions of discriminating variables for events passing the 1-lepton preselection criteria, after applying the kinematic reweighting to the m_{eff} distribution described in the text. The statistical and experimental systematic uncertainties (as defined in Sect. 7.7) are included in the uncertainty band. The last bin includes overflow events. The lower part of each figure shows the ratio of data to the background prediction. All backgrounds (including $t\bar{t}$) are normalized using the best available theoretical calculation described in Sect. 4.4. The background category $t\bar{t}+X$ includes $t\bar{t}W/Z$, $t\bar{t}h$ and $t\bar{t}t\bar{t}$ events. Example signal models with cross-sections enhanced by a factor of 50 are overlaid for comparison. **(a)** N_{jet}. **(b)** $N_{b\text{-jets}}$. **(c)** $E_{\mathrm{T}}^{\mathrm{miss}}$ **(d)** m_{eff}. **(e)** M_J^{Σ}. **(f)** m_{T}

7.4.1 Analysis Strategy and Background Treatment

In order to enhance the sensitivity to the various signal benchmarks described in Sect. 7.1.1, multiple SRs are defined. The SRs are defined to probe the existence of a signal or to assess model-independent upper limits on the number of signal events. The main background in all these regions is the production of a $t\bar{t}$ pair in association with heavy-flavor and light-flavor jets. A normalization factor ($\mu_{t\bar{t}}$) for this background is extracted for each individual SR from a data control region (CR) that has comparable background composition and kinematics. This is ensured by keeping the kinematic requirements similar in the two regions. The CRs and SRs are defined to be mutually exclusive. Signal contributions in the CRs are suppressed by inverting or relaxing some requirements on the kinematic variables (e.g., m_{T} or $m_{\mathrm{T,\,min}}^{b\text{-jets}}$), leading to a signal contamination in the CRs of 6% at most.

The $t\bar{t}$ normalization is cross-checked in validation regions (VRs) that share similar background composition, i.e., jet and lepton flavors, with the SR. The signal contamination in the VRs is found to be lower than 30% for benchmark signal mass points above the already excluded mass range. The $t\bar{t}$ purity is superior to 73 and 53% in the CRs and VRs, respectively. All of this is summarized in Fig. 7.7.

The non-$t\bar{t}$ backgrounds mainly consist of single-top, W + jets, Z + jets, $t\bar{t}$ $+W/Z/h$, $t\bar{t}t\bar{t}$, and diboson events. Their normalization is taken from the simulation normalized using the best available theory prediction. The multijet background is found to be very small or negligible in all regions. It is estimated using a procedure described in [39], in which the jet response is determined from simulated dijet events. This response function is then used to smear the jet response in low-$E_{\mathrm{T}}^{\mathrm{miss}}$ events. The jet response is cross-checked with data where the $E_{\mathrm{T}}^{\mathrm{miss}}$ can be unambiguously attributed to the mismeasurement of one of the jets.

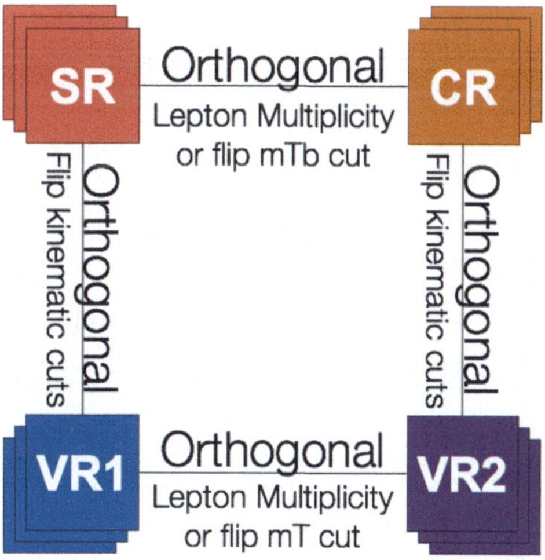

Fig. 7.7 A diagrammatic overview of how the different regions SR, CR, and VR are defined for the cut-and-count analysis

7.4.2 Optimization Strategy

This section describes the optimization of the SRs, CRs, and VRs. For the SRs, the figure of merit is the expected significance, defined by the function `BinomialExpZ` of `RooStats` https://www.slac.stanford.edu/grp/eg/minos/rootsys/cvs/roofit/roostats/src/NumberCountingUtils.cxx, https://root.cern.ch/root/html526/RooStats__NumberCountingUtils.html assuming 35 fb^{-1} of data.[9] This function estimates a significance for a signal above SM background, given the Poisson likelihood, with an estimate of the impact of systematic uncertainties.[10] We require that the MC statistical uncertainty on the background is less than 30% to ensure that the optimized selection is reliable. The optimization then motivates defining a number of SRs. The goal of these SRs is to provide robust regions capable of discovery of SUSY signatures, and for ease of reinterpretation by theorists [149].

Two categories of these discovery regions are defined: Gtt-0L and Gtt-1L in Sects. 7.4.3 and 7.4.4. Gtt-0L targets the zero lepton channel of the $\tilde{g} \to t\bar{t}$ process and Gtt-1L targets the one lepton channel. Determination of the signal and CRs for the cut-and-count analysis is carried out with several metrics in mind:

1. The expected significance is used as the primary metric to evaluate the performance of SRs.
2. The expected number of signal events must be $N_S > 2$ in order to ensure possibility of an observation.
3. The total background must always be $N_B > 0.5$.
4. The background composition within SRs and CRs is required to be dominated by $t\bar{t}$. The presence of other backgrounds, whose contributions are evaluated directly from simulation, is minimized.[11]
5. The statistical uncertainty on $t\bar{t}$ backgrounds must never be larger than 30%.
6. The signal contamination in validation and CRs is minimized.

I wrote a package, an optimization framework, that automates and streamlines this entire process in a very generic way. See Appendix A for more about the technical implementation.

7.4.3 Gtt-0L Optimization

The Gtt 0L optimization used the following variables which are summarized in Table 7.2.

---

[9]2015 + 2016 data accounted for 36.1 fb^{-1} at the end of 2016, so this is not too far off.

[10]This is the standard significance optimization technique within the SUSY group in ATLAS.

[11]The post-fit $t\bar{t}$ purity in all regions is >50%, and typically >70%.

Table 7.2 The 0-lepton optimization phase space that was scanned over

	Inclusive		Step
	Start	Stop	Size
N_{jet}	6	10	1
$N_{b\text{-jets}}$	3	6	1
$E_{\text{T}}^{\text{miss}}$	200 GeV	1000 GeV	50 GeV
m_{eff}	800 GeV	3000 GeV	100 GeV
M_J^{Σ}	0 GeV	500 GeV	100 GeV
$m_{\text{T, min}}^{b\text{-jets}}$	0 GeV	200 GeV	4 GeV0

This corresponded to 16,560 cuts. Each variable was scanned from "start" to "stop" in discrete step sizes. The starting and stopping values of the scan are inclusive

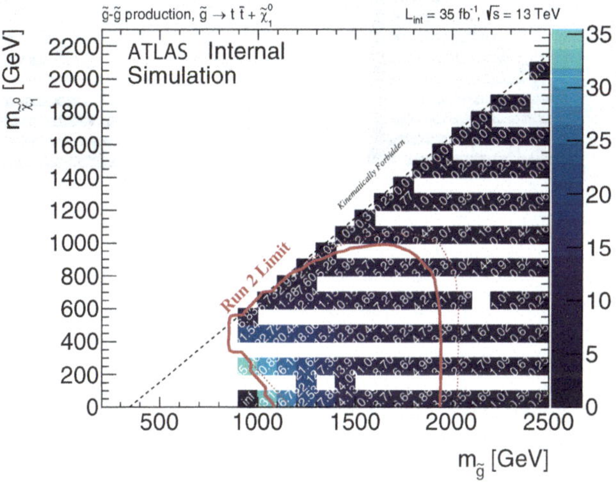

Fig. 7.8 Significance of optimal cuts (Fig. 7.9) for each grid point in the Gtt 0-lepton channel. The grid represents all the mass points for the $\tilde{g}$-$\tilde{\chi}_1^0$ pair with the mass of the gluino on the x-axis in GeV and the mass of the neutralino on the y-axis in GeV. This optimization was performed at an assumed total integrated luminosity of 35 fb^{-1}. The Run-2 limit from my first paper on this search at 3.6 fb^{-1} is overlaid in red [79]. The significance reported is an estimate of the CLs method using `BinomialExpZ` of `RooStats` https://www.slac.stanford.edu/grp/eg/minos/rootsys/cvs/roofit/roostats/src/NumberCountingUtils.cxx, https://root.cern.ch/root/html526/RooStats__NumberCountingUtils.html. This significance represents the discriminating power the optimal cut has in each mass point

Since the masses of the new particles are not known yet, the grid plotted in Figs. 7.8 and 7.9 represents all the mass points for the $\tilde{g}$-$\tilde{\chi}_1^0$ pair with the mass of the gluino on the x-axis in GeV and the mass of the neutralino on the y-axis in GeV. Figure 7.8 shows the optimal significance possible with the best selected variables at each grid point. Figure 7.9 shows the optimal cuts on various variables selected at each mass point. These were produced using the optimization framework

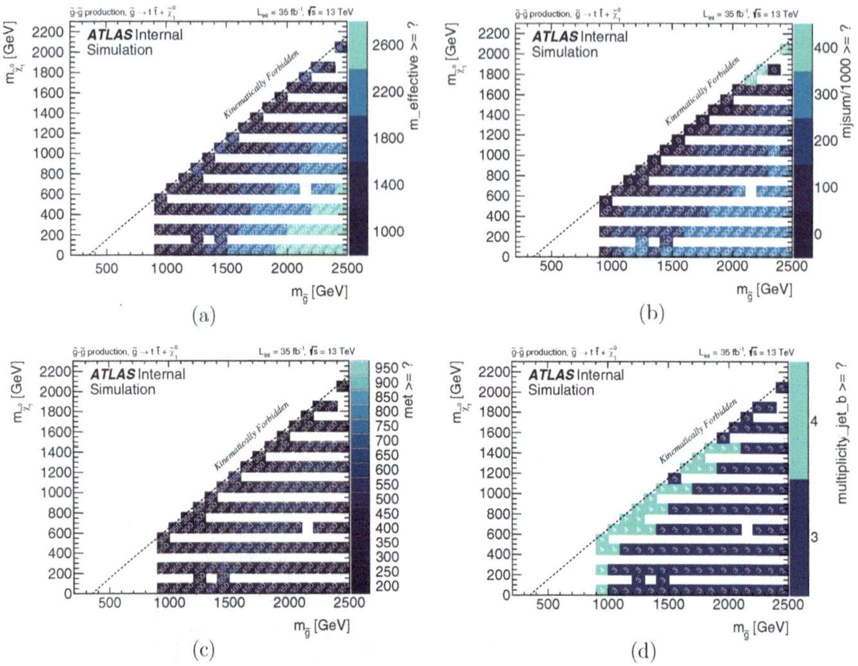

Fig. 7.9 Optimal cut values for each grid point in the Gtt 0-lepton channel. The grid represents all the mass points for the $\tilde{g}$-$\tilde{\chi}_1^0$ pair with the mass of the gluino on the x-axis in GeV and the mass of the neutralino on the y-axis in GeV. This optimization was performed at an assumed total integrated luminosity of 35 fb^{-1}. The kinematic variables were scanned over the values defined in Table 7.2, a significance for each combination of selections was computed, and the selection that provided maximum discriminating power in each mass point is plotted. Each plot represents the cut applied to the kinematic variable **(a)** m_{eff}, **(b)** M_J^{Σ}, **(c)** $E_{\text{T}}^{\text{miss}}$, **(d)** $N_{b\text{-jets}}$

that I wrote for this analysis, but is general enough to be applied to similar types of analyses doing a search and trying to understand the phase space of their signal model. Being able to visualize the change in the optimal cut for each variable across the different gluino-neutralino masses influences our choices for signal region definitions. For example, we can see two regions that appear based on the b-jet multiplicity where most mass points prefer requiring $N_{b\text{-jets}} \geq 3$ except for the compressed region which requires a tighter selection of $N_{b\text{-jets}} \geq 4$. Meanwhile, it is also nice to see a reinforcement of physics that we expect to see from this signal model as in the behavior of M_J^{Σ} where the boosted regions towards the bottom-right (a larger mass splitting) prefer a tighter cut on the mass of the four leading re-clustered jets than the compressed regions towards the top-left (a smaller mass splitting). This is consistent with re-clustered jets containing more merged decays and accidental substructure in a boosted topology.

7.4.3.1 Signal Regions

From these optimization results, three SRs have been designed for the 0-lepton channel of the Gtt analysis, summarized in Table 7.4. All 0-lepton SRs veto events with one or more reconstructed signal leptons, and require at least 3 b-tagged small-R jets. SR-B[12] targets the dramatic events in the region of high mass splitting between the $\tilde{g}$ and $\tilde{\chi}_1^0$. This region places large requirements on the m_{eff} and the largest requirement on M_J^{Σ} out of all Gtt SRs. SR-M[13] targets the bulk region of the signal grid, requiring large amounts of $E_{\text{T}}^{\text{miss}}$, m_{eff}, and M_J^{Σ}. SR-C[14] targets the "near-diagonal" region of the signal grid, where the small mass splitting between the $\tilde{g}$ and $\tilde{\chi}_1^0$ results in final states with jets which are typically softer than those found in the other regions of the signal plane. SR-C also has an additional b-tagged jet to further reject background in this portion of phase space.

The significance at each signal grid point for the optimal SR is plotted in Fig. 7.10b. The three SRs each target distinct regions of phase space (Fig. 7.10a), with little degradation in performance from the optimal set of cuts at each mass point (Fig. 7.11).

7.4.3.2 Control Regions

CRs are also defined for each SR in Table 7.4. As the background is expected to be dominated entirely by semi-lepton $t\bar{t}$ events, due to the b-jet requirement and the $E_{\text{T}}^{\text{miss}}$ requirement, following the strategy of the 2015 analysis we define only a single $t\bar{t}$ CR for the analysis and will take all the remaining backgrounds directly from simulation [20]. Since the background is expected to be semi-leptonic $t\bar{t}$, a single lepton CR strategy is used. An upper cut on m_{T} is enforced to prevent overlap with the Gtt SRs. All jet number related cuts are lowered by 1 between each SR and the corresponding CR (since a lepton replaces a jet), and cuts on $E_{\text{T}}^{\text{miss}}$ and m_{eff} and $m_{\text{T, min}}^{b\text{-jets}}$ are lowered to ensure sufficient statistics (a minimum of 15 expected events) in the CRs. The signal contamination in each CR is expected to be very small, and is shown in Fig. 7.12.

7.4.3.3 Validation Regions

Table 7.4 also defines one VR for each SR. The VR category validates the extrapolation from the 1L CR to the 0L SR. This is done by inverting the cut on M_J^{Σ}, which is shown to be well modeled in Sect. 7.3. Signal contamination plots for the VRs are shown in Figs. 7.13 and 7.14. Both are rather low.

[12]"B" for boosted.

[13]"M" for moderate-boost.

[14]"C" for compressed.

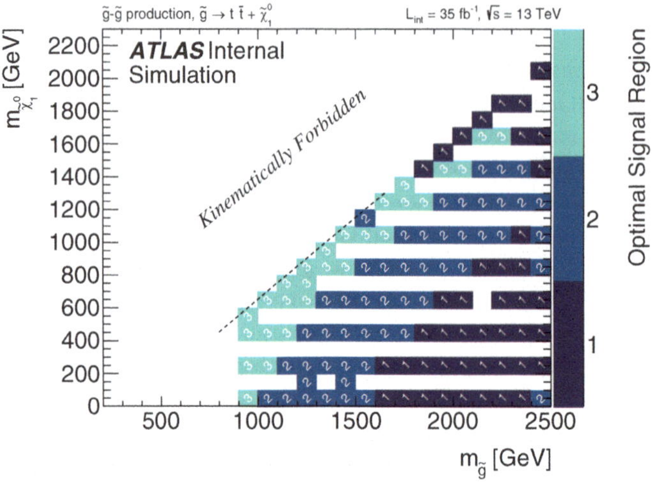

(a) Optimal SR for each grid point in the Gtt 0-lepton channel. In this figure, regions 1-2-3 correspond to B-M-C in the text.

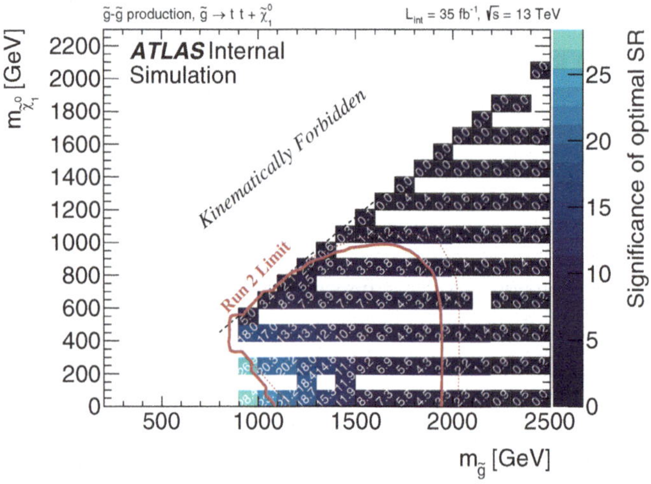

(b) Significance of optimal SR for each grid point in the Gtt 0-lepton channel.

Fig. 7.10 (**a**) Optimal SR and (**b**) significance for the optimal region for all points of the grid in the Gtt 0-lepton analysis

7.4.3.4 Background Composition

The composition of the background in the optimized Gtt-0L regions is all shown in Appendix E. Three of these composition plots are shown for the 0-lepton boosted region in Fig. 7.15 showing the heavy-flavor composition of the $t\bar{t}$ background. In all cases, the SR is dominated by $t\bar{t}$ as expected, as are the CRs and VRs. The expected

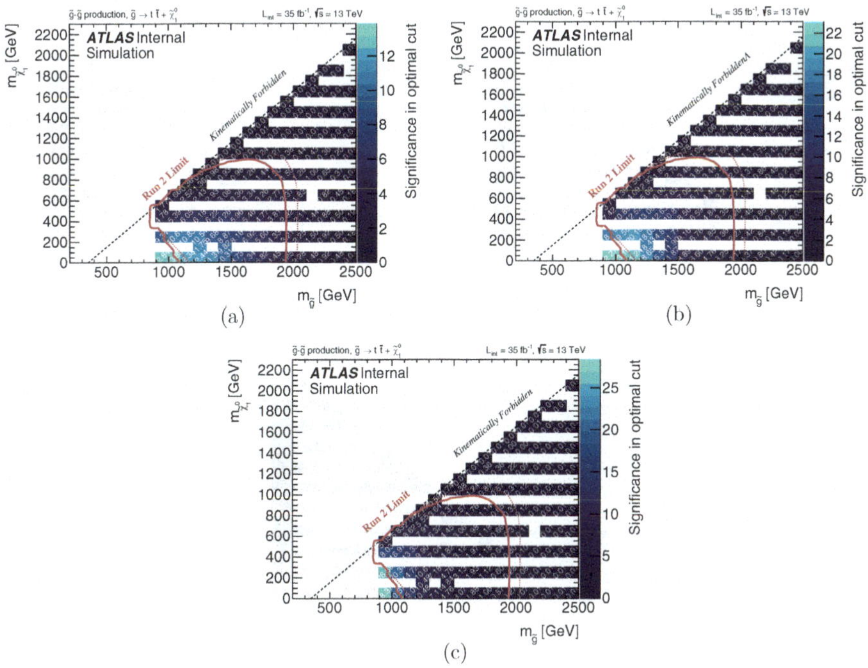

Fig. 7.11 Significance of each SR at each grid point of the Gtt 0-lepton analysis. (**a**) SR-B. (**b**) SR-M. (**c**) SR-C

yields of the CRs and VRs are also sufficient. Note all the composition plots are pre-fit (Table 7.3).

7.4.3.5 *N*-1 Plots

Distributions of *N*-1 plots for the kinematic variables used in the optimization of signal regions are all shown in Appendix D. Each of the variables shows significant discrimination power for the signals shown. An example plot is shown for the 0-lepton boosted signal region in Fig. 7.16 for the total jet mass variable. This plot is made by applying all of the *N* selections in the 0-lepton boosted signal region described in Table 7.4 except for the total jet mass variable M_J^{Σ}, hence the name *N*-1 plot.

7.4.4 Gtt-1L Optimization

The Gtt 1-lepton optimization is very similar to 0-lepton. The main change is the lowering of the N_{jet} requirement (accounting for the extra lepton) and the addition of

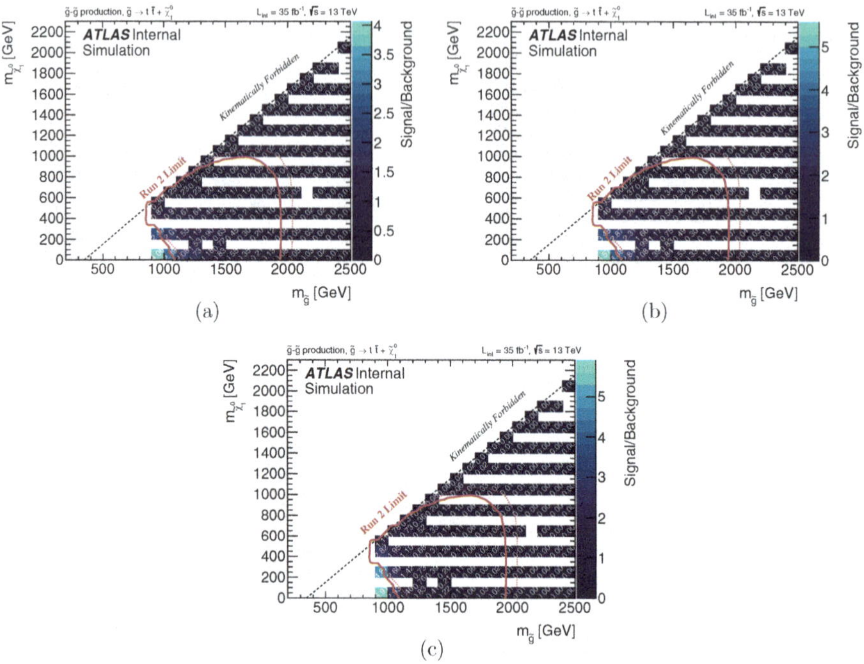

Fig. 7.12 Signal contamination of each CR at each grid point of the Gtt 0-lepton analysis. (**a**) CR-B. (**b**) CR-M. (**c**) CR-C

another transverse mass cut m_{T} which is inverted to create the CRs. To summarize, the variables used in this optimization are summarized in Table 7.3.

Figure 7.17 shows the optimal significance possible with the best selected variables at each grid point; Fig. 7.18 shows the optimal cuts on various variables selected at each mass point.

7.4.4.1 Signal Regions

From these optimization results, three SRs have been designed for the 1-lepton channel of the Gtt analysis, summarized in Table 7.5. All 1-lepton SRs require events with one or more reconstructed signal leptons,[15] and require at least 3 b-tagged small-R jets.

SR-B[16] targets the dramatic events in the region of high mass splitting between the $\tilde{g}$ and $\tilde{\chi}_1^0$. This region places large requirements on the m_{eff} and the largest

[15] In the first iteration of the analysis in 2015 [79], a study was done to determine whether baseline leptons or signal leptons would be used, and signal leptons were found to improve the sensitivity of the cut-and-count analysis.

[16] "B" for boosted.

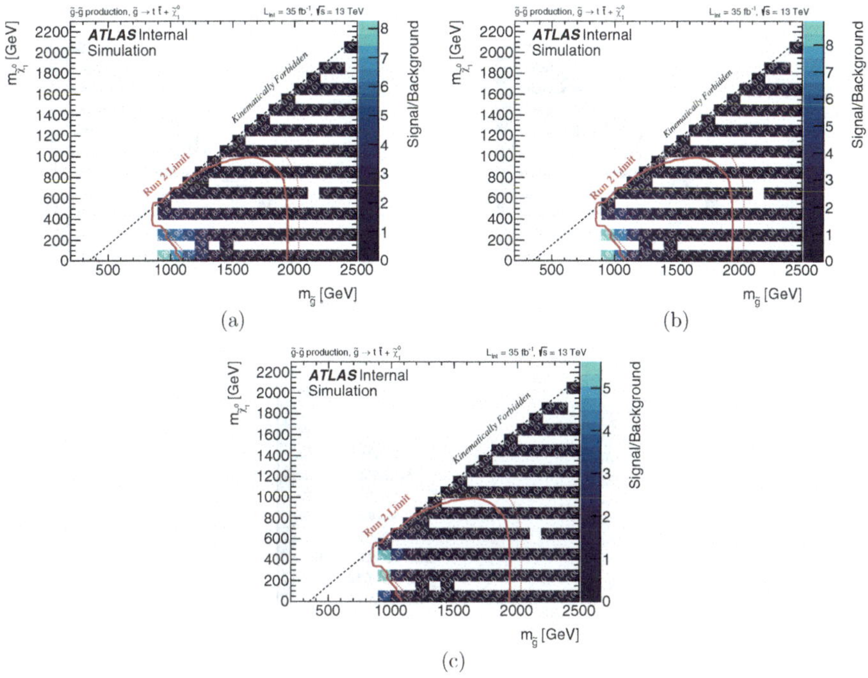

Fig. 7.13 Signal contamination of each VR at each grid point of the Gtt 0-lepton analysis. (**a**) VR-B. (**b**) VR-M. (**c**) VR-C

requirement on M_J^Σ out of all Gtt SRs. The large energy splitting between the $\tilde{g}$ and $\tilde{\chi}_1^0$ means that most of the energy goes into the visible decay products of the event, the boosted top quarks. SR-M[17] targets the bulk region of the signal grid, requiring large amounts of $E_{\mathrm{T}}^{\mathrm{miss}}$, m_{eff}, and M_J^Σ. SR-C[18] targets the "near-diagonal" region of the signal grid, where the small mass splitting between the $\tilde{g}$ and $\tilde{\chi}_1^0$ results in final states with jets which are typically softer than those found in the other regions of the signal plane as a larger fraction of the energy will be in the neutralinos, the invisible decay products, and so the visible decay products are less boosted and more resolved. SR-C also has an additional b-tagged jet to further reject background in this portion of phase space as the top quarks are more resolved and is more likely to have four isolated b-jets, rather than in a more boosted region where two of the b-jets might merge into a single jet.

The significance at each signal grid point for the optimal SR is plotted in Fig. 7.19b. The three SRs each target distinct regions of phase space (Fig. 7.19a),

[17]"M" for moderate-boost.

[18]"C" for compressed.

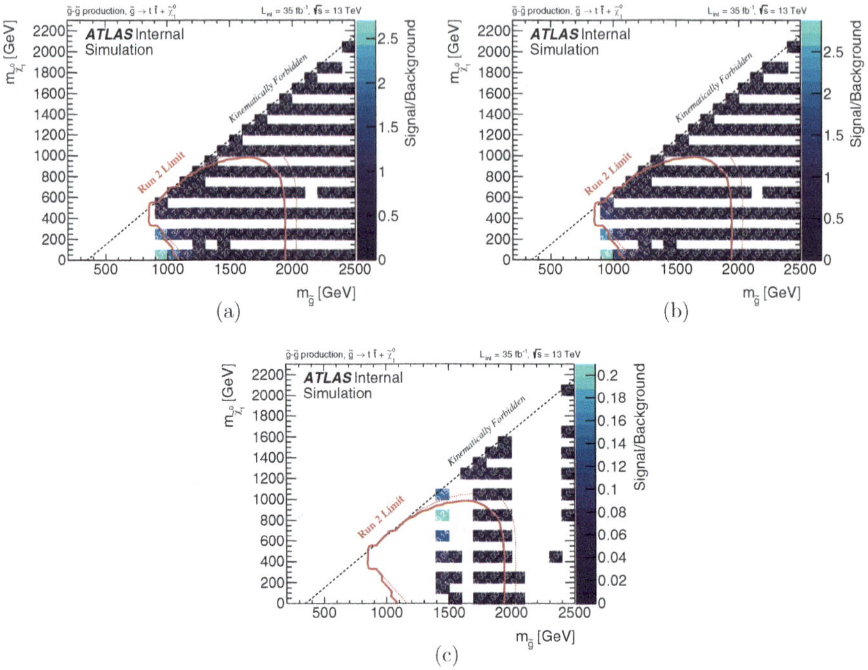

Fig. 7.14 Signal contamination of each VR at each grid point of the Gtt 0-lepton analysis. (**a**) VR-B. (**b**) VR-M. (**c**) VR-C

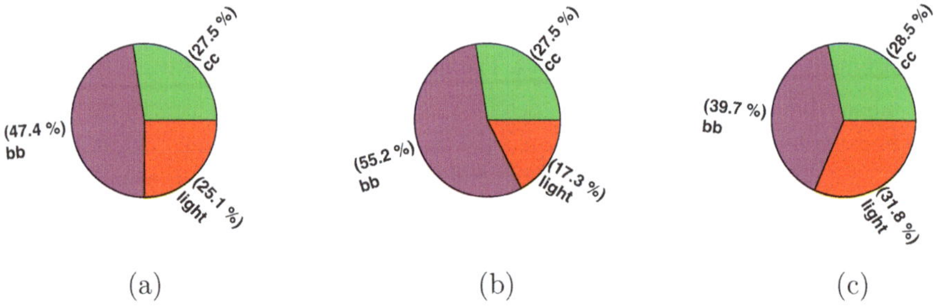

Fig. 7.15 Heavy-flavor composition of the $t\bar{t}$ component of the background in the optimized Gtt-0L boosted regions [21]. (**a**) CR-0L-B. (**b**) SR-0L-B. (**c**) VR-0L-B

with little degradation in performance from the optimal set of cuts at each mass point (Fig. 7.20).

7.4.4.2 Control Regions

CRs are also defined for each SR in Table 7.5. As the background is expected to be dominated entirely by semi-leptonic $t\bar{t}$ events, due to the b-jet requirement and the $E_{\mathrm{T}}^{\mathrm{miss}}$ requirement, following the strategy of the 2015 analysis we define only a

Table 7.3 The 1-lepton optimization phase space that was scanned over

	Inclusive		Step
	Start	Stop	Size
N_{jet}	5	9	1
$N_{b\text{-jets}}$	3	6	1
$E_{\text{T}}^{\text{miss}}$	200 GeV	1000 GeV	50 GeV
m_{eff}	800 GeV	3000 GeV	100 GeV
M_{J}^{Σ}	0 GeV	500 GeV	100 GeV
$m_{\text{T, min}}^{b\text{-jets}}$	0 GeV	200 GeV	4 GeV0
m_{T}	0 GeV	300 GeV	50 GeV

This corresponded to 115,920 cuts. Each variable was scanned from "start" to "stop" in discrete step sizes. The starting and stopping values of the scan are inclusive

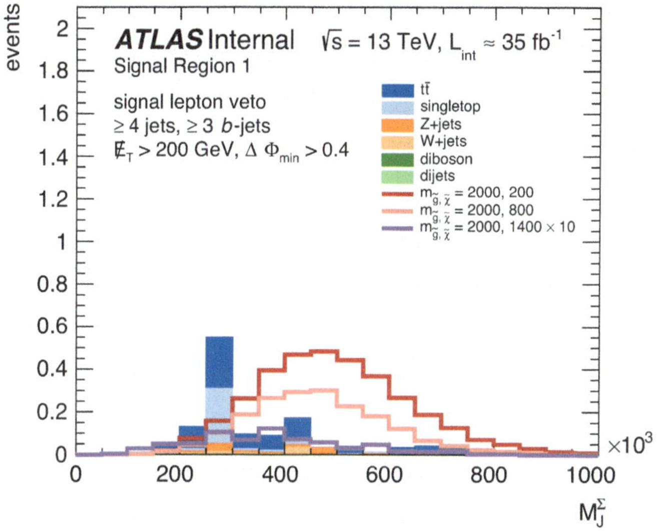

Fig. 7.16 N-1 plot of the total jet mass variable in the 0-lepton boosted region. The background is stacked in the histogram and three signal points are overlaid to show the shape comparisons, all normalized to 35 fb^{-1} except the highest mass signal curve which has been scaled up by a factor of 10

single $t\bar{t}$ CR for the analysis and will take all the remaining backgrounds directly from simulation [20]. Since the background is expected to be semi-leptonic $t\bar{t}$, a single lepton CR strategy is used. The cut on m_{T} is inverted in order to find a region with similar background composition but low signal contamination, while preventing overlap with the Gtt SRs. Cuts on $E_{\text{T}}^{\text{miss}}$ and m_{eff} and $m_{\text{T, min}}^{b\text{-jets}}$ are lowered to ensure sufficient statistics (a minimum of 15 expected events) in the CRs. The signal contamination in each CR is expected to be very small, and is shown in Fig. 7.21.

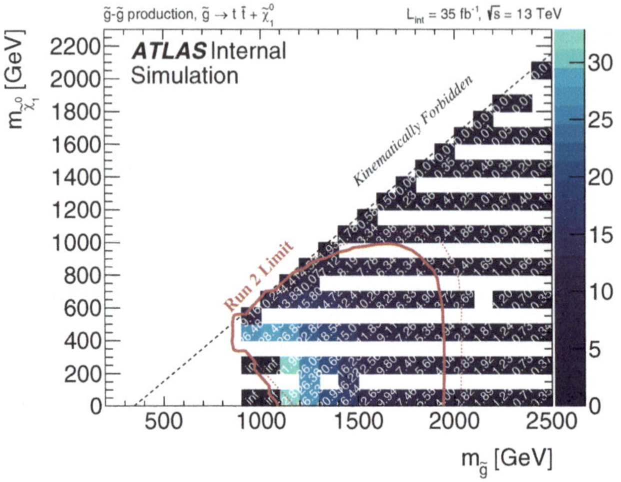

Fig. 7.17 Significance of optimal cuts for each grid point in the Gtt 0-lepton channel

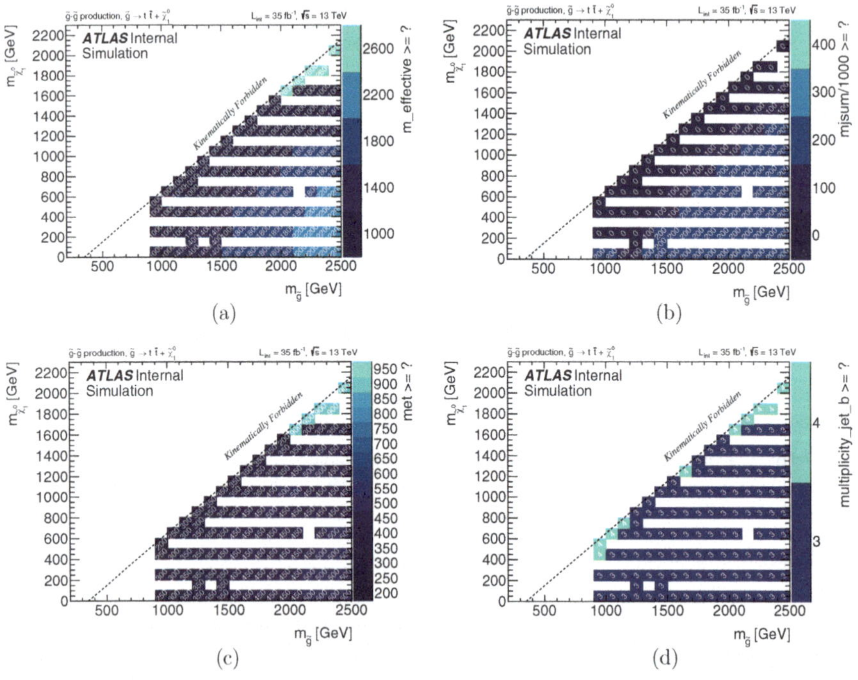

Fig. 7.18 Optimal cut values for each grid point in the Gtt 1-lepton channel. **(a)** m_{eff}. **(b)** M_J^{Σ}. **(c)** $E_{\text{T}}^{\text{miss}}$. **(d)** $N_{b\text{-jets}}$

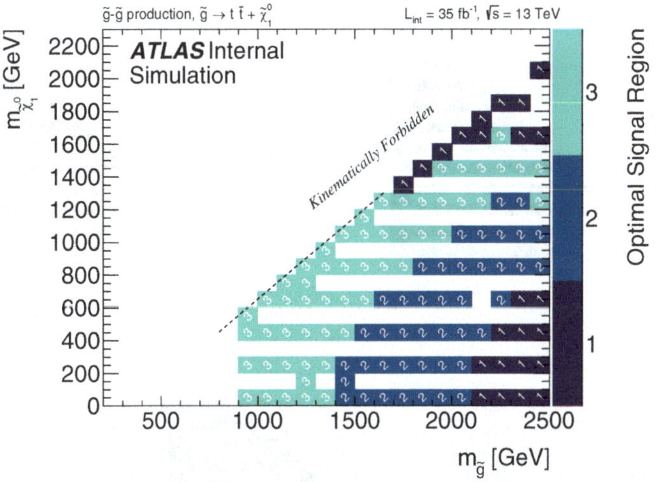

(a) Optimal SR for each grid point in the Gtt 1-lepton channel. In this figure, regions 1-2-3 correspond to B-M-C in the text.

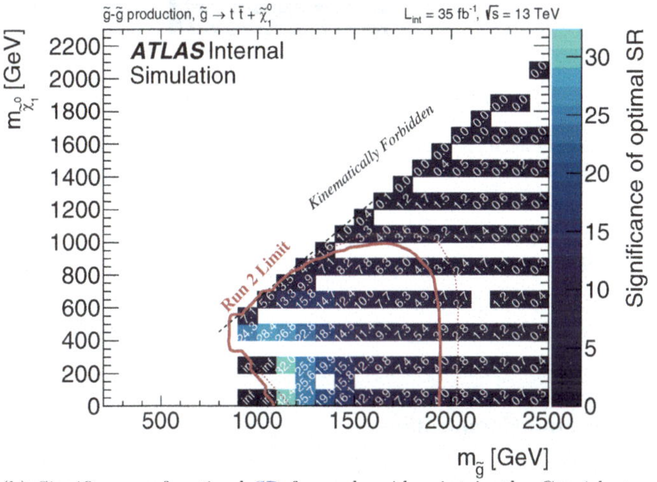

(b) Significance of optimal SR for each grid point in the Gtt 1-lepton channel.

Fig. 7.19 (**a**) Optimal SR and (**b**) significance for the optimal region for all points of the grid in the Gtt 1-lepton analysis

7.4.4.3 Validation Regions

Table 7.5 also defines two VRs for each SR. The VR-m_{T} category validates the extrapolation from low to high m_{T}; orthogonality is enforced with the SR by inverting the M_J^{Σ} (in the case of regions B and M) or $m_{\mathrm{T,\ min}}^{b\text{-jets}}$ cut (in the case of region C). The VR-$m_{\mathrm{T,\ min}}^{b\text{-jets}}$ category validates the extrapolation from no cut on

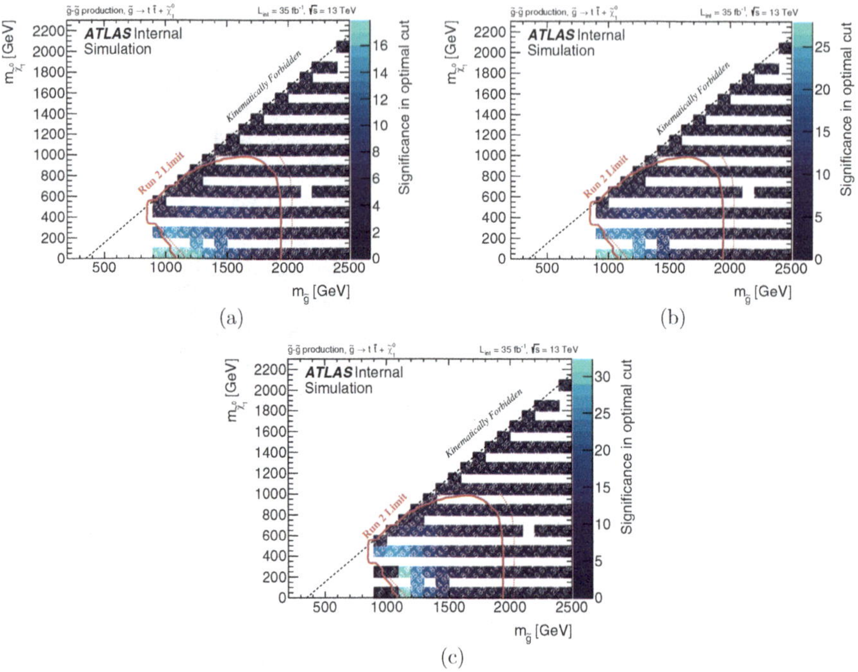

Fig. 7.20 Significance of each SR at each grid point of the Gtt 1-lepton analysis. (**a**) SR-1L-B. (**b**) SR-1L-M. (**c**) SR-1L-C

$m_{\mathrm{T,\ min}}^{b\text{-jets}}$ in the CR to the optimized cut on $m_{\mathrm{T,\ min}}^{b\text{-jets}}$ in the SR. Orthogonality with the CR is enforced by requiring $>N_{\mathrm{jet}}$, instead of exactly $= N_{\mathrm{jet}}$. Signal contamination plots for the VRs are shown in Figs. 7.22 and 7.23. Both are rather low.

7.4.4.4 Background Composition

The composition of the background in the optimized Gtt-1L regions is all shown in Appendix E. Three of these composition plots are shown for the 1-lepton boosted region in Fig. 7.24 showing the heavy-flavor composition of the $t\bar{t}$ background. In all cases, the signal region is dominated by $t\bar{t}$ as expected, as are the CRs and VRs. The expected yields of the CRs and VRs are also sufficient. Note all the results for composition are pre-fit.

7.4.4.5 *N*-1 Plots

Distributions of *N*-1 plots for the kinematic variables used in the optimization of signal regions are all shown in Appendix D. Each of the variables shows significant

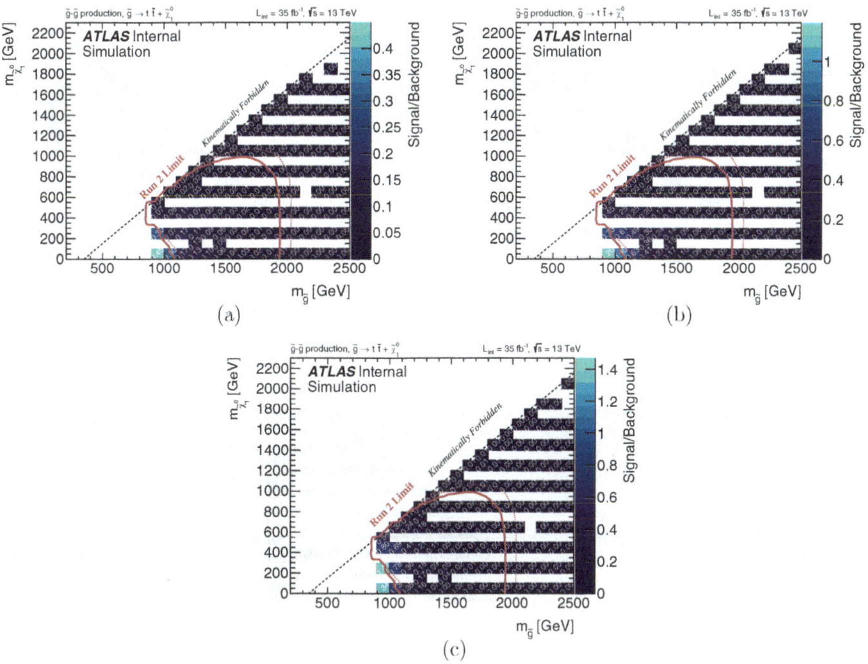

Fig. 7.21 Signal contamination of each CR at each grid point of the Gtt 1-lepton analysis. (**a**) CR-1L-B. (**b**) CR-1L-M. (**c**) CR-1L-C

discrimination power for the signals shown. An example plot is shown for the 0-lepton boosted signal region in Fig. 7.25 for the total jet mass variable. This plot is made by applying all of the N selections in the 0-lepton boosted signal region described in Table 7.5 except for the total jet mass variable M_J^{Σ}, hence the name N-1 plot.

7.5 Region Definitions for Cut-and-Count Analysis

The SRs are named in the form SR-X-YL-Z, where X indicates the target model, Y indicates the number of leptons, and Z labels the type of region targeted. The cut-and-count regions labelled B (for "boosted") are optimized for signals with a large mass difference between the gluino and the neutralino ($\Delta m \gtrsim$ 1.5 TeV), possibly leading to highly boosted objects in the final state. Conversely, regions C (for "compressed") primarily focus on signals for which the gluino decay products are softer due to the small Δm ($\Delta m \lesssim$ 300 GeV). Regions M (for "moderate") target intermediate values of Δm. SRs targeting the Gtt model in the 0-lepton and 1-lepton channels are presented in Tables 7.4 and 7.5.

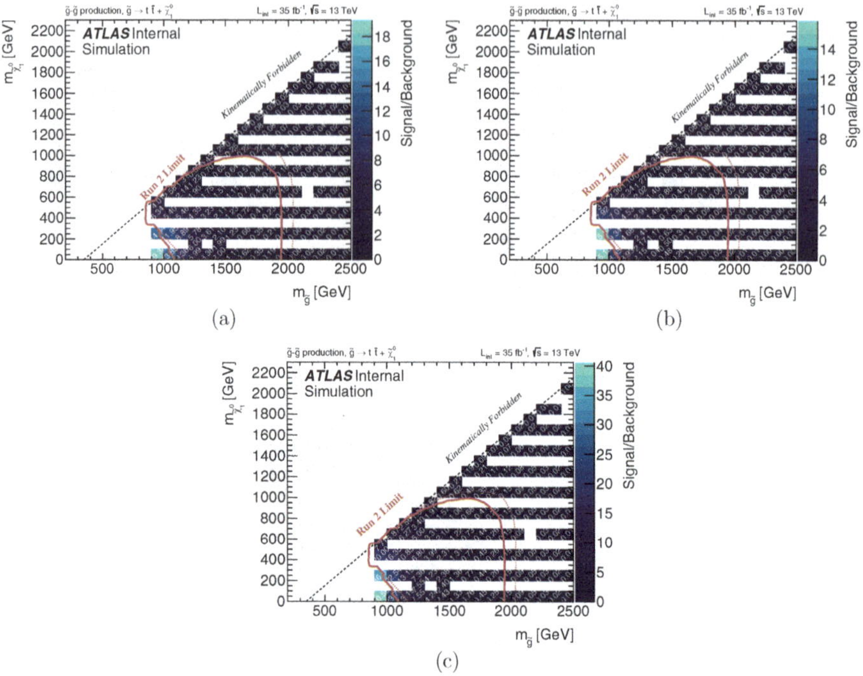

Fig. 7.22 Signal contamination of each VR-m_{T} at each grid point of the Gtt 1-lepton analysis. (**a**) VR-m_{T}-1L-B. (**b**) VR-m_{T}-1L-M. (**c**) VR-m_{T}-1L-C

In the 1-lepton channel, these regions differ mainly in their kinematic selections thresholds: m_{eff}, $E_{\mathrm{T}}^{\mathrm{miss}}$ and M_J^{Σ} selections are relaxed when going from region B to C to improve the acceptance for softer signals. The resulting background increase is compensated for by tightening the requirements on the number of (b-tagged) jets or $m_{\mathrm{T,\ min}}^{b\text{-jets}}$. CRs constraining the $t\bar{t}$ background are defined in the low-m_{T} region to remove overlaps with the SRs. The requirements on $m_{\mathrm{T,\ min}}^{b\text{-jets}}$ are removed, and the selections on kinematic variables are relaxed to ensure at least about 10 events in each CR.

The requirement of an exclusive jet multiplicity permits the definition of VRs kinematically close to the SRs and mutually exclusive to both the CRs and SRs. VR-m_{T} validates the background prediction in the high-m_{T} region. It is kept mutually exclusive with the SR by an inverted selection on M_J^{Σ} or $m_{\mathrm{T,\ min}}^{b\text{-jets}}$. VR-$m_{\mathrm{T,\ min}}^{b\text{-jets}}$ checks the background prediction in the high-$m_{\mathrm{T,\ min}}^{b\text{-jets}}$ regime, with an upper bound on m_{T} to keep the region mutually exclusive with the corresponding SR. The other kinematic requirements are kept as close as possible to those of the SRs to ensure that the event kinematics are similar and allow sufficiently large yields.

The SRs of the 0-lepton channel follow a similar strategy to the 1-lepton channel. Background composition studies performed on simulated event samples show that

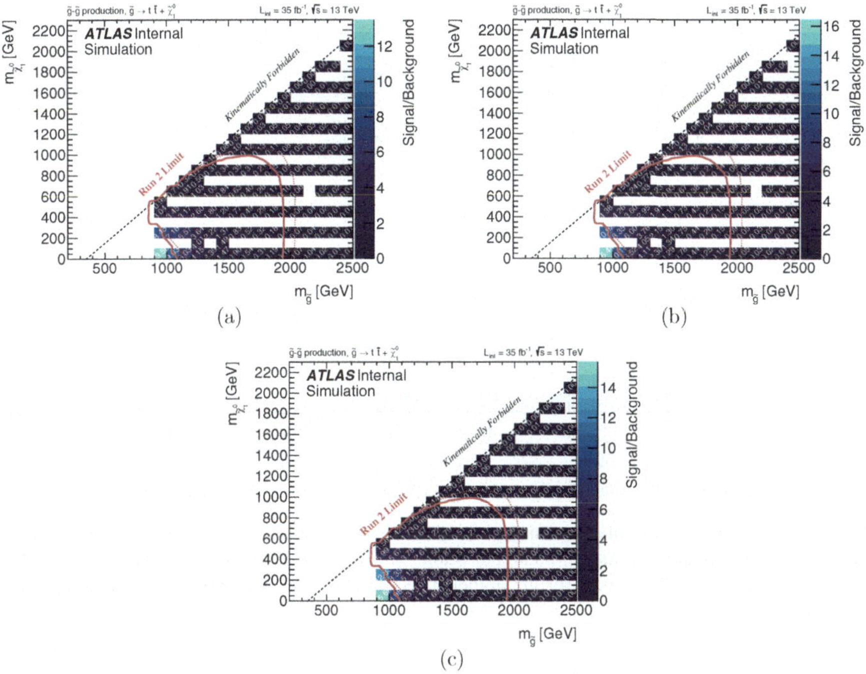

Fig. 7.23 Signal contamination of each VR-$m_{\mathrm{T,\,min}}^{b\text{-jets}}$ at each grid point of the Gtt 1-lepton analysis. (**a**) VR-$m_{\mathrm{T,\,min}}^{b\text{-jets}}$-1L-B. (**b**) VR-$m_{\mathrm{T,\,min}}^{b\text{-jets}}$-1L-M. (**c**) VR-$m_{\mathrm{T,\,min}}^{b\text{-jets}}$-1L-C

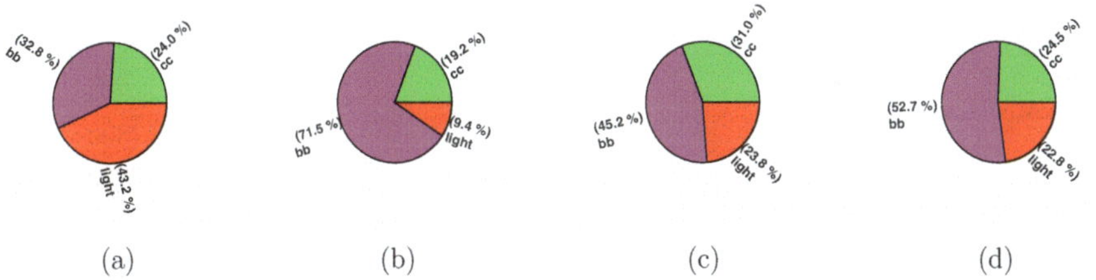

Fig. 7.24 Heavy-flavor composition of the $t\bar{t}$ component of the background in the optimized Gtt-1L regions. (**a**) CR-1L-B. (**b**) SR-1L-B. (**c**) VR-m_{T}-1L-B. (**d**) VR-$m_{\mathrm{T,\,min}}^{b\text{-jets}}$-1L-B

semi-leptonic $t\bar{t}$ events, for which the lepton is outside the acceptance or is a hadronically decaying τ-lepton, dominate in the SRs. Thus, CRs to normalize the $t\bar{t}$ + jets background make use of the 1-lepton channel, requiring the presence of exactly one signal lepton. An inverted selection on m_{T} is applied to suppress overlaps with the 1-lepton SRs. The background prediction is validated in a 0-lepton region, inverting the M_J^{Σ} selection to suppress any overlap with the SRs.

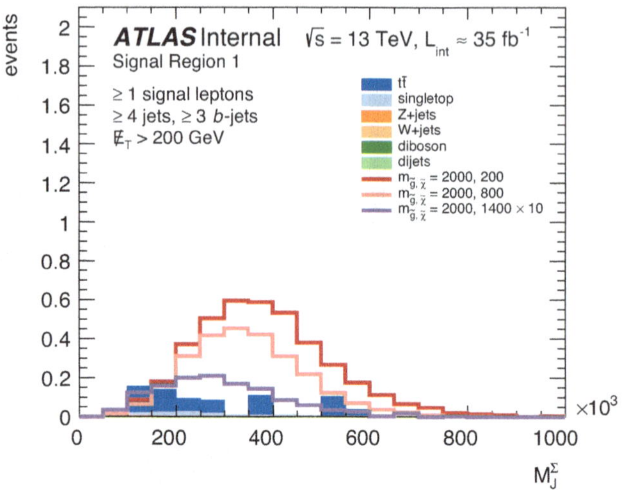

Fig. 7.25 N-1 plot of the total jet mass variable in the 1-lepton boosted region. The background is stacked in the histogram and three signal points are overlaid to show the shape comparisons, all normalized to 35 fb^{-1}, except for the highest mass signal curve which has been scaled up by a factor of 10

Table 7.4 Definitions of the Gtt 0-lepton SRs, CRs, and VRs of the cut-and-count analysis

Gtt 0-lepton										
Criteria common to all regions: $p_T{}^{jet} > 30$ GeV										
Targeted kinematics	Type	N_{lepton}	$N_{b\text{-jets}}$	N_{jet}	$\Delta\phi^{4j}_{\text{min}}$	m_T	$m^{b\text{-jets}}_{T,\text{min}}$	E_T^{miss}	$m_{\text{eff}}^{\text{incl}}$	M_J^Σ
Region B (Boosted, Large Δm)	SR	=0	≥3	≥7	>0.4	–	>60	>350	>2600	>300
	CR	=1	≥3	≥6	–	<150	–	>275	>1800	>300
	VR	=0	≥3	≥6	>0.4	–	–	>250	>2000	<300
Region M (Moderate Δm)	SR	=0	≥3	≥7	>0.4	–	>120	> 500	>1800	>200
	CR	=1	≥3	≥6	–	<150	–	>400	>1700	>200
	VR	=0	≥3	≥6	>0.4	–	–	>450	>1400	<200
Region C (Compressed, moderate Δm)	SR	=0	≥4	≥8	>0.4	–	>120	>250	>1000	>100
	CR	=1	≥4	≥7	–	<150	–	>250	>1000	>100
	VR	=0	≥4	≥7	>0.4	–	–	>250	>1000	<100

All kinematic variables are expressed in GeVexcept $\Delta\phi^{4j}_{\text{min}}$, which is in radians. The jet p_T requirement is also applied to b-tagged jets

Table 7.5 Definitions of the Gtt 1-lepton SRs, CRs, and VRs of the cut-and-count analysis

Gtt 1-lepton								
Criteria common to all regions: ≥ 1 signal lepton, $p_T^{\text{jet}} > 30$ GeV, $N_{b\text{-jets}} \geq 3$								
Targeted kinematics	Type	N_{jet}	m_T	$m_{T,\text{min}}^{b\text{-jets}}$	E_T^{miss}	$m_{\text{eff}}^{\text{incl}}$	M_J^{Σ}
Region B (Boosted, Large Δm)	SR	≥ 5	>150	>120	>500	>2200	>200
	CR	=5	<150	–	>300	>1700	>150
	VR-m_T	≥ 5	>150	–	>300	>1600	<200
	VR-$m_{T,\text{min}}^{b\text{-jets}}$	>5	<150	>120	>400	>1400	>200
Region M (Moderate Δm)	SR	≥ 6	>150	>160	>450	>1800	>200
	CR	=6	<150	–	>400	>1500	>100
	VR-m_T	≥ 6	>200	–	>250	>1200	<100
	VR-$m_{T,\text{min}}^{b\text{-jets}}$	>6	<150	>140	>350	>1200	>150
Region C (Compressed, small Δm)	SR	≥ 7	>150	>160	>350	>1000	–
	CR	=7	<150	–	>350	>1000	–
	VR-m_T	≥ 7	>150	<160	>300	>1000	–
	VR-$m_{T,\text{min}}^{b\text{-jets}}$	>7	<150	>160	>300	>1000	–

All kinematic variables are expressed in GeVexcept $\Delta\phi_{\text{min}}^{4j}$, which is in radians. The jet p_T requirement is also applied to b-tagged jets

7.6 Semi Data-Driven $t\bar{t}$ Normalization

The main source of background is the production of $t\bar{t}$ events. The third (and fourth) b-jet required in the region definitions (Sect. 7.5) can come from

- additional b-jets produced in association with a pair of top quarks
- from a c-jet
- a τ-lepton decaying to hadrons and a ν_τ is mistagged as a b-jet

The contribution from $t\bar{t}$ events with a light or gluon jet mistagged as a b-jet is subdominant but not negligible. In the 0-lepton channel, most of these $t\bar{t}$ events have a W boson decaying leptonically where the lepton is: not reconstructed, outside of acceptance, mis-identified as a jet, or a τ which decays hadronically. In the 1-lepton channel, the high m_T requirement enhances the contribution from dileptonic $t\bar{t}$ events with one hadronically decaying τ. Additional sources of background are single-top production, $t\bar{t}$+$W/Z/h$, W/Z+heavy-flavor jets, and diboson production; as already mention in Sect. 4.4.

The strategy used to estimate the $t\bar{t}$ is a semi-data-driven method which relies on the renormalization of $t\bar{t}$ simulated events in CRs enriched in $t\bar{t}$ background and with low expected yields from the targeted SUSY signals. The definition of these CRs is described in the Sect. 7.5. The extrapolation of the $t\bar{t}$ yield renormalized in the CR to the VRs and SRs is performed by a fit based on the profile likelihood method [99, 148]. Each signal region is fitted separately. The free parameter in each fit is the $t\bar{t}$ overall normalization scale, $\mu_{t\bar{t}}$, while the contributions from other background processes are set at the expected value and allowed to vary within their systematic uncertainties. The systematic uncertainties on the expected values are included in the likelihood as nuisance parameters with a Gaussian probability density function and the correlations are taken into account when appropriate. The likelihood function is built as the product of a Poisson probability density function, describing the event counts in each region, and the constraints on the nuisance parameters.

The entire $t\bar{t}$ background is normalized with one single scale factor. The fitted $t\bar{t}$ background is normalized in one CR with the same b-tag requirements as the corresponding SR. Three different likelihood fits can be performed to extract these results:

- Background-only fit: Only the CR is used to constrain the fit parameters. Any potential signal contamination is neglected and the number of observed events in the signal regions is not taken into account in the fit.
- Exclusion fit: Both CRs and SRs are used to constrain the fit parameters. The signal contribution as predicted by the tested model is taken into account in both regions using an additional free parameter for the non-SM signal strength, constrained to be non-negative, in the likelihood fit. Since the observed event yield in the signal region is used, the background prediction can differ from the prediction on the background-only fit. The exclusion fit configuration was used to produce all the model-dependent limits.
- Discovery fit: Both CRs and SRs are used to constrain the fit parameters. A potential signal contribution is considered in the signal regions but neglected in the CRs. This background prediction is conservative since any signal contribution in the CRs is attributed to background and thus yields a possible overestimate of the background in the signal regions. The discovery fit configuration is used to produce upper limits on the visible cross-sections.

Figure 7.26 shows the value of $\mu_{t\bar{t}}$ after the background-only fit in all of the SRs of the analysis described in Sect. 7.5. Chapter 8 shows the results of the background-only fit to the VRs and the unblinded SRs. The normalization factor is higher in the 1-lepton regions because of the rising data/MC slope corrected by a kinematic reweighting described in Sect. 7.3. Regions with 4 b-tags have a higher normalization because of the data/MC disagreement in the number of b-jets, primarily due to the poor modeling in MC.

7.7 Systematic Uncertainties

Figure 7.27 summarizes the relative systematic uncertainties in the background estimate for the cut-and-count analysis. These uncertainties arise from the extrapolation of the $t\bar{t}$ normalization obtained in the CRs to the SRs as well as from the yields of the minor backgrounds in the SRs, which are predicted by the simulation. The total systematic uncertainties range from approximately 20–80% in the various SRs.

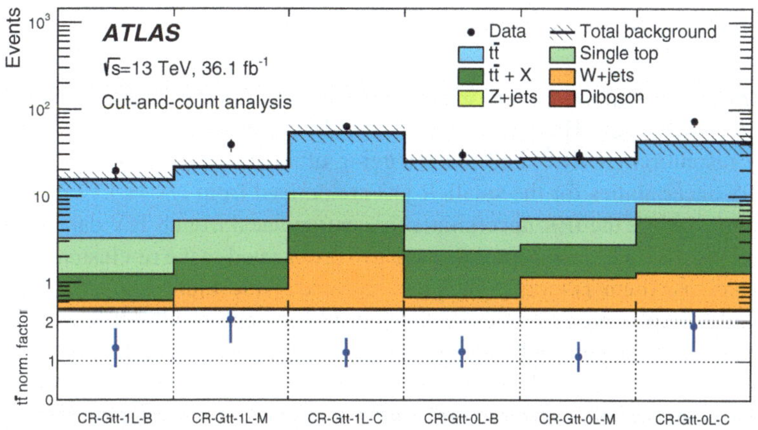

Fig. 7.26 The value of $\mu_{t\bar{t}}$, and uncertainty, after the background-only fit, for the regions defined in Sect. 7.5

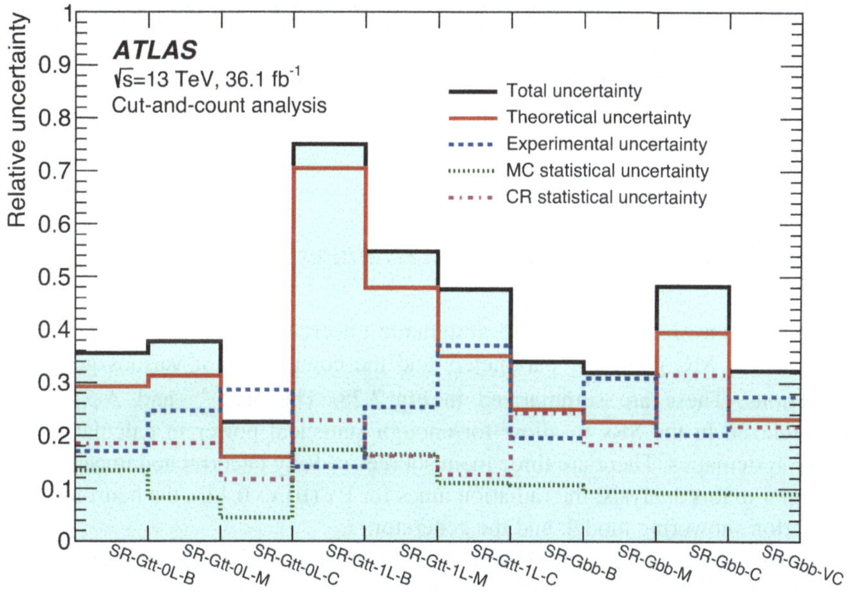

Fig. 7.27 Relative systematic uncertainty in the background estimate for the cut-and-count analysis. The individual uncertainties can be correlated, such that the total background uncertainty is not necessarily their sum in quadrature

7.7.1 Experimental Systematic Uncertainties

The detector-related systematic uncertainties affect both the background estimate and the signal yield. The largest sources in this analysis relate to the JES,[19] jet energy resolution (JER),[20] and the b-tagging efficiencies and mistagging rates.[21] The JES uncertainties for the small-R jets are derived from $\sqrt{s} = 13$ TeV data and simulations, while the JER uncertainties are extrapolated from 8 TeV data using MC simulations [61]. These uncertainties are also propagated to the re-clustered large-R jets, which use them as inputs. The jet mass scale and resolution uncertainties have a negligible impact on the re-clustered jet mass. The impact of the JES uncertainties on the expected background yields is between 4 and 35%, while JER uncertainties affect the background yields by approximately 0–26% in the various regions.

Uncertainties in the measured b-tagging efficiencies and mistagging rates are the subleading sources of experimental uncertainty. The impact of these uncertainties on the expected background yields is 3–24% depending on the considered region. The uncertainties associated with lepton reconstruction and energy measurements have a negligible impact on the final results. All lepton and jet measurement uncertainties are propagated to the calculation of $E_{\mathrm{T}}^{\mathrm{miss}}$, and additional uncertainties are included in the scale and resolution of the soft term. The overall impact of the $E_{\mathrm{T}}^{\mathrm{miss}}$ soft-term uncertainties is also small. Since the normalization of the $t\bar{t}$ background is fit to data in the CRs, uncertainties in the modeling of this background-only affect the extrapolation from the CRs to the SRs and VRs.

7.7.2 Theoretical Systematic Uncertainties on Background

The estimation of the background systematic uncertainties is performed by variations of the MC generator parameters and the comparison of various generator predictions. These are summarized in Fig. 7.28. The $m_{\mathrm{T,\,min}}^{b\text{-jets}}$ and $\Delta\phi_{\mathrm{min}}^{4\mathrm{j}}$ cuts were relaxed in the SRs to allow for enough statistical power in calculating the theory systematics. There are three main sources of fully uncorrelated uncertainties particular to this analysis: the radiation tunes for PYTHIA v6.428, the hadronization and parton showering model, and the generator.[22]

Hadronization and parton showering model uncertainties are estimated using a sample generated with POWHEG and showered by HERWIG++ v2.7.1 with the

---

[19]This is calculated by scaling up and down by 1σ the JES provided by the Jet/Etmiss group within ATLAS.

[20]This is calculated by smearing the $p_{\mathrm{T}}^{\mathrm{jet}}$ based on the $p_{\mathrm{T}}^{\mathrm{jet}}$ and η to account for a possible underestimate of the JER in MC.

[21]This is calculated by varying the η, $p_{\mathrm{T}}^{\mathrm{jet}}$, and flavor of each jet, separately for B-jets, C-jets, and light jets. This leads to three uncorrelated systematic uncertainties.

[22]For example, how the choice of MC generator affects our predictions.

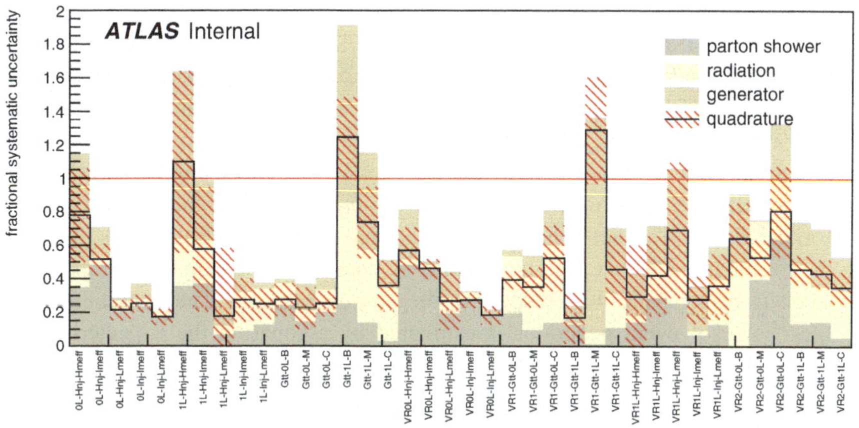

(a)

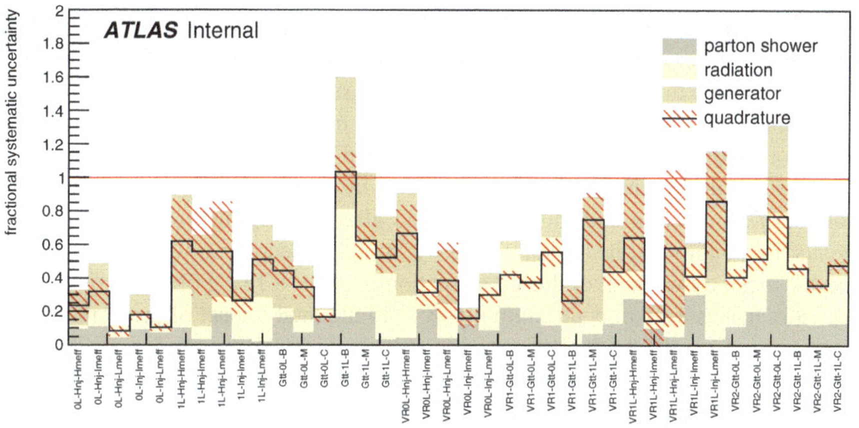

(b)

Fig. 7.28 The summary of the calculated theory uncertainties by comparison of generator predictions is shown here. Two different versions are shown, without and with truth b-tagging. One of the main problems was obtaining enough statistical power in truth generator samples. Instead of vetoing truth events that do not have enough b-tags, one uses truth b-tagging which weights each event by the probability for it to have the given $N_{b\text{-jets}}$ requirement (both inclusively and exclusively). Each component of the uncertainty is shown as stacked and the sum in quadrature is overlaid in black. The uncertainty of the systematic is shaded red. The unit is the full size of the uncertainty (i.e., 2 corresponds to a 200% uncertainty). (**a**) Without truth b-tagging. (**b**) With truth b-tagging

UEEE5 underlying-event tune. Systematic uncertainties in the modeling of initial-state radiation (ISR) and final-state radiation (FSR) are explored with POWHEG samples showered with two alternative settings of PYTHIA v6.428. The first of these uses the PERUGIA2012radHi tune [243] and has the renormalization and factorization scales set to twice the nominal value, resulting in more radiation in the final state. In addition, it has h_{damp} set to $2m_{\mathrm{top}}$. The second sample, using the PERUGIA2012radLo tune, has $h_{\mathrm{damp}} = m_{\mathrm{top}}$ and the renormalization and factorization scales are set to half of their nominal values, resulting in less radiation in the event. In each case, the uncertainty is taken as the change in the expected yield of $t\bar{t}$ background with respect to the nominal sample. The uncertainty due to the choice of event generator is estimated by comparing the expected yields obtained using a $t\bar{t}$ sample generated with MADGRAPH5_aMC@NLO and one that is generated with POWHEG. Both of these samples are showered with HERWIG++ v2.7.1. The total theoretical uncertainty in the inclusive $t\bar{t}$ background is taken as the sum in quadrature of these individual components.

An additional uncertainty is assigned to the fraction of $t\bar{t}$ events produced in association with additional heavy-flavor jets (i.e., $t\bar{t}+\geq 1b$ and $t\bar{t}+\geq 1c$), a process which suffers from large theoretical uncertainties. Simulation studies show that the heavy-flavor fractions in each set of SR, CR, and VR, which have almost identical b-tagged jets requirements, are similar. Therefore, the theoretical uncertainties in this fraction affect these regions in a similar way, and thus largely cancel out in the semi-data-driven $t\bar{t}$ normalization based on the observed CR yields. The residual uncertainty in the $t\bar{t}$ prediction is taken as the difference between the nominal $t\bar{t}$ prediction and the one obtained after varying the cross-section of $t\bar{t}$ events with additional heavy-flavor jets by 30%, in accordance with the results of the ATLAS measurement of this cross-section at $\sqrt{s} = 8$ TeV [80]. This component typically makes a small contribution (0–8%) to the total impact of the $t\bar{t}$ modeling uncertainties on the background yields, which ranges between 5 and 76% for the various regions. The statistical uncertainty of the CRs used to extract the $t\bar{t}$ normalization factors, which is included in the systematic uncertainties, ranges from 10 to 30% depending on the SR.

Modeling uncertainties affecting the single-top process arise especially from the interference between the $t\bar{t}$ and Wt processes. This uncertainty in Fig. 7.29 is estimated using inclusive $WWbb$ events, generated using MADGRAPH5_aMC@NLO, which are compared with the sum of $t\bar{t}$ and Wt processes. Furthermore, as in the $t\bar{t}$ modeling uncertainties, variations of PYTHIA v6.428 settings increasing or decreasing the amount of radiation are also used. An additional 5% uncertainty is included in the cross-section of single-top processes [189].

Overall, the modeling uncertainties affecting the single-top process lead to changes of approximately 0–11% in the total yields in the various regions.

Uncertainties in the W/Z + jets backgrounds are estimated by varying independently the scales for factorization, renormalization, and resummation by factors of 0.5 and 2. The scale used for the matching between jets originating from the matrix element and the parton shower is also varied. The resulting uncertainties in the total yield range from approximately 0 to 50% in the various regions. A 50%

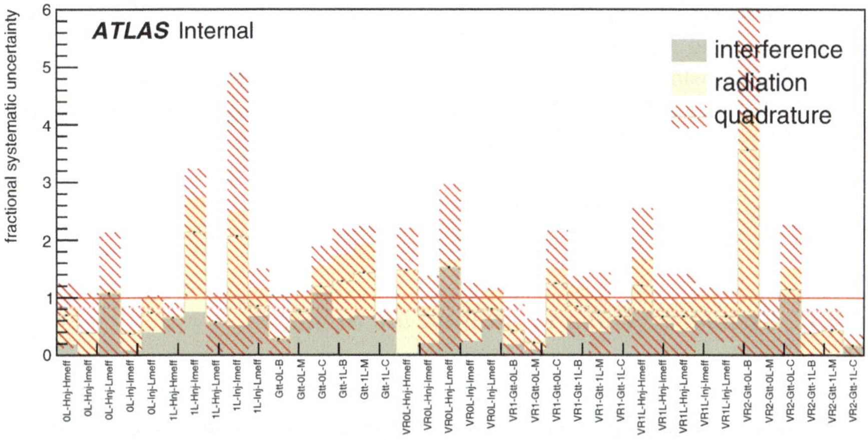

(a)

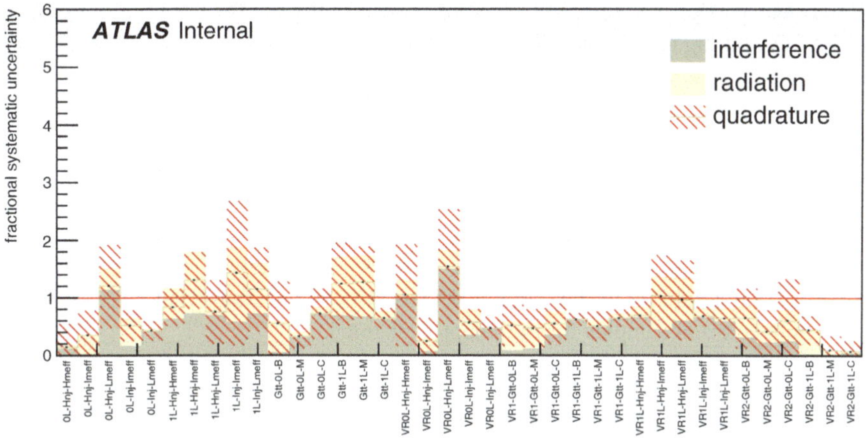

(b)

Fig. 7.29 The summary of the calculated theory uncertainties for single-top is shown here. Two different versions are shown, without and with truth b-tagging. One of the main problems was obtaining enough statistical power in truth generator samples. Instead of vetoing truth events that do not have enough b-tags, one uses truth b-tagging which weights each event by the probability for it to have the given $N_{b\text{-jets}}$ requirement (both inclusively and exclusively). Each component of the uncertainty is shown as stacked and the sum in quadrature is overlaid in black. The uncertainty of the systematic is shaded red. The unit is the full size of the uncertainty (i.e., 2 corresponds to a 200% uncertainty). (**a**) Without truth b-tagging. (**b**) With truth b-tagging normalization uncertainty is assigned to $t\bar{t}+W/Z/h$, $t\bar{t}t\bar{t}$ and diboson backgrounds and are found to have no significant impact on the sensitivity of this analysis. Uncertainties arising from variations of the parton distribution functions were found to affect background yields by less than 2%, and therefore these uncertainties are neglected here. Uncertainties due to the limited number of events in the MC background samples are included if above 5%. They reach approximately 20% in regions targeting large mass splitting.

7.7.3 Systematic Uncertainties on the Signal

The signal samples are normalized using the best cross-section calculations at NLO in the strong coupling constant, adding the resummation of soft gluon emission at NLL accuracy [111–113, 197, 198]. The nominal cross-section and the uncertainty are taken from an envelope of cross-section predictions using different PDF sets and factorization and renormalization scales, as described in [119]. The cross-section of gluino pair-production in these simplified models is $14 \pm 3\,\text{fb}^{-1}$ for a gluino mass of 1.5 TeV, falling to $1.0 \pm 0.3\,\text{fb}^{-1}$ for 2 TeV mass gluino. This is also summarized in Table 4.1.

7.7.4 Other Systematic Uncertainties

A systematic uncertainty is also assigned to the kinematic correction described in Sect. 7.3. The total size of the correction is used as an uncertainty and is applied to all simulated event samples for the 1-lepton channel.

Chapter 8
Results

This chapter presents the results and interpretation of the analysis performed in Chap. 7. Once a theory has been described (Chap. 2), the data collected (Chaps. 3 and 4) and reconstructed (Chap. 5), analyzed and with systematics accounted for (Chap. 7), the search regions can be applied to data, compared to Monte Carlo signal and background models, and hypothesis testing performed to determine if new physics has been found.

Once the Monte Carlo is normalized using a data-driven method (Sect. 7.6), the results need to be first validated in a background-only fit described in Sect. 8.2.1. If there are no significant deviations in the modeling of data in the validation regions, **unblinding**, Sect. 8.2.2, can happen where data is allowed to fill in the signal regions. Finally, as there is no significant excess observed in the cut-and-count analysis, exclusion limits are set using the CLs method in Sect. 8.3. To wrap up the interpretation, truth acceptances and signal region selection efficiencies are described in Sect. 8.4 for theorists and users who wish to reinterpret the results of the analysis.

8.1 General Likelihood

Here, I try to use consistent notation so that $n_{\text{subscript}}$ refers to observed events, a number corresponding to data yields in the given region, while $s_{\text{subscript}}$ and $b_{\text{subscript}}$ refer to predicted yields for signal and background, respectively, corresponding to yields from Monte Carlo simulation in the given region. The exception is for n_{pred} which refers to the total predicted yield of backgrounds and signal.

© Springer Nature Switzerland AG 2020

G. Stark, *The Search for Supersymmetry in Hadronic Final States Using Boosted Object Reconstruction*, Springer Theses,
https://doi.org/10.1007/978-3-030-34548-8_8

$$L(\boldsymbol{n}, \boldsymbol{\theta}^{\mathbf{0}}|\mu_{\rm sig}, \boldsymbol{b}, \boldsymbol{\theta}) = \underbrace{P(n_S|\lambda_S(\mu_{\rm sig}, \boldsymbol{b}, \boldsymbol{\theta}))}_{P_{\rm SR}} \times \underbrace{P(n_{t\bar{t}}|\lambda_{t\bar{t}}(\mu_{\rm sig}, \boldsymbol{b}, \boldsymbol{\theta}))}_{P_{\rm CR}} \times \underbrace{C_{\rm syst}(\boldsymbol{\theta}^0, \boldsymbol{\theta})}_{C_{\rm syst}}. \quad (8.1)$$

The general likelihood[1] L (Eq. (8.1)) fit of analyses is a product of the Poissonian distributions of event yields in the signal regions and control regions of interest [99, 151]. As the analysis strategy was designed to minimize the signal contamination in the control region, the normalization $\mu_{t\bar{t}}$ will depend on the amount of signal, $\mu_{\rm sig}$. Similarly, because the signal regions are designed to maximize SUSY discovery, this will depend on the amount of background present in the signal regions. This likelihood is used to perform hypothesis testing on the predicted number of events in the signal region described with Monte Carlo simulation, $n_{\rm pred}$, compared to the observed number of events seen by data. Equation (8.2) describes the number of predicted events $n_{\rm pred}$ as a function of number of predicted signal events s, number of predicted $t\bar{t}$ events $b_{t\bar{t}}$ scaled by the transfer factor $\mu_{t\bar{t}}$, and the predicted yields of other backgrounds $\boldsymbol{b}_{\rm other}$. $\mu_{\rm sig}$ is a binary parameter with $\mu_{\rm sig} = 0$ for a background-only fit (no signal models are used) and $\mu_{\rm sig} = 1$ for a total fit for discovery or setting exclusion limits. The likelihood describes the normalization factors for background processes such as $t\bar{t}$ (using data-driven techniques) as well as the **nuisance parameters**, $\boldsymbol{\theta}$, that parameterize the systematic uncertainties, $\boldsymbol{\theta} \sim (\theta_0, \theta_1, \ldots, \theta_m) \equiv \theta_i$ for the m systematic uncertainties. Each θ_i is a nuisance parameter that continuously interpolates from nominal variation to a systematic variable, such as from $\theta_i = \theta^0 \equiv 0 \mapsto \pm 1$ for $\pm 1\sigma$ variations.[2] Systematics can kill an experimental observation if they are not under control. In the limit of large numbers, the significance of an observation is $S/\sqrt{B}$. In the presence of systematics, this significance is smaller. For example, if there is a systematic uncertainty θ_B on the background, the significance becomes something like $S/\sqrt{B(1+\theta_B^2 B)} \rightarrow S/B\theta_B$ in the limit of large background, and so a large systematic uncertainty can make it difficult if not impossible to achieve a significant observation. And finally, λ_i are part of Poissonian expectations that are functions of the background predictions and the systematics, the normalization factor for background processes, and $\mu_{\rm sig}$ [151] as shown in Eq. (8.3).

$$n_{\rm pred} = \mu_{\rm sig} s + \underbrace{\mu_{t\bar{t}} b_{t\bar{t}} + \boldsymbol{b}_{\rm other}}_{\boldsymbol{b}} \quad (8.2)$$

$$P(k|\lambda) = \frac{\lambda^k e^{-\lambda}}{k!}. \quad (8.3)$$

[1] Also referred to as the "extended maximum likelihood fit."

[2] One standard deviation.

Systematic uncertainties are included in this likelihood through the probability density function C_{syst} which describes the variations of $\boldsymbol{\theta}$ around $\boldsymbol{\theta^0}$. Changes in the nuisance parameters are described by λ_S and λ_i, the functions that predict signal and background. If each systematic is described by a Gaussian with $\sigma = 1$, the probability density function C_{syst} can be written as in Eq. (8.4)

$$C_{\text{syst}}\left(\boldsymbol{\theta^0}, \boldsymbol{\theta}\right) = \prod_{\text{all systematics}} \frac{1}{\sqrt{2\pi}} e^{-\frac{1}{2}\left(\theta_i^0 - \theta_i\right)^2}. \tag{8.4}$$

8.2 Background-Only Fit

8.2.1 Validation

In order to verify that our normalization in Sect. 7.6 is well-modeled, validation regions are defined in a way to minimize the contribution from signal while remaining orthogonal to control regions to suppress any effects of cross-correlation. This procedure is done using something called a **background-only fit** [99] which propagates the estimate of the $t\bar{t}$ normalization factor, $\mu_{t\bar{t}}$, to the corresponding validation and signal regions, allowing us to predict the background event yields in the validation regions and signal regions. As the background-only fit only uses the control regions in the fit, this allows for external groups to use the background-only fit results to perform hypothesis testing on an entirely different untested signal model not studied by the current analysis. This procedure is known as **reinterpretation** [149].

Figure 8.1 shows the results of the background-only fit to the control regions, extrapolated to the validation regions for the cut-and-count analysis. The number of events predicted by the background-only fit is compared to data in the upper panel. The pull, χ, (Eq. (8.5a)), defined by the difference between the observed number of events, n_{obs}, and the predicted background yield, n_{pred}, divided by the total uncertainty, σ_{tot}, is shown for each region in the lower panel. The total uncertainty is the total systematic uncertainty on the background prediction, σ_{pred} (described in Sect. 7.7), added in quadrature to the Poissonian variation on the expected number of background events, $\sigma_{\text{stat, exp}}$. On average, if the pulls for all validation regions are negative (positive), the data is overestimated (underestimated) and the background model needs to be corrected.

$$\chi \equiv \text{pull} = \frac{n_{\text{obs}} - n_{\text{pred}}}{\sigma_{\text{tot}}} \tag{8.5a}$$

$$\sigma_{\text{tot}} = \sigma_{\text{pred}} \otimes \sigma_{\text{stat, exp}}. \tag{8.5b}$$

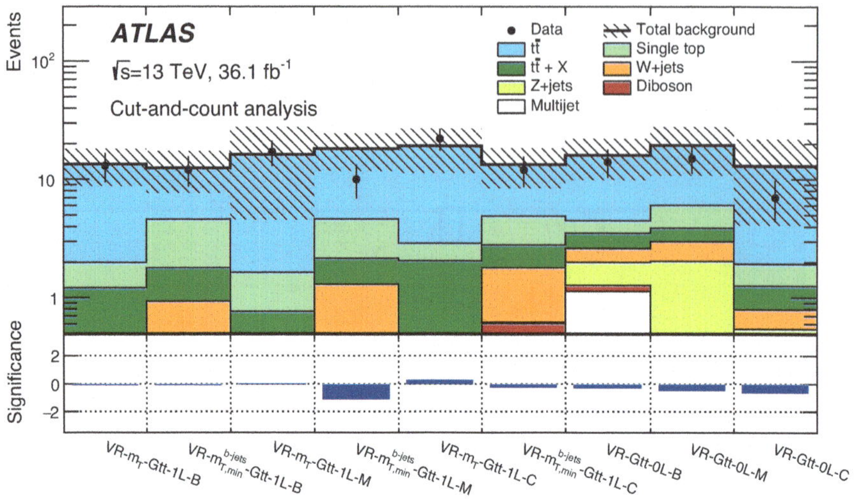

Fig. 8.1 Results of the background-only fit extrapolated to the validation regions of the cut-and-count analysis. The $t\bar{t}$ normalization, $\mu_{t\bar{t}}$, is obtained from the fit to the control regions shown in Fig. 7.26. The upper panel shows the observed number of events and the predicted background yield. All uncertainties are included in the uncertainty band. The background category $t\bar{t}$ + X includes $t\bar{t}$ + W/Z, $t\bar{t}$ + H, and $t\bar{t}\,t\bar{t}$ events. The lower panel shows the pulls in each validation region

8.2.2 *Unblinding*

As seen in Sect. 8.2.1, none of the pulls exceed 2σ[3] which indicates no significant mismodeling. Given the successful validation, one proceeds to look at the background-only fit in the signal regions, unblinded, which means to include observations (data) as well. The analysis is initially blinded to minimize bias towards region definitions, and through a formal procedure within ATLAS, an analysis can go through approval to unblind. The event yields in the unblinded signal regions for the cut-and-count analysis is shown in Fig. 8.2 where the pull (Eq. (8.5a)) is shown for each region in the lower panel. The background is dominated by $t\bar{t}$ in all signal regions. For 0-lepton regions, the subdominant background is $Z(\rightarrow \nu\nu)$+jets and $W(\rightarrow \ell\nu)$ + jets.[4] For the 1-lepton regions, the subdominant background is single-top, $t\bar{t}W$, and $t\bar{t}Z$. No significant excess is found above the predicted background. Table 8.1 shows the observed and predicted number of events from the background-only fit in the Gtt 0-lepton and 1-lepton regions for the cut-and-count analysis. The central value of the fitted background is, in general, larger than

[3]Remember, 95% CL.

[4]Because these are subdominant in the 0-lepton regions, the lepton is either an unidentified electron, muon, or a hadronically decaying τ lepton.

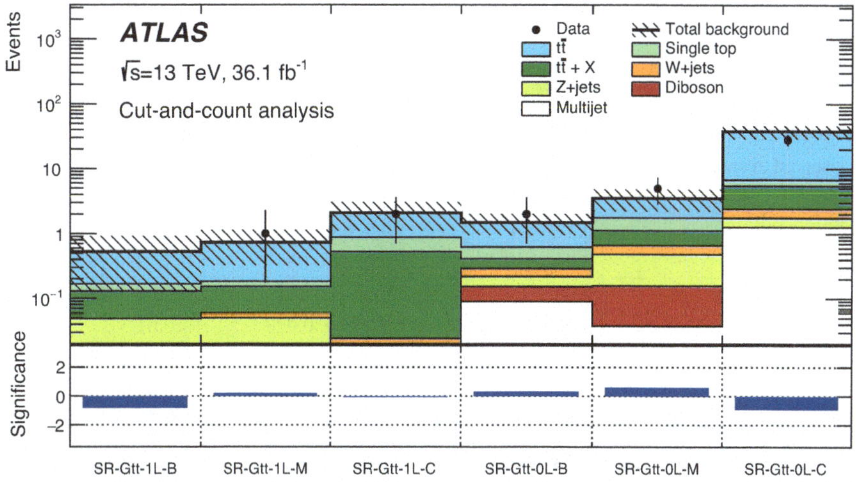

Fig. 8.2 Results of the background-only fit extrapolated to the unblinded signal regions of the cut-and-count analysis. The $t\bar{t}$ normalization, $\mu_{t\bar{t}}$, is obtained from the fit to the control regions shown in Fig. 7.26. The data in the signal regions are not included in the fit. The upper panel shows the observed number of events and the predicted background yield. All uncertainties are included in the uncertainty band. The background category $t\bar{t}$ +X includes $t\bar{t}$ +W/Z, $t\bar{t}$ +H, and $t\bar{t}$ $t\bar{t}$ events. The lower panel shows the pulls in each signal region

the MC-only prediction. This is primarily due to an underestimation of the cross-section of $t\bar{t}+\geq 1b$ and $t\bar{t} \geq 1c$ [82].

8.3 Limits

Since there is no significant excess over the expected background from Standard Model processes, the data is included in an exclusion fit to derive one-sided upper limits at 95% Confidence Level (CL). The limits are calculated using the CL_s[5] prescription [148]. In particular the CL_s method is derived from the probability density functions of $-2\ln(Q)$ with Q being the ratio of likelihoods (Eq. (8.1)) for the two hypotheses of interests for the exclusion test [228] in Eq. (8.6). As a consequence of the Neyman–Pearson lemma [216], if H_0 is the null hypothesis (background-only) and H_1 is the alternate hypothesis (signal + background), then the most powerful statistic one can construct is the likelihood ratio in Eq. (8.6). Both hypotheses can be parameterized by μ_{sig} such as $h = \mu_{\mathrm{sig}}s + b$ where a background-only hypothesis corresponds to $\mu_{\mathrm{sig}} = 0$, and a signal + background hypothesis corresponds to $\mu_{\mathrm{sig}} = 1$.

[5]This has an unfortunate name.

Table 8.1 Results of the background-only fit extrapolated to the Gtt zero and one lepton signal regions in the cut-and-count analysis, for the total background prediction and breakdown of the main background sources

	SR-Gtt-1L		
Targeted kinematics	B	M	C
Observed events	0	1	2
Fitted background	0.5 ± 0.4	0.7 ± 0.4	2.1 ± 1.0
$t\bar{t}$	0.4 ± 0.4	0.5 ± 0.4	1.2 ± 0.8
Single-top	0.04 ± 0.05	0.03 ± 0.06	0.35 ± 0.28
$t\bar{t} + X$	0.08 ± 0.05	0.09 ± 0.06	0.50 ± 0.28
Z + jets	0.049 ± 0.023	0.050 ± 0.023	<0.01
W + jets	<0.01	<0.01	0.024 ± 0.026
Diboson	<0.01	<0.01	<0.01
MC-only background	0.43	0.45	1.9
	SR-Gtt-0L		
Targeted kinematics	B	M	C
Observed events	2	5	28
Fitted background	1.5 ± 0.5	3.5 ± 1.3	38 ± 8
$t\bar{t}$	0.9 ± 0.4	1.8 ± 0.7	31 ± 8
Single-top	0.21 ± 0.14	0.6 ± 0.4	1.3 ± 1.1
$t\bar{t} + X$	0.12 ± 0.07	0.45 ± 0.25	3.0 ± 1.6
Z + jets	0.06 ± 0.10	0.3 ± 0.9	0.49 ± 0.31
W + jets	0.07 ± 0.06	0.18 ± 0.15	0.67 ± 0.22
Diboson	0.06 ± 0.07	0.12 ± 0.07	<0.01
Multijet	0.09 ± 0.11	0.04 ± 0.05	1.3 ± 2.1
MC-only background	1.3	3.3	23

The uncertainties shown include all systematic uncertainties. The data in the signal regions are not included in the fit. The background category $t\bar{t} + X$ includes $t\bar{t}W/Z$, $t\bar{t}H$, $t\bar{t}t\bar{t}$ events. The row "MC-only background" provides the total background prediction when the $t\bar{t}$ normalization is obtained from a theoretical calculation [153]

$$Q \equiv \frac{L(s+b)}{L(b)}. \tag{8.6}$$

To illustrate how the CL$_s$ method works, I refer to Fig. 8.3 where blue corresponds to a background and red corresponds to the background+signal, both are Poisson distributed with the given means. In the context of this example, a likelihood for observing n_{obs} events in data with a hypothesis is defined as $p(n_{\text{obs}}|\text{hypothesis})$. $p(10|b) = 0.014$ (shaded blue). A p-value, p_b, corresponding to this is the probability of a future measurement with $n \geq n_{\text{obs}}$ for the background-only hypothesis ($\mu_{\text{sig}} = 0$) used to quantify a **discovery**, an excess of events over the background expectation. For the signal + background hypothesis $p(10|s + b) = 0.583$ (shaded red). Similarly, a p-value, p_{s+b}, corresponding to this is the

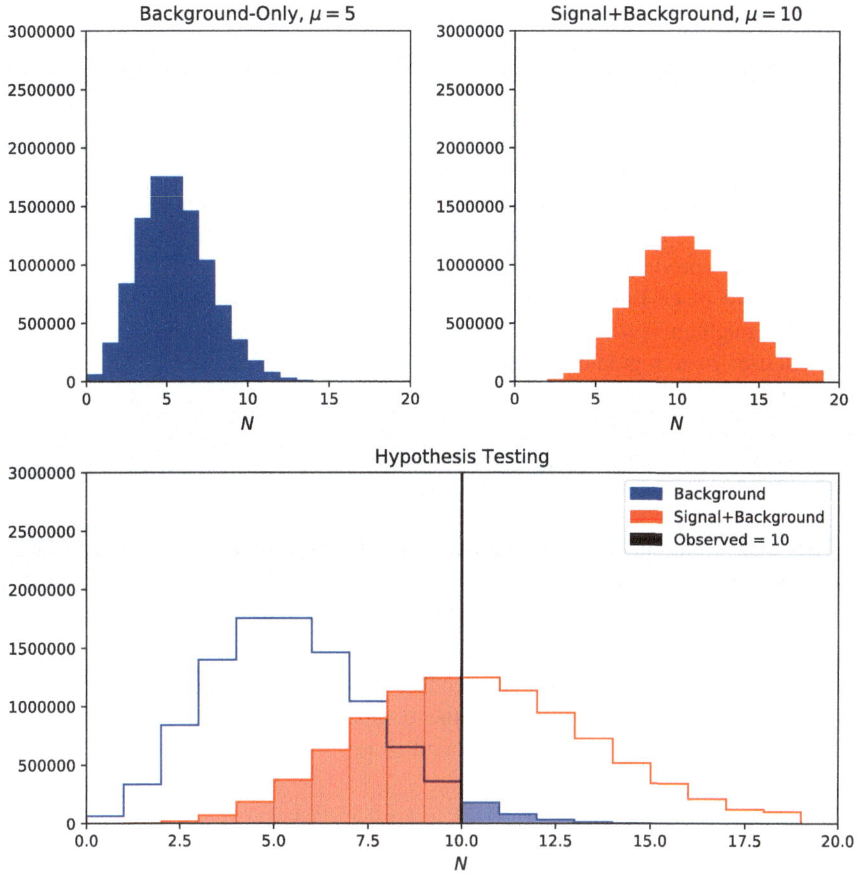

Fig. 8.3 Example Poissonian probability density functions for background (blue) and signal + background (red) hypotheses for 10 observed events (black line). In this example, $p(10|b) = 0.018$ and $p(10|s + b) = 0.125$. The observed data is more likely under the $s + b$ hypothesis than background-only

probability of a future measurement with $n \leq n_{\text{obs}}$ for the signal + background hypothesis.[6] These are formally summarized in Eq. (8.7a).

$$p_{s+b} = \int_{-\infty}^{N_{\text{obs}}} \text{Poisson}(q(\mu_{\text{sig}} = 1))\text{d}N \tag{8.7a}$$

$$p_b = \int_{N_{\text{obs}}}^{\infty} \text{Poisson}(q(\mu_{\text{sig}} = 0))\text{d}N \tag{8.7b}$$

[6]Note the inverted integral here!

The convention in ATLAS is to then define the upper limit as the point at which $p_{s+b} = 0.05$. Signal models[7] with $p_{s+b} > 0.05$ are excluded at the 95% CL while signal models with $p_{s+b} < 0.05$ are not excluded. This definition of p_{s+b} is reasonable for setting limits, given the assumption that observed data is at least consistent with background. However, if observed data has a downward fluctuation with respect to the background expectation, then one can exclude a signal model with $\mu_{\text{sig}} = 0$, as well as all signal models for $0 < \mu_{\text{sig}} < 1$! Going back to Fig. 8.3, but instead setting $n_{\text{obs}} = 1$ (below background), one will find that $p_{s+b} = 0.04$ for $\mu_{\text{sig}} \sim 0$ which produces artificially strong limits at 95% CL!

In order to place limits on new physics and solve the problem of data with a downward fluctuation with respect to data, a new quantifier, called CL_s, incorporates p_{s+b} but uses p_b to regulate the behavior of n_{pred} compared to predicted b as shown in Eq. (8.8)

$$\text{CL}_s = \frac{p_{s+b}}{1 - p_b} = \frac{\text{CL}_{s+b}}{\text{CL}_b}. \tag{8.8}$$

The CL_s value is used to set exclusion limits on specific models [228]. Equation (8.6) is a simplified form of this likelihood. The LHC standard is the profile likelihood from the Neyman–Pearson lemma [216] in Eq. (8.9) where $\hat{\mu}, \hat{\theta}$ are computed to maximize the likelihood function, $\hat{\hat{\theta}}$ is calculated to maximize the likelihood function for the particular μ. This new definition allows us to consider the hypothesis testing as comparing the compatibility of data with signal and background with respect to just background alone.

$$q(\mu_{\text{sig}}) = -2\ln\left(\frac{L\left(\mu, \hat{\hat{\theta}}\right)}{L\left(\hat{\mu}, \hat{\theta}\right)}\right). \tag{8.9}$$

Note that the Neyman construction of parameter estimation can be rather cumbersome. There are two observables, n and b, and two possible true values, $s\mu$ and $\hat{b}$. For each μ, the maximum likelihood estimator of $\boldsymbol{\theta}$ is found, $\hat{\hat{\boldsymbol{\theta}}}(s\mu, n)$. However, as this procedure is remarkably computationally intensive, an approximation can be done by fixing $\hat{\hat{\boldsymbol{\theta}}}(s\mu, n_{\text{obs}})$ to reduce the dimensionality. Another useful feature of Eq. (8.9) is that in the limit of large N (asymptotic or Asimov approximation), this can be evaluated analytically [148]. If this was not the case, this sort of determination would be computationally unfeasible due to the large number of

[7]One can imagine, for the provided example here, signal models as being more Poisson distributions with $\mu > 10$, such as for $\mu = 15, 20, \ldots$. As these get further and further away from the observed value, it becomes a matter of identifying which value of μ (by interpolation) is the limit at which you cannot exclude the distribution anymore; e.g., to fail to reject the null hypothesis.

systematics and signal model variables one needs to evaluate. Within ATLAS, it is convention to use the 95% CL, so that we exclude signal models with $CL_s < \alpha = 0.05$. For each the signal model being considered (Fig. 7.1) and as described in Sect. 2.2.2, each signal point is parameterized by the mass of the gluino, $\tilde{g}$, and the lightest supersymmetric particle, $\tilde{\chi}_1^0$. For each signal point, an exclusion fit is performed and the CL_s is calculated. Those with $CL_s < 0.05$ are excluded at the 95% CL. The last point of contention is to statistically combine the 0-lepton and 1-lepton regions to maximize the sensitivity. Each signal point will have six regions of CL_s computed, three for 0-lepton (boosted, moderate, compressed) and three for 1-lepton. For each region, model-dependent limits have been drawn and are shown in Appendix G. The signal region that provides the smallest CL_s at each signal point is chosen and collected into a "Gtt combination" plot shown in Fig. 8.4. The $\pm 1\sigma^{\text{SUSY}}_{\text{theory}}$ lines around the observed limits are obtained by changing the SUSY cross-section by one standard deviation. The yellow band around the expected limit shows the $\pm 1\sigma$ uncertainty, including all statistical and systematic uncertainties except the theoretical uncertainties in the SUSY cross-section. Compared to the previous results [70], the gluino mass sensitivities of the current search (assuming massless $\tilde{\chi}_1^0$) have improved by 450 GeV. Gluinos with masses below 1.97 TeV are excluded at the 95% CL for $\tilde{\chi}_1^0$ masses below 300 GeV.

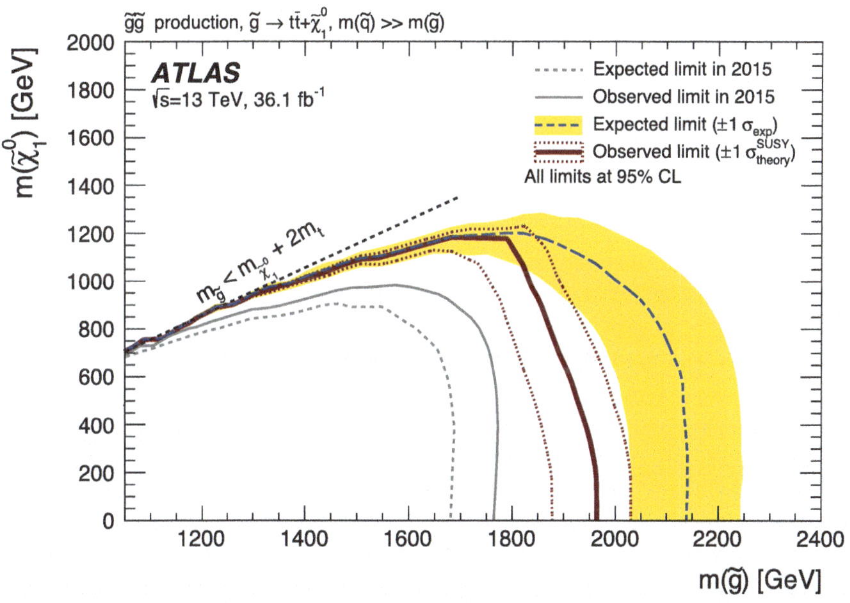

Fig. 8.4 Exclusion limits in the $\tilde{\chi}_1^0$ and $\tilde{g}$ mass plane for the Gtt model obtained in the context of the cut-and-count analysis. The dashed and solid bold lines show the 95% CL expected and observed limits, respectively. The shaded bands around the expected limits show the impact of the experimental and background uncertainties. The dotted lines show the impact on the observed limit of the variation of the nominal signal cross-section by $\pm 1\sigma$ of its theoretical uncertainty. The 95% CL expected and observed limits from the ATLAS search based on 2015 data [70] are also shown

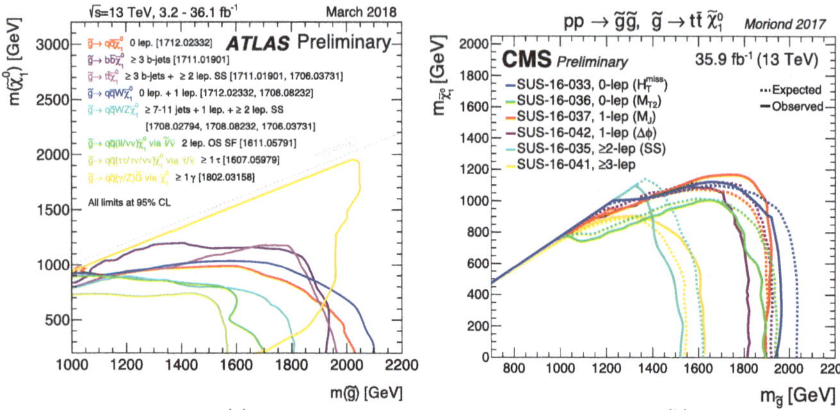

(a) (b)

Fig. 8.5 Exclusion limits at 95% CL based on 13 TeV data for (**a**) ATLAS [55] and (**b**) CMS [136] in the ($\tilde{g}$, $\tilde{\chi}_1^0$) mass plane for different simplified models featuring the decay of the gluino to the lightest supersymmetric particle (lightest neutralino or gravitino) either directly or through a cascade chain featuring other SUSY particles with intermediate masses. For each line, the gluino decay mode is reported in the legend, along with the arXiv reference, and it is assumed to proceed with 100% branching ratio. Some limits depend on additional assumptions on the mass of the intermediate states, as described in the references provided in the plot (ATLAS [1–6, 82]; CMS [137–142]). The search presented in this thesis is shown in magenta for ATLAS

The simplified model does make some explicit assumptions that may end up not being physical, however, it is useful to frame the results of the search and its exclusion limits in the context of other searches performed in Fig. 8.5 for both the ATLAS and CMS collaborations that set limits in the ($\tilde{g}$, $\tilde{\chi}_1^0$) mass plane. The results of the search described in this analysis are shown in magenta for the ATLAS summary plot.

8.4 Signal Acceptances and Experimental Efficiencies

These last set of plots I have made for HEPData [205] allows theorists to perform reinterpretation of existing searches. This is done by providing the measurements, cutflows, and efficiencies for the selections in the analysis . Two such examples of this data are the reconstruction efficiencies and the detector acceptance of the signal models that we considered for this analysis. At a truth level, the signal acceptance provides a way to quantify what portion of the signal one can measure with a perfect detector for our selections, which is evaluated at a truth level using truth information in Monte Carlo simulations. At a reconstruction level, the efficiency, really a detector acceptance ⊗ selection efficiency, quantifies the impact of the detector and our selections on the signal efficiency. So by evaluating these two forms, one can calculate the signal acceptance (truth), signal acceptance ⊗ effi-

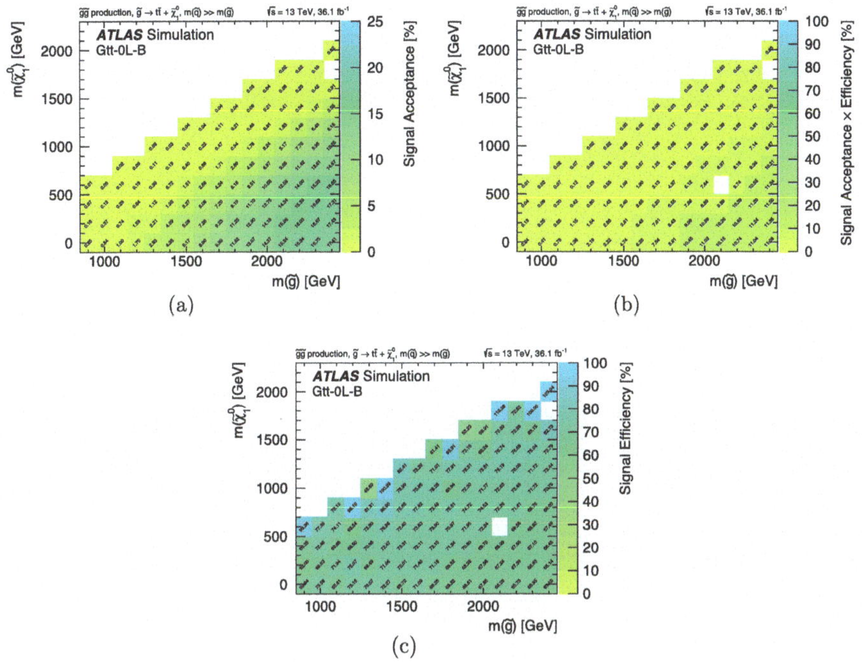

Fig. 8.6 For the Gtt 0-lepton boosted region, (**a**) signal acceptance at truth level, (**b**) signal acceptance ⊗ efficiency at the reconstruction level, and (**c**) calculated signal efficiency is shown in the ($\tilde{g}$, $\tilde{\chi}_1^0$) mass plane. The z-axis represents the value of each bin in units % with 0% being yellow, and 100% (25% for acceptance) being green

ciency (reconstruction), and divide the two to get a signal efficiency (incorporation reconstruction and truth) which quantifies the impact of the detector on our signal models. Figure 8.6 shows the three different plots for the 0-lepton boosted signal region selection. As seen towards the bottom right of the signal efficiency, the boosted regime suffers from a lower detector efficiency than in a more resolved region, motivating the need for detector-level improvements for boosted objects, such as the gFEX trigger upgrade I work on. See Appendix H for the full set of plots.

One potential complication is the fluctuations that are present because the analysis imposes very tight selections which can cause signal efficiencies > 100%. For example, the b-tagging algorithm described in Sect. 5.2 does mistag light jets and c-jets as b-jets, but at a very reduced rate. Because our analysis requires at least three b-jets and the tight selections push it into extreme regions of the phase space, the impact of mistagged jets is more amplified. In this case, the mistagging at the reconstruction level will artificially inflate the acceptance ⊗ efficiency compared to the truth level, where there is no mistagging. Dividing the truth level from the reconstruction level to get a signal efficiency can create an efficiency > 100%.

Chapter 9
Upgrade Studies

This chapter provides a summary of the preliminary upgrade studies I have performed on gFEX, an upgrade project introduced in Sect. 4.2.1.

The gFEX [114] subsystem of the ATLAS L1Calo trigger is one of several modules designed as part of the Phase-I upgrade [203] to maintain trigger acceptance against increasing LHC luminosity in Run 3 (2021) and beyond. It is designed to enhance the selectivity of the L1 trigger and increase sensitivity to key physics channels, such as identifying boosted tops in the final state, a focus of this thesis analysis. A key feature of gFEX is that the entire calorimeter is available in a single module, which enables the use of algorithms that can scan the entire η range of the calorimeter, especially for calculating event-level observables. One of these full-scan algorithms can identify boosted hadronic topologies that are characteristic of new physics scenarios. For example, a gFEX trigger algorithm can capture the entire decay of a top quark which can, under a Lorentz-boosted topology, shower over a large area without any significant local energy deposits in a limited region of interest. Finally, the architecture of gFEX permits event-by-event local pile-up suppression, providing robust observables which reduced pile-up dependencies.

This chapter provides an overview of trigger analysis studies performed with the instrumentation upgrade that I have been involved with, gFEX. First, a strong physics motivation is described and how gFEX can potentially contribute to the trigger upgrades in Sect. 9.1. The gFEX reconstruction algorithm used in the following studies is described in Sect. 9.2. Next, Sect. 9.3 discusses the necessary background to read and understand **turn-on curves**. A turn-on curve allows us to parameterize a given trigger in terms of the efficiency of selecting offline reconstructed objects. Once the necessary background is in place, a few preliminary studies (Sect. 9.4) are shown, with lots of room for future improvement and continuation.

© Springer Nature Switzerland AG 2020

G. Stark, *The Search for Supersymmetry in Hadronic Final States Using Boosted Object Reconstruction*, Springer Theses,
https://doi.org/10.1007/978-3-030-34548-8_9

9.1 Motivating gFEX

Let us suppose, for example, we are performing a search for $Z'(\rightarrow t\bar{t})$ by identifying its decay products as shown in Fig. 9.1. A jet is clustered from calorimeter clusters shown as black dots in the event display. Subjets are formed from topoclusters associated with the jet using the C/A algorithm with $R = 0.2$. The initial state radiation (blue) can contribute significantly to the amount of pile-up energy in this jet and reducing the resolution of measuring the jet. Finally, a black circle is drawn to help visualize the size of the $R = 1.0$ anti-k_t jet for the event while a purple rectangular box shows the size of the sliding window used in the Level-1 trigger algorithms online. As the Z' has a large amount of mass, the decay products ($t\bar{t}$) will have a significant Lorentz boost which makes it possible to observe the full top quark decay within a large-R jet.

As you have read about in Chaps. 2 and 7, high p_{T} Lorentz-boosted top quarks, W/Z/h bosons, and exotics are critical elements of the ATLAS physics program. As described in Chap. 4, moving to an environment with more luminosity and more pile-up energy density will cause the trigger thresholds to go up to manage rates. gFEX is one of a series of instrumentation upgrades that will enable us to control the rates, while still being able to maintain an efficient trigger for such programs. As it has been made clear in Fig. 9.1, the current Level-1 trigger uses a small sliding window which becomes inefficient for jets that decay over a larger area, exactly like

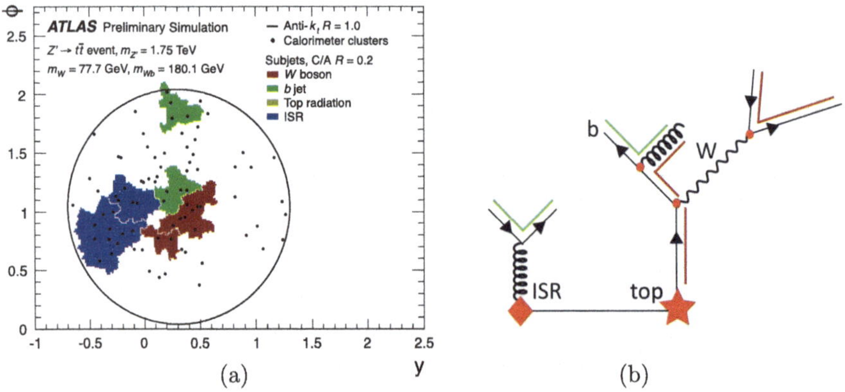

Fig. 9.1 Illustration of a simulated large radius anti-k_t jet, with $R = 1.0$ from a top quark produced in a $Z' \rightarrow t\bar{t}$ decay with $m_{Z'} = 1.75$ TeV [45]. The (**a**) event display and (**b**) parton shower history for an example decay. Subjets are identified by a particular color in the event display: W boson (red), b-jet (green), top radiation (yellow), and initial state radiation (blue). Shown is a black circle representing the size of the $R = 1.0$ anti-k_t offline jet that is clustered and can capture the full information of the top decay and a dashed, purple rectangular window of size 0.8×0.8 representing the L1 trigger sliding window algorithm for identifying energy above a fixed threshold

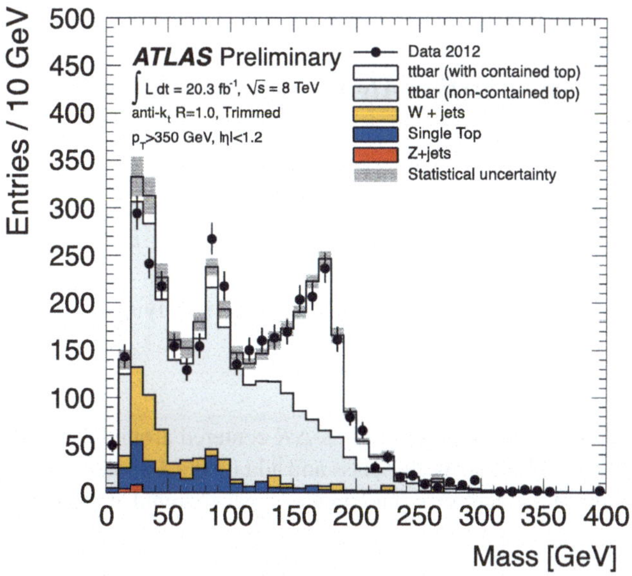

Fig. 9.2 Jet mass for leading p_T anti-k_t trimmed jets with $R = 1.0$, $|\eta| < 1.2$, and $p_T > 350\,\text{GeV}$. Here, "contained" refers to events having a hadronically decaying top quark t with collimated daughter particles at the truth level (all three daughter quarks $q_i \equiv bq\bar{q}$ satisfy $\Delta R(q_i, t) < 1.0$) [40]. The shaded band represents the bin-by-bin statistical uncertainty in Monte Carlo simulation

the reconstructed objects I search for in this thesis analysis. Figure 9.2 shows the jet mass distribution for different physics processes: $t\bar{t}$, W/Z + jets and single top for Lorentz-boosted jets. A top quark, with $p_T > 350\,\text{GeV}$ will have approximately a size parameter $R \sim 2\frac{m}{p_T} < 1.0$ and can be fully captured in a large-R jet. The white shows a fully contained top inside an $R = 1.0$ anti-k_t jet that peaks around the mass of the top quark. A non-fully contained top has peaks around the invariant mass of the two quarks from the hadronic W-boson decay and the invariant mass of the b-jet and one of the quarks from the hadronic W-boson decay.

This gives rise to the concept of substructure, which talks about a hadronic top quark being a three-pronged decay, while a hadronic W-boson is a two-pronged decay. As seen from the leading jet mass in Fig. 9.2, the mass of a jet indicates a measure of the amount of substructure inside. So if a jet with sufficient energy decays over too large of an area, the Level-1 trigger will not fire, it cannot capture the full energy of the jet, and this sliding window algorithm is therefore **inefficient for jets with significant substructure**. In the next sections, we will explore trigger studies and quantify the efficiency of the gFEX trigger compared to the Level-1 trigger for jets with substructure.

9.2 gFEX Algorithms

9.2.1 *The Reconstruction Algorithm*

The reconstruction algorithm is very primitive right now. Below, we provide a step-by-step process of how the jet is reconstructed. For each event,

1. Filter the gTowers to only use those that pass a fixed E_{T} threshold. These are used to seed the algorithm.
2. For each seed, grab the gTowers around the seed satisfying

$$\Delta R \leq X \tag{9.1}$$

   or, in other words, a circle of radius ΔR centered around the seed.
3. We use ROOT's TLorentzVector class and add up the 4-vectors for all gTowers to the seed and use this to create our trigger object centered at the seed's geometric η, ϕ.

$$E_T^{\text{object}} = E_T^{\text{seed}} + \sum_{\text{towers around seed}} E_T^{\text{tower}} \tag{9.2}$$

Unless otherwise specified, large-R ($R = 1.0$) objects are the primary focus of this study, and the gFEX trigger. The area of a trigger jet is defined as the sum of the area of the gTowers included in the jet.

9.2.2 *The Offline-Trigger Object Pairing Algorithm*

In order to pair our objects, to help us understand how well our algorithm works with respect to a reference, such as offline reconstructed jets, we need to be able to match our reconstructed trigger objects with their corresponding offline reference object. For an event,

1. For each offline object—filter out the trigger objects so only trigger objects satisfying

$$\Delta R \leq X. \tag{9.3}$$

2. From the "distance"—filtered trigger objects, identify the object with the highest E_{T} and use this as the offline event's paired object.

A typical cut is to apply $\Delta R < 1.0$ for reconstructed, isolated offline jets as gTowers are often colocated with the energetic jets as seen in Fig. 9.3. A tighter ΔR cut can be applied if multiple energetic objects are within proximity of the leading offline jet, but these substructure-based studies will be discussed in Sect. 9.4.3.

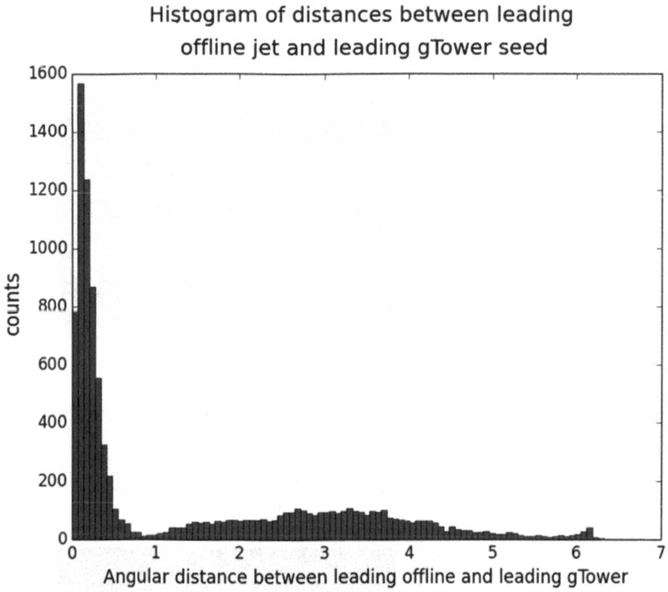

Fig. 9.3 A distribution of the angular variable ΔR between the leading jet in the event and the leading gTower in the event for Monte Carlo simulated $t\bar{t}$ samples with $\langle\mu\rangle = 80$ at a center-of-mass energy of $\sqrt{s} = 14\,\text{TeV}$. A majority of towers are found within $\Delta R < 1.0$ of the reconstructed, isolated offline jet in the event

9.2.3 Event Displays

Figures 9.4, 9.5, and 9.6 show example event displays for the (a) offline, reconstructed jets in the event, the (b) trigger jets formed from the reconstruction algorithm, and (c) the gTowers in the event for a $t\bar{t}$ Monte Carlo simulated sample with center-of-mass energy $\sqrt{s} = 13\,\text{TeV}$.

9.3 Efficiency of Triggers

Turn-on curves are one of the fundamental ways to quantify a trigger. One of the primary goals of the trigger is to maintain a high efficiency for offline reconstructed objects, based on how the trigger itself was designed. The calculation follows the formula Eq. (9.4) where N^{offline} describes the distribution of offline objects before a trigger selection (subscript naught) and after a trigger selection (subscript $\mathcal{T}$).

$$\varepsilon_{\mathcal{T}} = \frac{N_{\mathcal{T}}^{\text{offline}}}{N_0^{\text{offline}}} \tag{9.4}$$

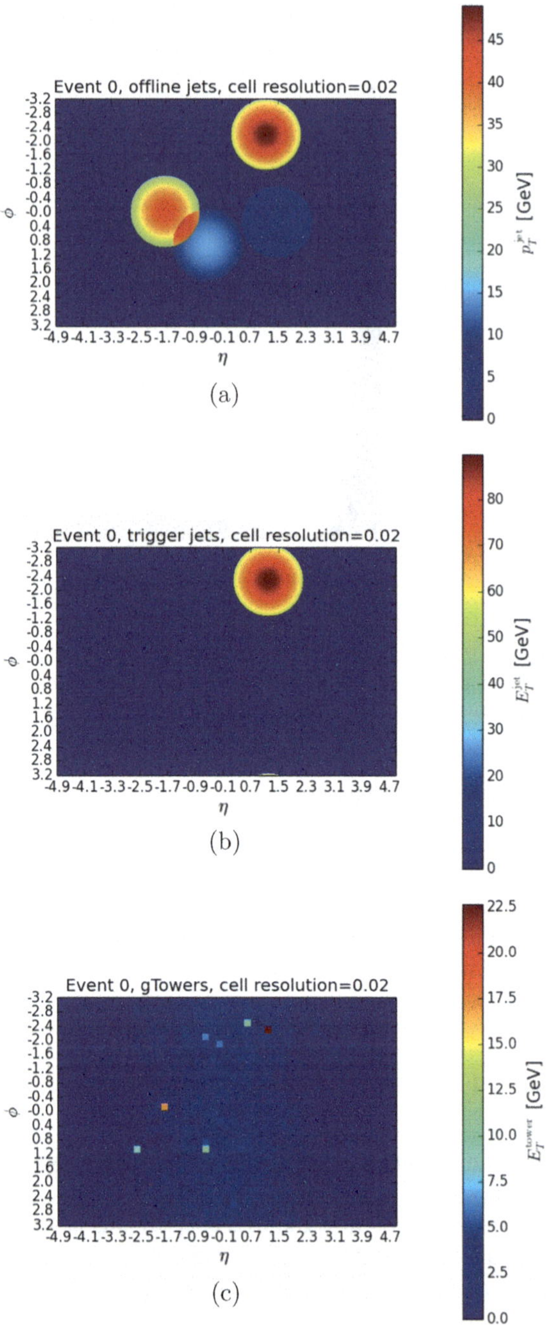

Fig. 9.4 A canonical example that demonstrates the algorithms in the preceding subsections. (**a**) Offline jets. (**b**) Trigger jets. (**c**) gTowers

Fig. 9.5 No gTowers were found for the given threshold of 20 GeV. (**a**) Offline jets. (**b**) Trigger jets. (**c**) gTowers

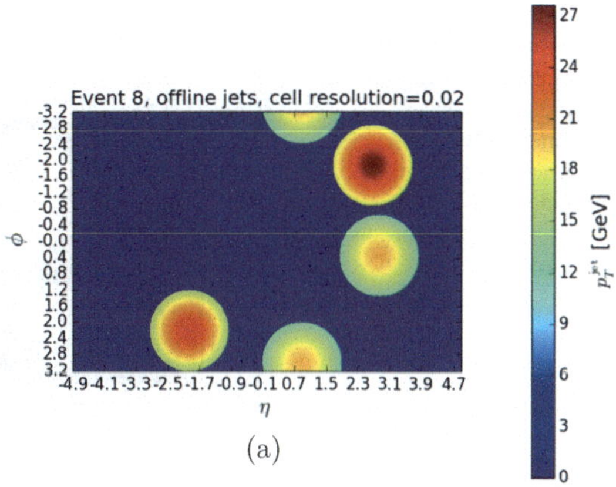

(a)

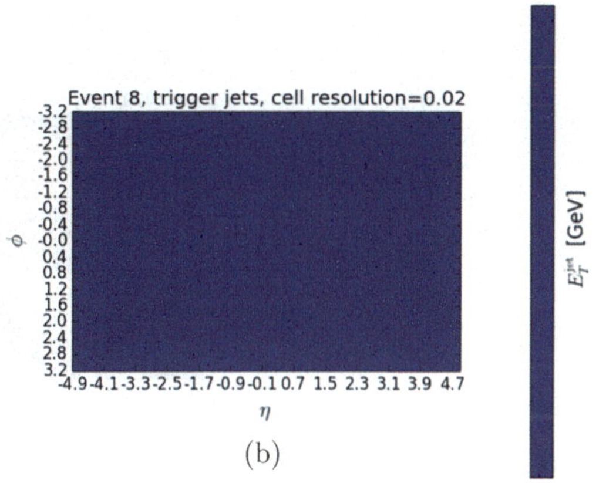

(b)

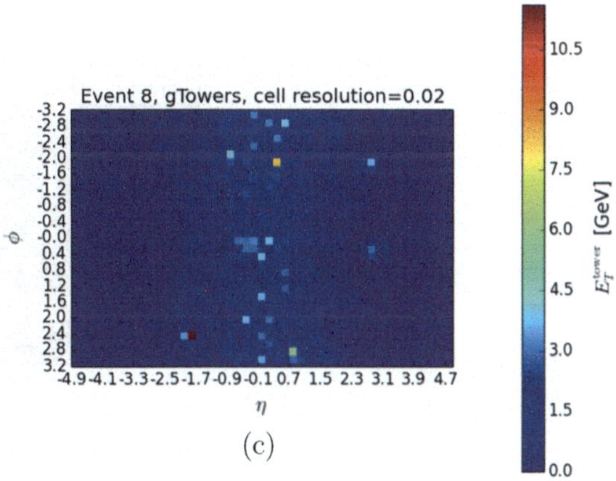

(c)

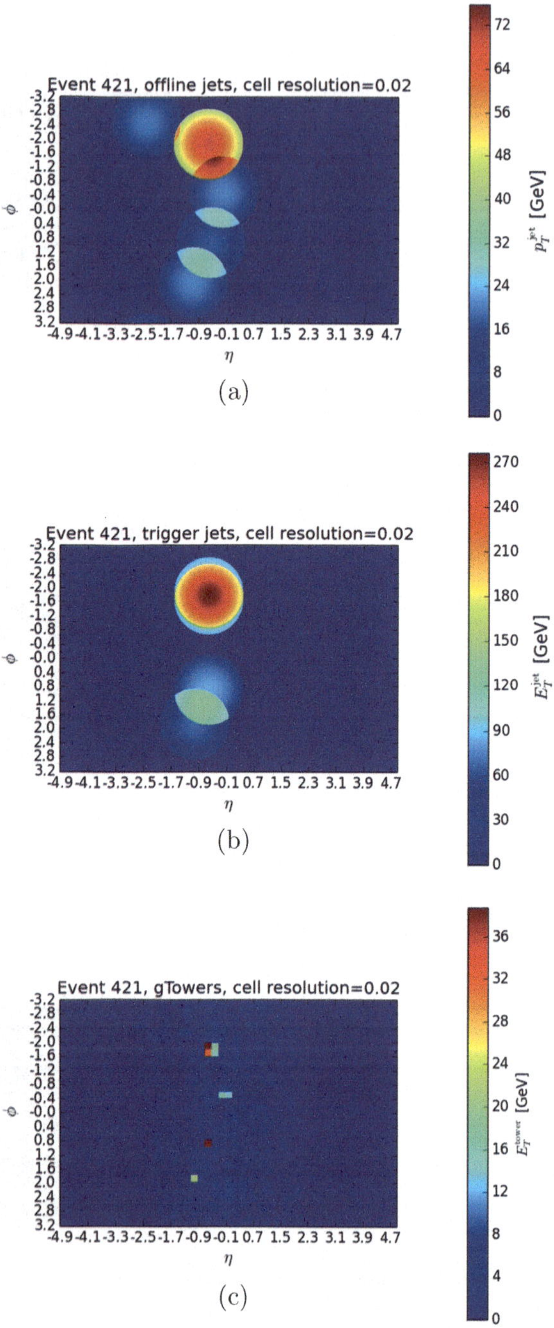

Fig. 9.6 Overlapping gTowers with $E_{\mathrm{T}} > 20\,\mathrm{GeV}$ showing that even a crude version of a clustering algorithm is still able to identify at least two jets at almost identical locations. (**a**) Offline jets. (**b**) Trigger jets. (**c**) gTowers

In terms of a technical implementation, you can generate a histogram of offline, reconstructed, leading jet p_T and then apply the trigger selection, and divide the two distributions bin-by-bin. This approach is known as the differential approach as it divides in exclusive bins. An integral approach, as the name suggests, divides the cumulative versions of the histograms, bin-by-bin. In gFEX, a typical turn-on curve compares a selection using trigger objects such as gTowers, gBlocks against the offline, reconstructed, jets. Let us take an illustrative example with distributions shown in Fig. 9.7, where dividing each trigger-selected distribution (b–d) by the one with no trigger selection (a) produces familiar turn-on curves Fig. 9.8.

Each turn-on curve can be parameterized, or quantified, by two numbers: the resolution and the plateau location. The resolution of a turn-on curve is a measure of how sharply it "turns on." A trigger with better resolution turns on more sharply; therefore, the width of the turn-on region is smaller, and this is a good proxy for the resolution. The plateau location is the position along the x-axis where the turn-on reaches large efficiency, typically 95% efficiency. The quantification of turn-on curves allows one to be able to quickly visualize the performance of different trigger selections with two numbers that fully describe the turn-on. Figure 9.9 shows an example of three different turn-on curves, part of a study described more later in Sect. 9.4, to demonstrate how the quantification of the resolution and the plateau can help compare trigger efficiencies. This quantification is useful as one is not easily able to see which of the three curves (red, blue, green) have the best resolution.

So now that the foundation is laid out, Fig. 9.10 shows the gFEX trigger using a $t\bar{t}$ Monte Carlo simulated sample in a center-of-mass energy $\sqrt{s} = 14$ TeV which is expected for Run 3. There are two colors representing the two different triggers, red for the inclusive gFEX jet trigger and blue for the L1 inclusive jet trigger. For each color, there are three curves with different marker shapes representing the number of reconstructed subjets on the isolated, offline reconstructed jet matched to the corresponding trigger jet. Circles are offline jets with a single subjet, squares are offline jets with two subjets, and triangles are offline jets with three or more subjets. As also shown in Fig. 9.2, the number of subjets in the offline reconstructed jet corresponds very well with the particular physics process, such as a top quark hadronic decay or a hadronic W-boson decay. Therefore, one can think of circle markers representing dijet events, square markers representing hadronic W-bosons, and triangles representing top quarks. For an offline jet with a single subjet, both the L1 trigger and gFEX triggers have similar resolution[1] with a similar 99% plateau point[2] around 200 GeV. So the gFEX matches the behavior of the L1 trigger. When requiring that the offline jet has two or more subjets, the gFEX trigger is able to maintain the same resolution with the same plateau location, while the L1 trigger

[1]The resolution is determined by the width of the slope of the turn-on curve. A sharper turn-on curve indicates better offline resolution.

[2]On a turn-on curve, one identifies the plateau by the x-value such that the efficiency is approximately 99%. A trigger with a lower plateau is more preferred as it reduces the selections needed to be applied on the offline object.

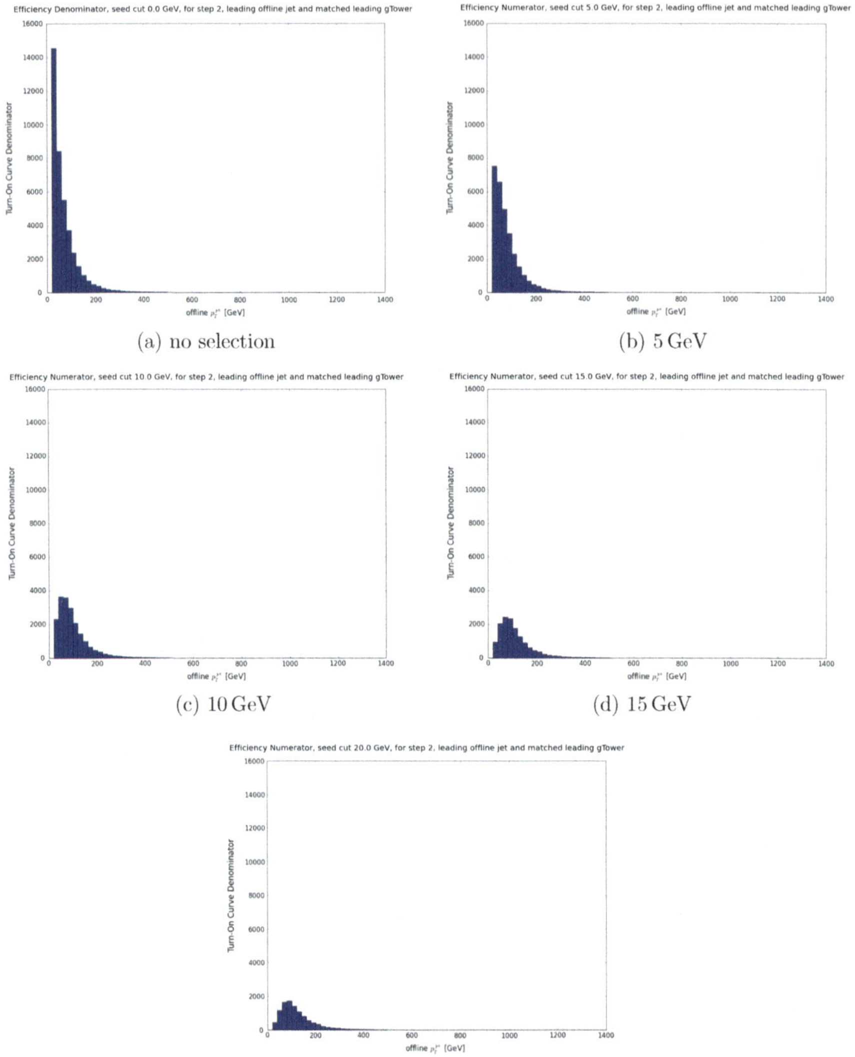

(a) no selection (b) 5 GeV (c) 10 GeV (d) 15 GeV (e) 20 GeV

Fig. 9.7 Example distributions of reconstructed, uncalibrated, leading, anti-k_t $R = 1.0$ offline jet p_{T} (**a**) without a trigger selection applied and (**b–d**) a requirement on the leading gTower E_{T}. Offline jets are matched to the leading gTower in an event, so this amounts to an event-level trigger selection. The y-axis is the number of events. Turn-on curves from dividing each trigger-selected distribution by the denominator (no selection) is shown in Fig. 9.8

sees degraded performance. To put this in context for a physics analysis that depends on an inclusive jet trigger, using the L1 trigger `L1_J100` would require the offline, $R = 1.0$ anti-k_t jets to have $p_{\mathrm{T}} > 200\,\mathrm{GeV}$ to stay in the region of a fully efficient trigger, provided that the jet has one subjet. If the jet has more substructure,

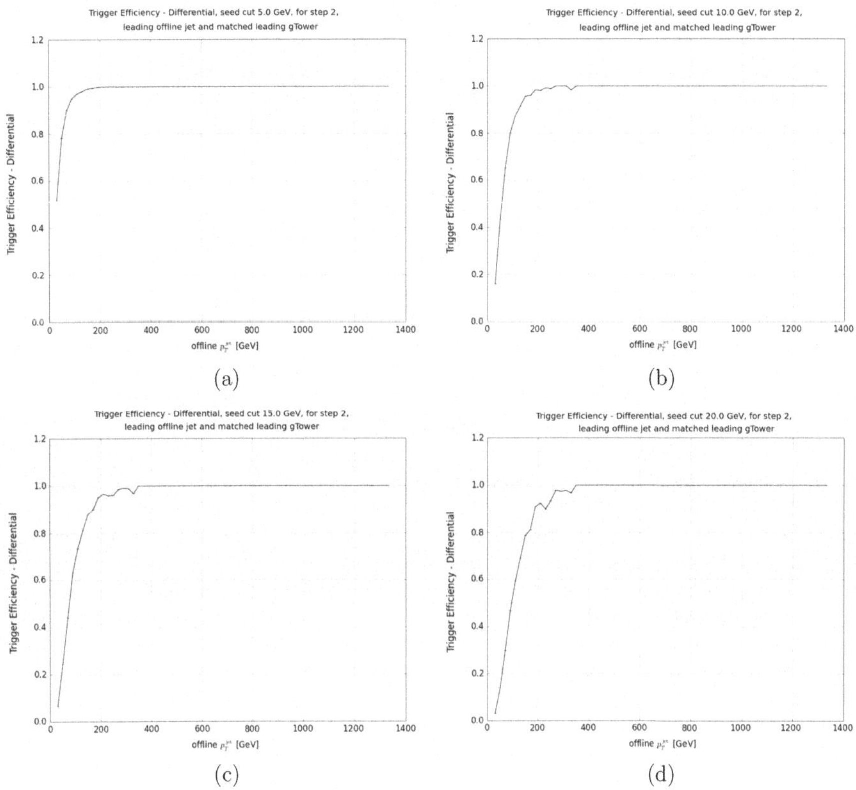

Fig. 9.8 Example turn-on curves of reconstructed, uncalibrated, leading, anti-k_t $R = 1.0$ offline jet p_{T} with a requirement on the leading gTower E_{T}. Offline jets are matched to the leading gTower in an event, so this amounts to an event-level trigger selection. The y-axis is the efficiency of the trigger. These curves were calculated from distributions in Fig. 9.7. (**a**) 5 GeV. (**b**) 10 GeV. (**c**) 15 GeV. (**d**) 20 GeV

a significantly tighter cut of $p_{\mathrm{T}} > 500$ GeV needs to be applied just to stay in the region of a fully efficient trigger. Jets in a moderately boosted regime, from $p_{\mathrm{T}} > 200$–500 GeV would not be efficiently selected by the current L1 trigger, but would be covered by the proposed gFEX trigger. These jets are crucial to physics programs, such as the analysis search presented in this thesis.

9.4 gFEX Studies

9.4.1 Pile-Up Energy Density Calculations

As described in Sect. 5.1.3.2, Eq. (5.6), pile-up energy density is an important quantity to estimate for events at the LHC where one can expect to see many

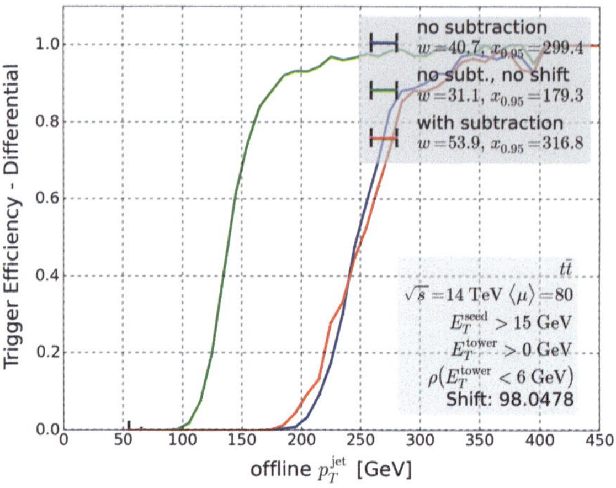

Fig. 9.9 Example turn-on curves of different triggers to understand the impact of changes in the gFEX jet reconstruction algorithms for Monte Carlo simulated $t\bar{t}$ events with center-of-mass energy $\sqrt{s} = 14\,\text{TeV}$, requiring the leading trigger jet to have $p_T > 140\,\text{GeV}$. The gFEX jet reconstruction algorithm is seeded by a 15 GeV gTower to form gFEX trigger jets. The green curve represents the efficiency of this standard reconstruction. The red curve represents the efficiency of the trigger jets, after they have been corrected for estimated pile-up energy density in the event, described more in Sect. 9.4.1. Because there was an observed shift in the location of the trigger curve, the blue curve represents the efficiency using uncorrected trigger jets, but tightens the trigger selection from 140 to 240 GeV (a shift of 98 GeV), to closely match the location of the red curve and understand the impact on the resolution of the trigger. Each turn-on curve is parameterized by the resolution w and the plateau $x_{0.95}$: green is $(w, x_{0.95}) = (31.1, 179.3)$, blue is $(w, x_{0.95}) = (40.7, 299.4)$, and red is $(w, x_{0.95}) = (53.9, 316.8)$

interactions per bunch crossing as well as in the future of the LHC program, the HL-LHC, which could see up to 200 interactions! Being able to estimate pile-up at the trigger level is crucial for gFEX jet reconstruction and having this quantity calculated as early as possible in the collision helps downstream algorithms, such as those in the HLT perform more efficiently. This section describes a series of studies performed in order to estimate the offline pile-up in gFEX, which I will call **online pile-up**.

A hardware consideration is the latency constraints described in Table 4.2 which only allow up to five bunch crossings to run algorithms on gFEX. While the pile-up calculation done offline uses a median-based approach, gFEX does not have enough time to sort 1284 gTowers in order to compute the pile-up. It also does not have enough time to form trigger jets and calculate pile-up using those jets. However, all is not lost, as a truncated-mean-based approach can work. This is described in Eq. (9.5) which only considers pile-up by summing gTowers below a particular energy threshold.

$$\rho_{\text{online}} = \text{mean}\left\{\frac{E_{\text{T}}^{i}}{\text{area}_i}\right\}, \forall i \in \text{gTowers with } E_{\text{T}} < X\,\text{GeV} \tag{9.5}$$

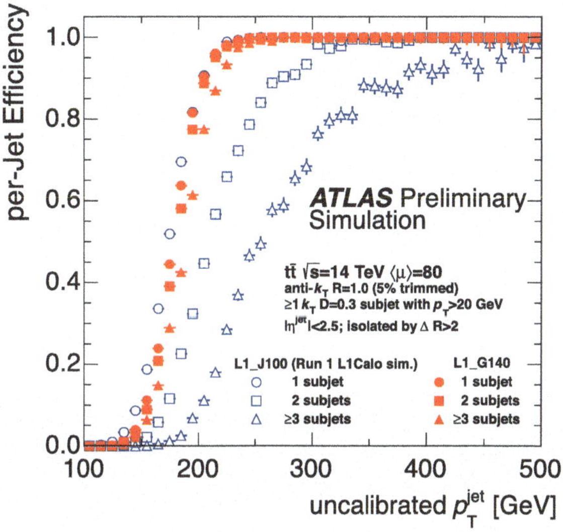

Fig. 9.10 Per-jet efficiency turn-on curves in Monte Carlo (MC) simulation for multiple Phase I upgrade Level-1 jet trigger options [56]. A global feature extraction (gFEX) reconstruction algorithm (closed red markers, left) from the TDAQ Phase I Upgrade Technical Design Report (TDR) [12] with a 140 GeV threshold is compared to full simulation of the Run I Level-1 calorimeter jet trigger (open blue markers, left and right) with a 100 GeV threshold. The gFEX reconstruction implements a simple seeded cone algorithm with a nominal radius of $R = 1.0$ and with a seed selection of 15 GeV applied to calorimeter towers with area 0.2×0.2 in $\eta \times \phi$. The 140 GeV gFEX trigger threshold is chosen to match the `L1_J100` single subjet turn-on curve. Pair-produced top quark Monte Carlo simulation samples are simulated with a pile-up level equivalent to an average number of interactions per bunch crossing $\langle \mu \rangle = 80$. For each algorithm, the efficiency curves are shown as a function of the offline trimmed anti-k_t $R = 1.0$ jet p_T with different offline subjet multiplicities. The trimming parameters specify that any subjets with a p_T fraction of the original jet less than 5% are to be discarded. The subjets are defined using the k_t-clustering algorithm with a nominal radius parameter of $D = 0.3$. For subjet counting, the subjets are required to have a subjet $p_T > 20$ GeV. The offline trimmed jets are required to be isolated from any other offline jet by at least a radial distance of $\Delta R > 2.0$ rad and to be within the pseudorapidity range $|\eta| < 2.5$. The turn-on curves measure per-jet efficiencies after requiring that the Level-1 gFEX jet be within $\Delta R < 1.0$ of the offline trimmed jet

Figure 9.11 shows some distributions made for a study to explore the truncated-mean-based approach described in Eq. (9.5) with two upper thresholds at 3 GeV and 6 GeV. Various upper thresholds were studied in steps of 1 GeV from 1 GeV to 15 GeV and two are chosen to be shown. The red curve corresponds to using all gTowers, the teal and magenta curves are for the negative and positive central region, respectively, and the gold and black are for the negative and positive forward region, respectively.

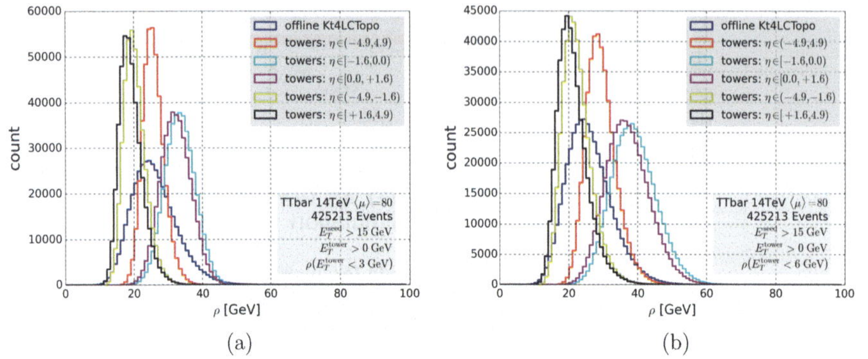

Fig. 9.11 Distributions of truncated-mean-based online pile-up calculation using gTowers across different η ranges compared to offline pile-up (blue) for Monte Carlo simulated $t\bar{t}$ events with $\langle\mu\rangle = 80$ at center-of-mass energy $\sqrt{s} = 14\,\text{TeV}$. While the scale is not important as this can be calibrated later, the width of each distribution is and how well it corresponds to offline. The upper threshold, X, also labeled on the plot as $\rho(E_T^{\text{tower}} < X\text{GeV})$, is also specified as (**a**) 3 GeV and (**b**) 6 GeV

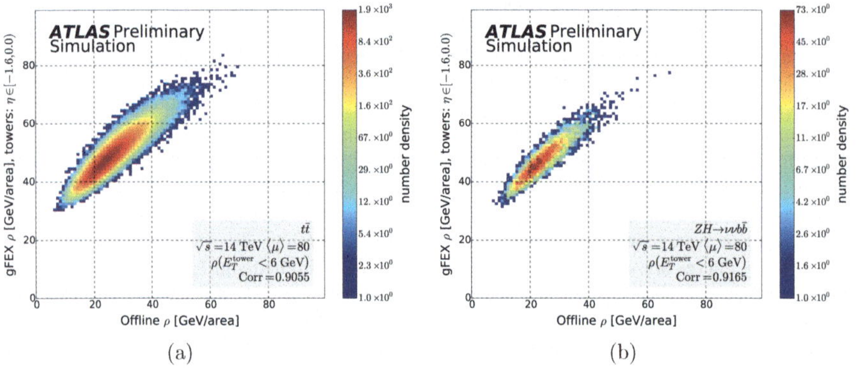

Fig. 9.12 Correlation between the offline event energy density ρ [37] on the horizontal axis and a simplified calculation of the event energy density in the L1Calo trigger using gFEX with a truncated-mean-based approach using gTowers with $E_T^{\text{tower}} < 6\,\text{GeV}$ and $-1.6 < \eta < 0.0$ [56]. The correlation for (**a**) $t\bar{t}$ and (**b**) $ZH \rightarrow \nu\nu b\bar{b}$ events is greater than 90%. Both Monte Carlo simulation samples are simulated with average number of interactions $\langle\mu\rangle = 80$ at a center-of-mass energy $\sqrt{s} = 14\,\text{TeV}$. In each case, the strong correlation means that the average value of ρ measured by the gFEX trigger for a given offline ρ is similar

Figure 9.12 shows the first of these successful studies demonstrating the efficacy of the truncated-mean-based calculation of online ρ using central[3] gTowers for two different Monte Carlo simulated samples. Both samples have an average interaction

[3]The gTowers selected correspond to an η range of the original proposal of gFEX. These studies will need to be updated again using current Monte Carlo simulation upgrade samples and an updated η range.

per crossing $\langle\mu\rangle = 80$ at center-of-mass energy $\sqrt{s} = 14\,\text{TeV}$. The online pile-up is shown to be strongly correlated to the offline pile-up, ignoring a difference in the scales which is expected as the gTowers have not been calibrated. The very nice conclusion is that the online calculation of ρ is independent of the physics processes we are studying. By definition, ρ should not depend on the hard scatter and it does not! Upper thresholds were explored in steps of 1 GeV from 1 GeV to 15 GeV; $X = 6\,\text{GeV}$ was found to have the strongest correlation.

9.4.2 Pile-Up Mitigation Studies

The next set of studies is a natural continuation from the pile-up calculation studies in Sect. 9.4.1 by focusing on the effects of incorporating pile-up calculations in the gFEX trigger efficiency. Do we lose resolution? Do we see performance gains? Using the trigger jet reconstruction algorithm described in Sect. 9.2 to build gFEX trigger jets with $R = 1.0$, an area-based mitigation approach is taken to subtract pile-up from the jet. This is described in Eq. (9.6) which corrects the energy of a jet based on the area of the jet.

$$E_T^{\text{jet, corr.}} = E_T^{\text{jet}} - \rho_{\text{online}} \times \text{area}^{\text{jet}}. \tag{9.6}$$

Using this area-based pile-up mitigation, Fig. 9.13 shows the correlation between the offline jet and online jet energies before and after the correction. What is interesting to note is that while the scale of the trigger jets has expectedly and notably shifted downwards, the correlation between the offline jet energy and trigger

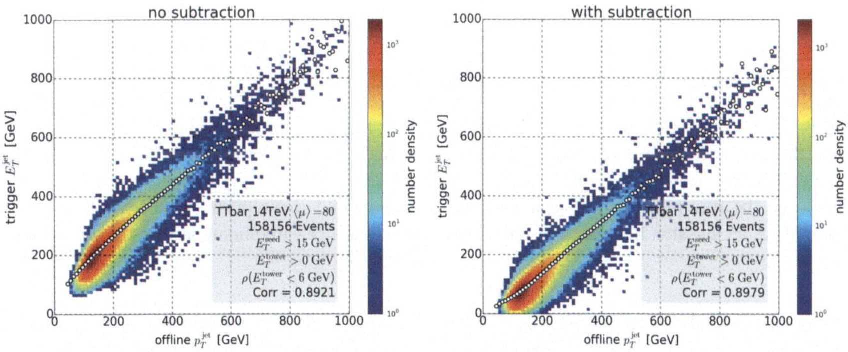

(a) no mitigation (b) with mitigation

Fig. 9.13 Correlation between isolated offline jet and matched gFEX trigger jet energies for $-1.6 < \eta < 0.0$ in a Monte Carlo simulated $t\bar{t}$ sample with $\langle\mu\rangle = 80$ at a center-of-mass energy $\sqrt{s} = 14\,\text{TeV}$. The correlations are shown (**a**) before and (**b**) after pile-up mitigation is applied. The trigger jets were seeded using towers with $E_\text{T} > 15\,\text{GeV}$ and the truncated-mean-based pile-up ρ_{online} was calculated using towers with $E_\text{T} < 6\,\text{GeV}$ which was optimized. The white circles represent the average trigger jet energy in each offline jet bin

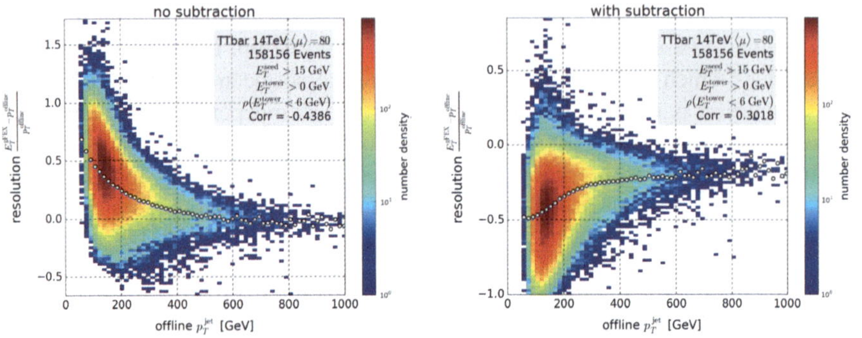

(a) no mitigation (b) with mitigation

Fig. 9.14 Correlation between isolated offline jet and the energy resolution of the matched gFEX trigger jet is shown for $-1.6 < \eta < 0.0$ in a Monte Carlo simulated $t\bar{t}$ sample with $\langle\mu\rangle = 80$ at a center-of-mass energy $\sqrt{s} = 14\,\text{TeV}$. The resolutions are shown (**a**) before and (**b**) after pile-up mitigation is applied. The trigger jets were seeded using towers with $E_T > 15\,\text{GeV}$ and the truncated-mean-based pile-up ρ_{online} was calculated using towers with $E_T < 6\,\text{GeV}$ which was optimized. The white circles represent the average resolution in each offline jet bin. The resolution of the trigger jet energy is defined as a measure of the difference with respect to the matched offline jet compared to the energy of the offline jet

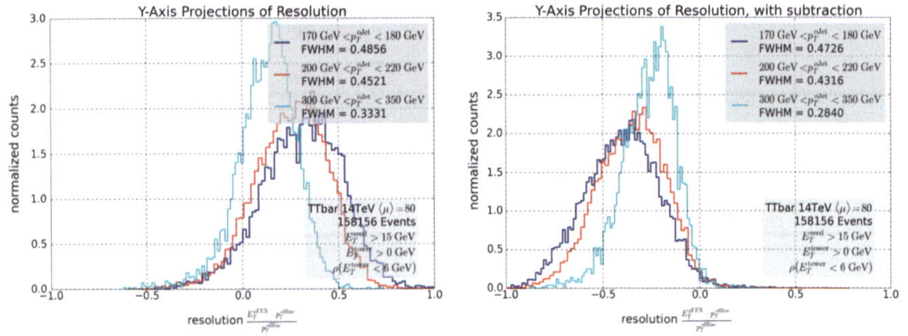

(a) no mitigation (b) with mitigation

Fig. 9.15 Y-axis projections of the resolution plots in Fig. 9.14 for selected offline jet p_T ranges: 170–180 GeV, 200–220 GeV, and 300–350 GeV. This was done on Monte Carlo simulated $t\bar{t}$ events with $\langle\mu\rangle = 80$ at a center-of-mass energy $\sqrt{s} = 14\,\text{TeV}$. The projections are shown (**a**) before and (**b**) after pile-up mitigation is applied. Each legend also reports the full-width half-max (FWHM) of a Gaussian fit to each of the projections, a smaller value being a stronger resolution

jet energy remains just as strong. At low energies, there was a slight non-linearity observed before the correction that seems to be partially linearized after the pile-up correction is applied. Further study is needed here.

The next study was to determine how the pile-up correction improved the resolution of the trigger jets with respect to the matched, isolated offline jet as shown in Fig. 9.14. The resolution calculation (Eq. (9.7)) describes how well the online trigger jet measures the matched, isolated, offline reconstructed jet. Figure 9.14

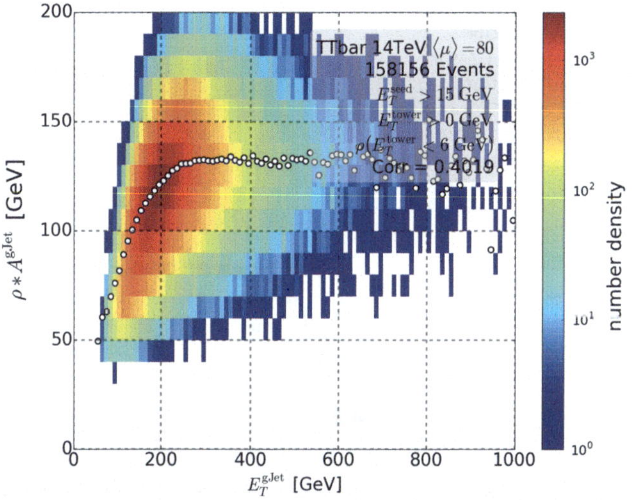

Fig. 9.16 Correlation showing the amount of pile-up energy density subtracted from a given trigger jet as a function of the matching, isolated offline reconstructed jet for $-1.6 < \eta < 0.0$. This was done on Monte Carlo simulated $t\bar{t}$ events with $\langle\mu\rangle = 80$ at a center-of-mass energy $\sqrt{s} = 14\,\text{TeV}$. For offline jets below 200 GeV, there are usually not enough energetic gTowers around the seeded tower with $E_\text{T} > 15\,\text{GeV}$ to be included in the reconstruction and so those associated, lower energy trigger jets tend to be smaller in area as they have less towers and so the correction falls off with energy. At a certain point, the correction is approximately the same which is the "full-occupancy" trigger jet with all gTowers within $\Delta R < 1.0$ of the seed participating in the reconstruction

shows correlation plots of the resolution of trigger jets as a function of the matched, isolated offline jet p_T before and after pile-up mitigation. While there is still significant non-linearity after pile-up mitigation, pile-up mitigation significantly improves the resolution of the trigger jets for offline jets with $p_\text{T} > 250\,\text{GeV}$ which is right in the region that gFEX is designed to specialize in. Moreover, there is no non-linearity before any pile-up mitigation which makes it very difficult to calibrate the energy of the trigger jets. Figure 9.15 shows y-projections of the trigger jet resolution in a few selected offline jet p_T ranges before and after the pile-up mitigation. The pile-up mitigation does not negatively affect the width of the distribution of trigger jets for lower energy offline jets with $p_\text{T} < 250\,\text{GeV}$ but does measurably narrow the distribution of trigger jets for offline jets with $p_\text{T} > 300\,\text{GeV}$.

$$\mathcal{R} = \frac{E_T^{\text{gFEX jet}} - p_\text{T}^{\text{offline jet}}}{p_\text{T}^{\text{offline jet}}} \tag{9.7}$$

Finally, the last study for pile-up mitigation is a question of how much energy is being subtracted from each trigger jet shown in Fig. 9.16.

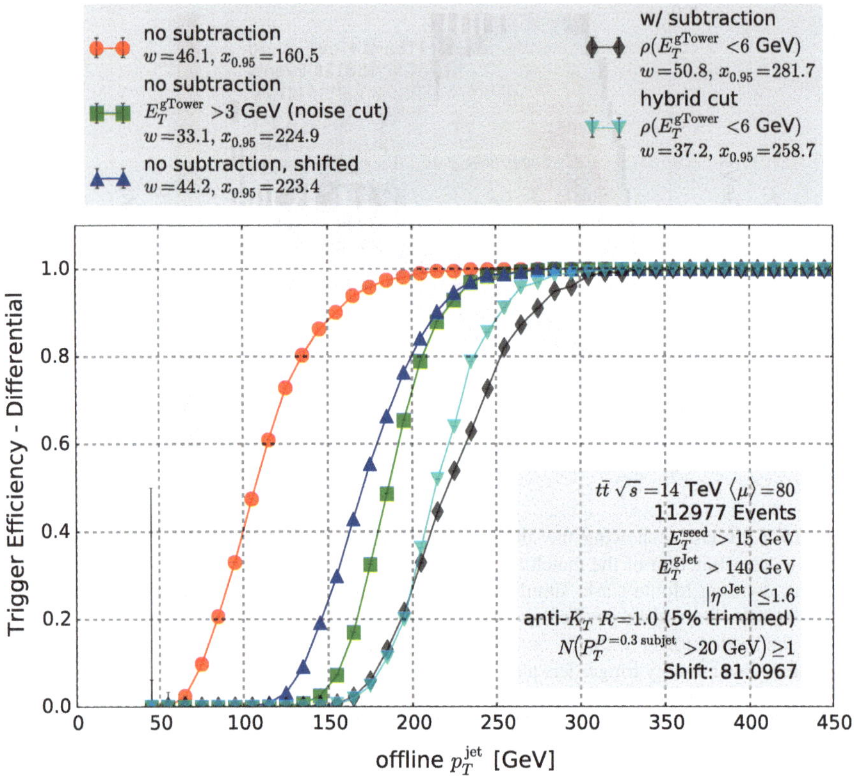

Fig. 9.17 Trigger efficiency curves for the five different pile-up mitigation techniques. These are: no subtraction, no subtraction but a noise cut applied, no subtraction and simply shifted, with pile-up subtraction, and a hybrid cut. The legend reports the width of the turn-on w and the location of the plateau at 95% $x_{0.95}$. The efficiency is reported as a function of the large-R $R = 1.0$ anti-k_t isolated offline jet matched to the given trigger jet for a trigger jet requirement of 140 GeV

9.4.2.1 Efficiency of Pile-Up Mitigation Techniques

Thus far, there have been demonstrated significant studies into the area-based pile-up mitigation techniques. However, up until now, no turn-on curves have been shown to demonstrate these. An area-based subtraction is indeed possible. Other choices for reducing pile-up are

- **noise cut**: a simple selection removing towers below a certain threshold
- **hybrid cut**: a combination of a noise cut at 3 GeV and applying 25% of the area-based pile-up subtraction ($0.25\rho_{\text{online}}$)

A study was proposed and done to study the impact of all of these different pile-up mitigation techniques on the trigger efficiency for a fixed trigger selection shown

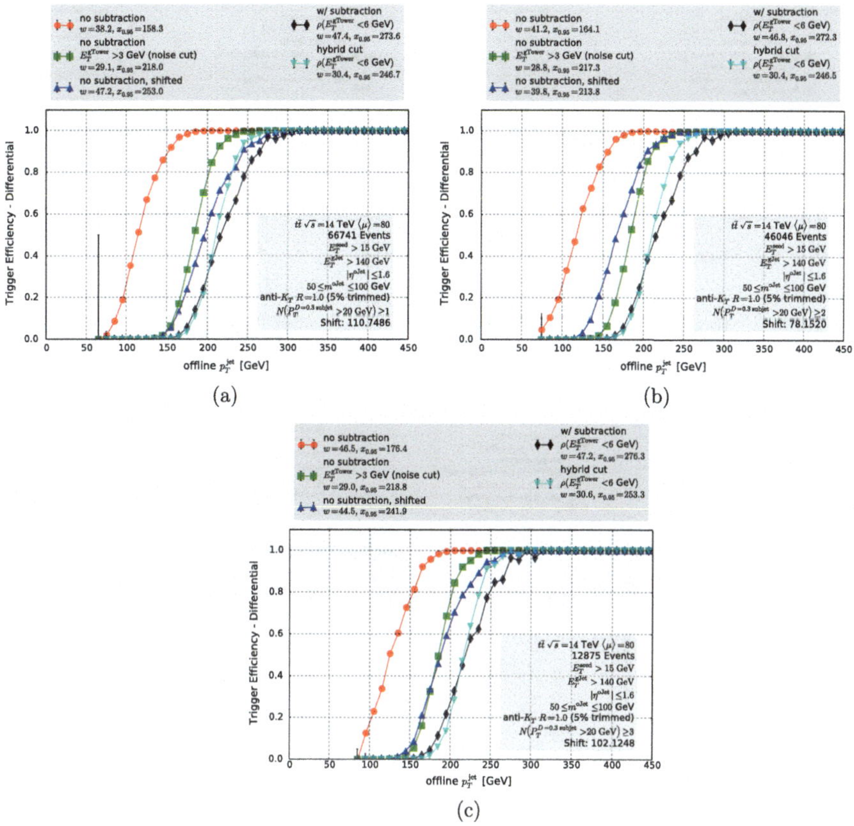

Fig. 9.18 Trigger efficiency curves for the five different pile-up mitigation techniques. These are: no subtraction, no subtraction but a noise cut applied, no subtraction and simply shifted, with pile-up subtraction, and a hybrid cut. The legend reports the width of the turn-on w and the location of the plateau at 95% $x_{0.95}$. The efficiency is reported as a function of the large-R $R = 1.0$ anti-k_t isolated offline jet matched to the given trigger jet for a trigger jet requirement of 140 GeV. An additional selection on the mass of the offline jet is required to be within 50–100 GeV to enhance hadronic W-bosons. (**a**) At least 1 subjet. (**b**) At least 2 subjets. (**c**) At least 3 subjets

in Fig. 9.17.[4] For each curve, the resolution (width of turn-on w) and the plateau location $x_{0.95}$ is reported. It is seen that a hybrid cut tends to have measurably better resolution than just an area-based subtraction alone, but the noise cut appears to have the largest impact on the resolution of a trigger jet. This is expected as the noisy towers can smear the resolution of a jet energy. Further selections applied on the offline jet mass to enhance the hadronic W-bosons (Fig. 9.18) and hadronic tops (Fig. 9.19) are also shown with very similar conclusions.

[4]These studies need to be redone with equal rate settings.

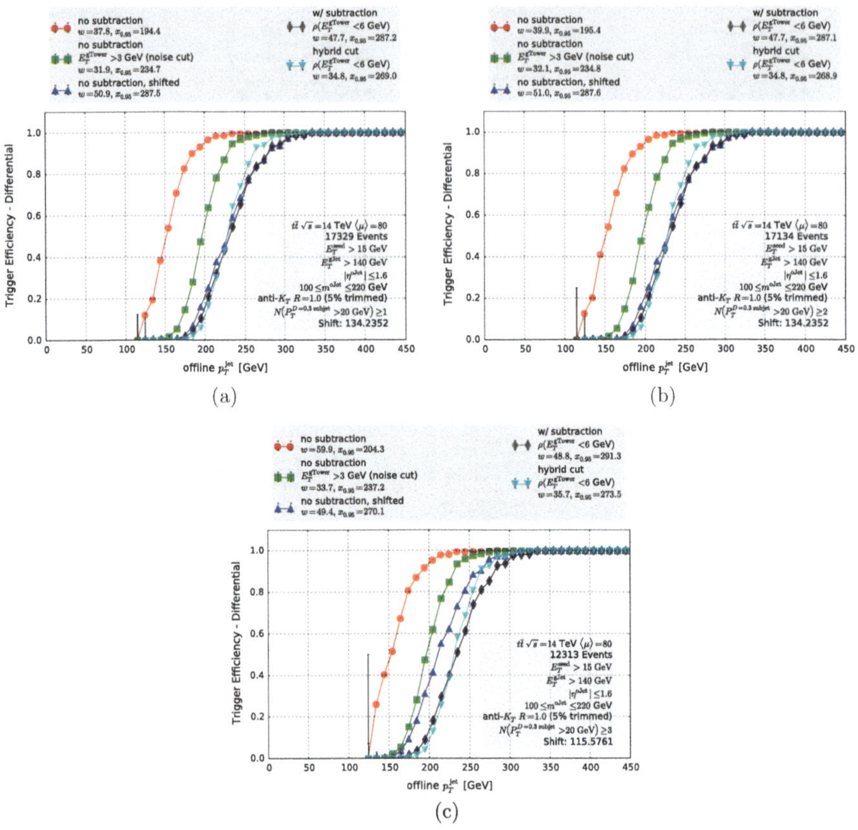

Fig. 9.19 Trigger efficiency curves for the five different pile-up mitigation techniques. These are: no subtraction, no subtraction but a noise cut applied, no subtraction and simply shifted, with pile-up subtraction, and a hybrid cut. The legend reports the width of the turn-on w and the location of the plateau at 95% $x_{0.95}$. The efficiency is reported as a function of the large-R $R = 1.0$ anti-k_t isolated offline jet matched to the given trigger jet for a trigger jet requirement of 140 GeV. An additional selection on the mass of the offline jet is required to be within 100–200 GeV to enhance hadronic tops. (**a**) At least 1 subjet. (**b**) At least 2 subjets. (**c**) At least 3 subjets

9.4.3 Substructure Studies

Now we are on the last series of upgrade studies that are presented in this thesis. All along, the jets that have been formed were often seeded by a single gTower above some particular E_{T} threshold, usually 15 GeV for the studies shown so far. However, if we think about the representation of a $t\bar{t}$ decay and its parton shower, it seems that the most energetic gTowers inside a reconstructed trigger jet should correspond with the subjets. Therefore, it is highly motivated to try and see if it is possible to identify trigger jets with significant substructure using the kinematics of the gTowers themselves. The first thing is to look at the correlation between leading

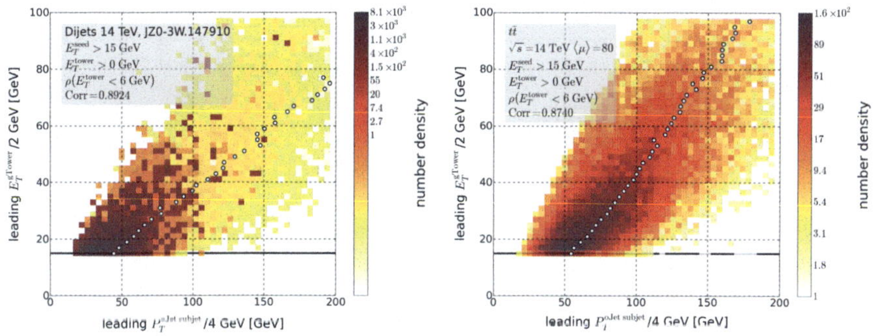

(a) (b)

Fig. 9.20 Correlations of the energy between leading gTower and the matched offline jet's leading subjet for (**a**) QCD multijet and (**b**) $t\bar{t}$ Monte Carlo simulated events with $\langle\mu\rangle = 80$ at a center-of-mass energy of $\sqrt{s} = 14$ TeV. The trigger jets are seeded with gTowers with $E_{\rm T} > 15$ GeV and pile-up corrected using an upper threshold of 6 GeV. Notice that there is a nice average linearity and a relatively strong correlation (>85%) for both Monte Carlo samples. For multijet events, the leading gTower contains a majority of the energy of the offline jet, while for $t\bar{t}$ this is lower as expected for jets with significant substructure and energy spread out more

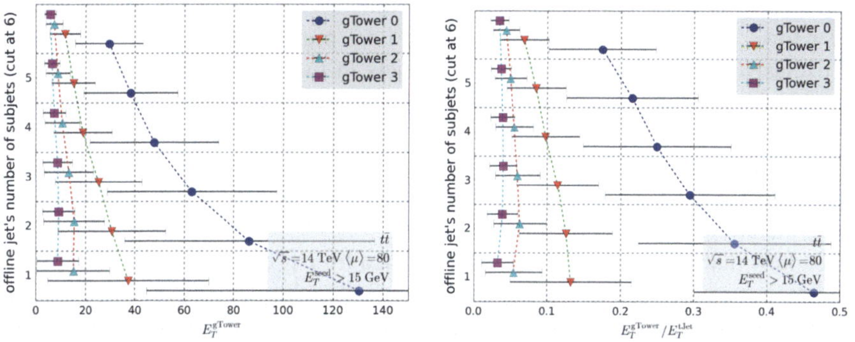

(a) (b)

Fig. 9.21 Distributions of the (**a**) gTower energy and (**b**) fraction of gFEX trigger jet energy carried by each of the leading towers in the given jet. The y-axis is exclusive binning in the number of subjets of the matched, isolated, reconstructed offline jet. Dashed lines connect points which map the same type of gTower to understand the trends of the leading, subleading, etc. gTowers as you require more and more substructure in the offline jet. Each distribution of gTower energies for an offline subjet selection is fitted to a Gaussian and the mean is extracted and drawn as the marker, while the standard deviation is extracted and drawn as error bars

gTower in each trigger jet that was reconstructed as shown in Fig. 9.20 which shows a particularly strong correlation between the subjets of an offline jet and the leading gTowers in the associated trigger jet. This indicates that there is potentially some leverage to construct a discriminating variable for identifying subjets.

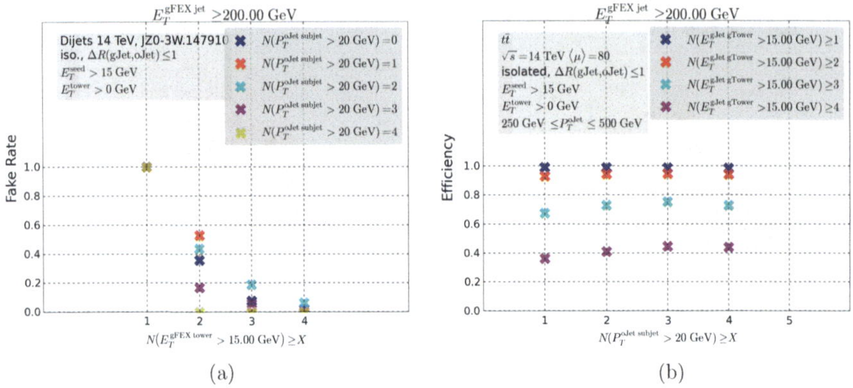

Fig. 9.22 This is a plot of the (**a**) fake rate of dijet samples and (**b**) efficiency of the $t\bar{t}$ samples for Monte Carlo simulated events with $\langle\mu\rangle = 80$ and a center-of-mass energy $\sqrt{s} = 14\,\text{TeV}$. (**a**) is the background fake rate where a gFEX trigger jet E_{T} selection is applied to the denominator and numerator and a "subtower" multiplicity is applied to the numerator. (**b**) is the signal efficiency where the offline selection detailed in the plot is applied to the numerator and denominator and the trigger selection listed is applied to the numerator

So given all this information, the next straightforward study is to understand the gTower energy and the fraction of the total trigger jet energy as a function of exclusive binning in the number of subjets in the offline jet. This is a bit of a mouthful and this is shown in Fig. 9.21. This is an enormously rich plot to breakdown. First, one can look at just the leading gTower in the event ("gTower 0") and can potentially apply a high, inverted selection on its energy to suppress dijet-like events where the offline jets only have one subjet. This is a preliminary study that needs to be explored further.

So now that the leading four gTowers, sorted by E_{T}, can be used as a proxy for subjets, can start to define a gFEX jet "subtower" multiplicity by $N(E_{\text{T}}^{\text{tower}} > X)$, the number of gTowers in a trigger jet is greater than an E_{T} threshold. If we then consider $t\bar{t}$ a signal, and multijet as a background, we can start exploring trigger selections on "subtower multiplicity" in a way to maximize signal over background. This requires studying the trigger efficiency on signal and the fake rate on background, which is shown in Fig. 9.22. In this study, isolated offline jets are matched to gFEX jets with $\Delta R < 1.0$, seeded with towers $E_{\text{T}} > 15\,\text{GeV}$ with no pile-up mitigation applied. The trigger selection applied is to require trigger jet $E_{\text{T}} > 200\,\text{GeV}$. In order to enhance the signal of $t\bar{t}$ to look purely at hadronic top decays that are fully captured, an offline jet mass cut is applied around the mass of the top quark $100\,\text{GeV} < m^{\text{offline jet}} < 220\,\text{GeV}$ and the p_{T} of the jet is required to be semi-boosted $250\,\text{GeV} < p_{\text{T}} < 500\,\text{GeV}$.

To test your understanding of Fig. 9.22, let us walk through a single point in both background and signal. Take signal for a second and let us think about requiring that the offline jet has one subjet with $p_{\text{T}} > 20\,\text{GeV}$. The probability of finding more than one gTowers with $E_{\text{T}} > 15\,\text{GeV}$ goes down as we require more gTowers,

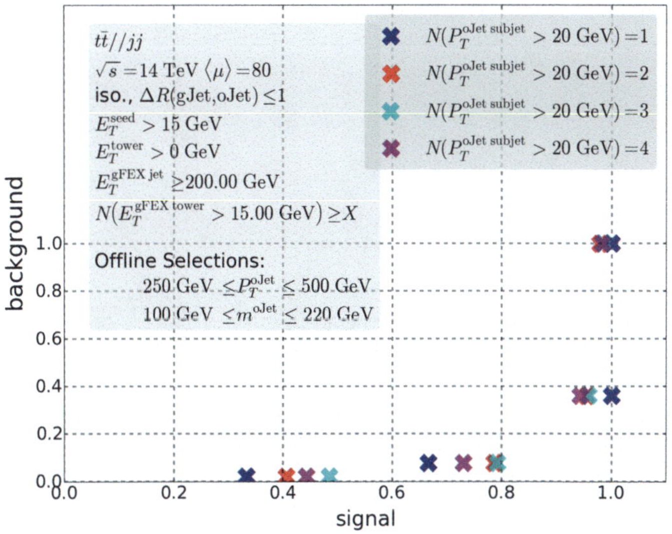

Fig. 9.23 The signal efficiency is shown as a function of the background "fake rate" from Fig. 9.22

and therefore the efficiency goes down as well in that particular column. Now take the background and again, think about requiring four subjets in the offline jet with $p_T > 20$ GeV. As we increase the "subtower multiplicity" trigger selection in the numerator, the rate falls down, as it becomes less and less likely to find enough gTowers with sufficient energy. In particular, for dijet events where there is very little substructure to begin with, a dijet event that has many offline subjets will have many low energy subjets, compared to a dijet with less subjets each with higher energy. This is why the rate of dijet events with more offline jets falls off more sharply.

So now that we have a fake rate and a signal efficiency, we can combine the two into a single plot shown in Fig. 9.23. This plot shows the signal efficiency as a function of the background "fake rate" (with no offline selection on the background). To test your understanding of this combination, a 6.3% background "fake rate" with 66% signal efficiency corresponds to $N(E_T^{tower} > 15\,\text{GeV}) \geq 3$ which is a point that corresponds with boosted top quark decays! Precisely the kind of signal that my thesis analysis is sensitive to, but with a designed trigger that enables me to enhance the efficiency of selecting events while suppressing background, or uninteresting, events. Further study needs to be done to quantify the rate of the trigger given a trigger selection such as specified.

Chapter 10
Conclusion

In the first operation of the LHC, an analysis searching for new physics using the same simplified model [15, 16], observed no significant excess of events over the expected SM background [46]. In 2014, the lower limit on the $\tilde{g}$ mass was set to be at 1.4 TeV at the 95% confidence level for this simplified signal model. The results presented in this thesis span the first 2 years of the second operational period of the Large Hadron Collider, during 2015 and 2016, utilizing new techniques of boosted object reconstruction to greatly extend the sensitivity to new physics and improve our understanding of the Standard Model. The lower limit of the gluino mass is now set at 1.95 TeV at the 95% confidence level using three different kinds of kinematic observables: missing energy-type, energy scale-type, and energy structure-type [143]. Even though no excess was observed and tighter limits were set, simplified models are useful to provide topology-based limits on searches to identify the boundaries of search sensitivity and derive limits on more general models by reinterpreting [150] the limits in the context of a different signal topology. Even though simplified models may not be motivated by realistic SUSY scenarios, they help to understand the limits of the detector technology. Experimentalists and theorists alike can identify kinematic ranges for which existing searches are not efficient or sensitive, and then define new search strategies to attempt to cover the gaps in the exploration of phase space. Experimentalists can use the results of searches being performed now to define goals and plans for covering regions of phase space that are particularly hardware-limited. The second portion of this thesis has a focus on the instrumentation upgrades to be ready for the third operation of the LHC physics program, to last for the foreseeable lifetime of the LHC which is around 2045. As the current trigger system in ATLAS is not efficient at detecting many of the boosted objects that are copiously produced in the high energy proton–proton collisions today, the gFEX module is being built and designed to recover this efficiency. This instrumentation upgrade will benefit countless analyses using boosted objects in the future, including those that have not yet been considered as a potential model of new physics. This thesis is but a chapter in the rich

© Springer Nature Switzerland AG 2020

G. Stark, *The Search for Supersymmetry in Hadronic Final States Using Boosted Object Reconstruction*, Springer Theses,
https://doi.org/10.1007/978-3-030-34548-8_10

story of boosted objects a significant advance in the attempt to use boosted object reconstruction to find supersymmetry.

But there are still many areas for improvement. When I first started in 2014, the LHC was set to start up in 2015. For the entirety of 2015 and 2016, it has been a rapid sprint to get preliminary results on these crucial simplified models out to the theorists and the public. But now, this sprint becomes a marathon. The LHC will be shutting down for upgrades at the end of the year. By the end of 2018, experimental particle physicists in the ATLAS collaboration will have $150\,\text{fb}^{-1}$ of data to play with for the next 4–5 years. The increased lower limits on the mass of gluinos may have weakened the case for naturalness, though the possibility of natural supersymmetry still has not been excluded. While the simplified model is certainly unrealistic, what with gluinos decaying through stop squarks 100% of the time, and those stop squarks decaying to top quarks 100% of the time, the analysis of this model is crucial to reinterpretation in other regions of phase space. I look forward to seeing what neighboring physics models we are able to contribute some sensitivity to. The techniques presented in this thesis are brand new and there remains the opportunity to refine them during the next iteration of the analysis. There are also many areas of improvement that have been uncovered during my time with this analysis, such as the non-perturbative modeling of QCD in order to better model the tails of the important kinematic observables. The systematic uncertainties associated with jets and flavor-tagging, both of which this flagship SUSY analysis is very sensitive to, are generally the dominating uncertainties by far and modulating these uncertainties can strengthen the reach of this analysis. Similarly, many physics studies of gFEX are now being done for the first time and have been shown in Chap. 9 and will need to be redone using the most advanced and latest Monte Carlo simulations and detector geometry definitions that exist. Just like substructure found its way into popularity of jet physics [121, 201, 239], I hope that the first studies of substructure in gFEX gain some momentum, enabling future physicists to design a trigger menu that provides sensitivity to the currently unobserved boosted Higgs decays [122] or the simultaneous production of four top quarks [110]. What more can we reveal of nature? Tune in and see!

Appendix A
Optimizing Optimizations

The most up-to-date documentation can be found with the source code at github:kratsg/optimization on GitHub. This tool allows you to take a series of ROOT ntuples, signal and background, apply a lot of cuts automatically, and figure out the most optimal selections to maximize significance. It comes packed with a lot of features

- generator script to create, what we call, a supercuts file containing all the rules to tell the script what cuts to apply and on which branches
- cut script which will take your signal, background, and supercuts; run them all; and output a series of files with the appropriate event counts for all cuts provided
- optimization script which will take your signal counts and background counts; run them all; and output a sorted list of optimal cuts
- hash look up script to reverse-engineer the cut for a given hash when you supply the supercuts file

Note as part of making the script run as fast as possible, I try to maintain a low memory profile. It will not store (or remember) the cut used to create a significance value. Instead, we compute a 32-bit hash which is used to encode the cuts, and a way to "decode" the hash is also provided.

A.1 Major Dependencies

- PyROOT (which technically requires ROOT)
- numpy
- root_numpy

© Springer Nature Switzerland AG 2020

G. Stark, *The Search for Supersymmetry in Hadronic Final States Using Boosted Object Reconstruction*, Springer Theses,
https://doi.org/10.1007/978-3-030-34548-8

A.2 Top-Level

```
usage: rooptimize [-h] [-a] {generate,cut,optimize,
hash,summary} ...

Author: Giordon Stark. vX.Y.Z

positional arguments:
  {generate,cut,optimize,hash,summary}
                              actions available
    generate                  Write supercuts template
    cut                       Apply the cuts
    optimize                  Calculate significances for a
                              series of computed cuts
    hash                      Translate hash to cut
    summary                   Summarize Optimization Results

optional arguments:
  -h, --help                  show this help message and exit
  -a, --allhelp               show this help message and all
                              subcommand help
                              messages and exit

This is the top-level. You have no power here.
```

A.2.1 Parameters

There is only one required position argument: the `action`. You can choose from

- generate—generate a supercuts template file from an input ROOT file for running the rest of the optimization framework
- cut—perform the first step of optimization by generating the cuts and then applying them to all samples
- optimize—perform the second step of optimization by calculating the significances for all signal samples for all cuts
- hash—look up the cut associated with a generated identifying hash
- summary—summarize the results of an optimization into a single JSON file for reproducibility and plotting

We also provide an optional argument `-a, --allhelp` which will print all the help documentation at once instead of just the top-level `-h, --help`.

Appendix B
xAODAnaHelpers

xAODAnaHelpers has its source code in GitHub: github:UCATLAS/xAODAna Helpers/. This is the largest analysis framework in ATLAS for Run 2 data-taking physics built for two different kinds of offline analysis release software: AnalysisBase and AnalysisTop.

This package is meant to be the minimal needed to use the Combined Physics (CP) tools properly to calibrate, select, and correct the physics objects used for most physics analyses. Each step of the analysis chain is done by an `EL::Algorithm` which utilizes `TStore` to pass information to the algorithms down the chain. The final product can be a `TTree`, histograms, or a mini xAOD. The philosophy adopted is not to remake the Event Data Model or to alter it but to make minimal wrapper around CP tools to help users configure them properly and connect the full chain without much hassle. To this end, some details are hidden for the user and set automatically in the tools. As much as possible we used the same names as is shipped with the xAOD objects or the CP tools themselves. The user is not meant to learn a new EDM but rather to learn the minimal needed to start doing the fun stuff—Physics!!

B.1 Background

An analysis is done in steps with an `EL::Algorithm` running for each. For example, one algorithm is used to calibrate the jet collection, another to apply some selection cuts, and then a third algorithm can contain your analysis code to calculate something with the jets or one of the general plotting algorithms that will fill a configurable set of plots. A second jet calibrator and selector can be added to have a second jet collection at the same time. A base class for a tree has also been created with some standard branches and a way for the user to add more as well. The tree is configurable with the same options as the histogramming classes—with a string of

© Springer Nature Switzerland AG 2020

G. Stark, *The Search for Supersymmetry in Hadronic Final States Using Boosted Object Reconstruction*, Springer Theses,
https://doi.org/10.1007/978-3-030-34548-8

categories of interest. Each algorithm can be configured via a text file read by TEnv. Example for all are given and one can look for the "configure" function to see what options are available (also noted below). Development changes and help requests can be obtained on the e-group "atlas-sw-xAODAnaHelpersFW" or directly here on GitHub.

A word on **systematics**. When the object itself is altered (i.e., JES calibration and JES systematics) a new collection is made and put into `TStore`. The name of the nominal collection after calibration is set from the config file. The name of the systematically varied collection uses the same name plus the name of the systematic directly from the CP tool. The next algorithm in the chain using these objects needs to know which collections were created. To avoid hardcoding things and all that, when the systematics are applied a vector is created containing the names of each systematic. Downstream algos pick up this vector then know which collections to run over. Each algorithm will loop over all collections before going to the next step. If selectors are configured with limits in the number of events passing the cuts, only collections passing the cuts will be passed to algos downstream. If none passes, the next event is analyzed.

The entire framework comes with a flexible runner script written in Python that allows the user to write a python configuration for their entire analysis, chaining/hooking up algorithms together in any particular order they would like. On top of this, the configuration is flexible enough to even run algorithms not provided by xAODAnaHelpers itself, such as perhaps algorithms the user has written themselves, or algorithms found in other smaller analysis frameworks. The idea is that xAODAnaHelpers is to be treated like a toolkit to be used when you need to speed up some development time without sacrificing a loss in physics or configurability. On top of this, because each algorithm is meant to stand on its own, a user could potentially run two analyses at the same time with a single configuration, and just keep everything together by chaining things and producing multiple outputs. The significant flexibility and complexity offered by the configuration allows xAODAnaHelpers to be used in a broad range of analyses within ATLAS.

Appendix C
Ironman: Slow-Control and Monitoring

C.1 IPBus

IPBus is a simplified transaction [202] used to communicate with the hardware using IP connections such as UDP and TCP/IP. In particular, it extends UDP with handshake capability to minimize the number of packets lost in transport while trying to minimize the overhead in implementing the communication on the boards.

At the time when IPBus was designed, the amount of space available on chips was very crucial and care was taken to minimize the amount of code put in. Now that embedded processors exist on the market and have been proven it is possible to avoid the unreliability of UDP and simply implement TCP/IP. Because the IP protocol is not fixed, and IPBus is not transport-neutral, this leads to the need for such an environment.

C.2 Ironman

The ironman software manages the logic of connecting inbound IPBus requests with outbound IPBus responses in a transport-neutral way (Fig. C.1). The full documentation for ironman can be found online at http://iron-man.readthedocs.io/en/latest/. The goals of ironman are:

- provide a wide array of standard networking protocols for reading and writing packets,
- allow for implementation of custom communication protocols for reading and writing the various hardware components,
- allow for definition of custom hardware maps which specify the layout of the entire board, and

© Springer Nature Switzerland AG 2020

G. Stark, *The Search for Supersymmetry in Hadronic Final States Using Boosted Object Reconstruction*, Springer Theses,
https://doi.org/10.1007/978-3-030-34548-8

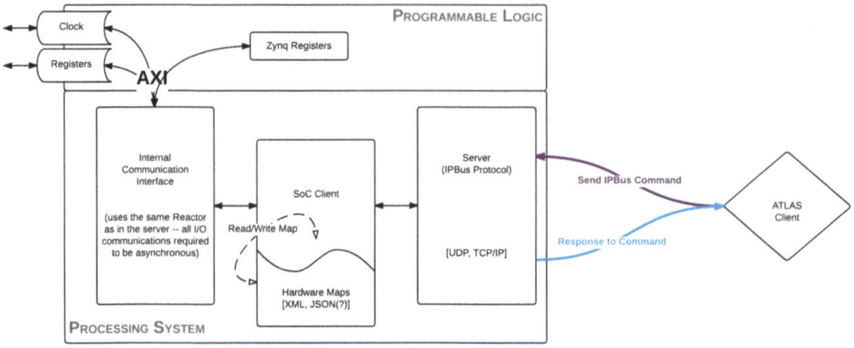

Fig. C.1 Overview of the ironman architecture

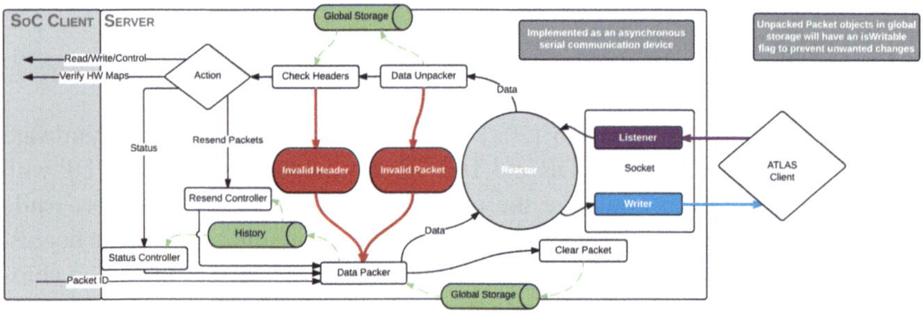

Fig. C.2 Overview of the ironman server

- use a single-threaded reactor model, an event-driven model, which is a global loop that fires listeners when certain events have triggered.

An external client (Fig. C.2) is tasked with the job of communicating a transaction request or a status query of a piece of hardware (ATLAS client). This is sent as an IPBus command to the board running ironman software. This request is received by a server actively listening and then dispatches this request to the SoC client (Jarvis). The SoC client is made self-aware using hardware definitions (Fig. C.3) and dispatches this request along a callback chain to the internal communications Interface (Fig. C.5). It is at this point that the software handles the customized communication to fulfill the request of the biological being controlling the ATLAS client (Fig. C.4). After this, the response is formed and transferred back to the server who will transmit the information back to the client.

C.2.1 Server

The server (Fig. C.2) has a few jobs to do upon receipt of a packet. As the server is actively listening, it is going to plug itself into the reactor and kick off the callback

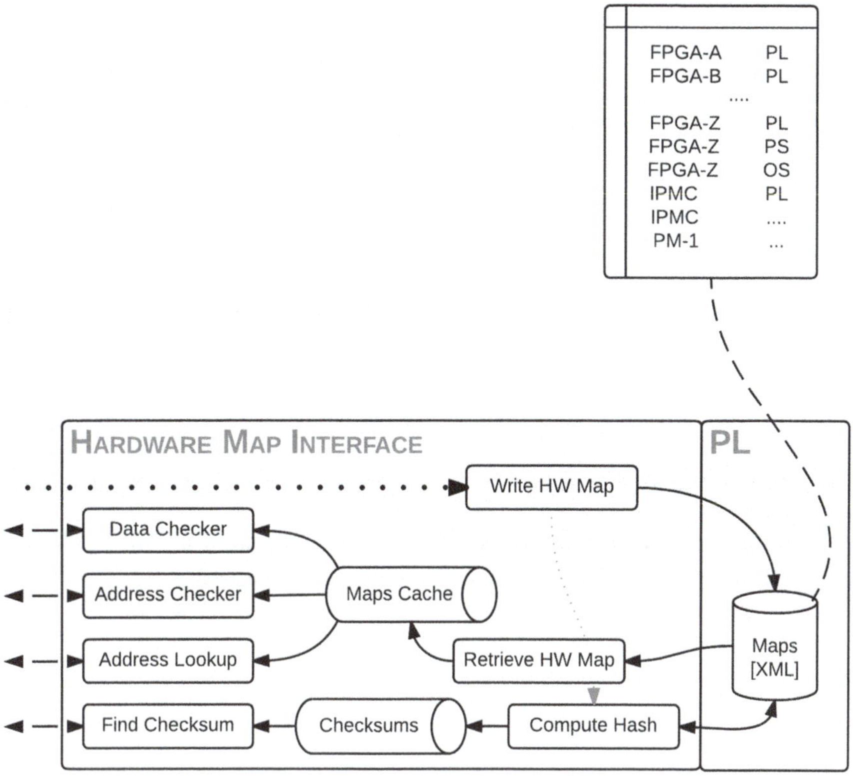

Fig. C.3 Overview of the ironman hardware interface

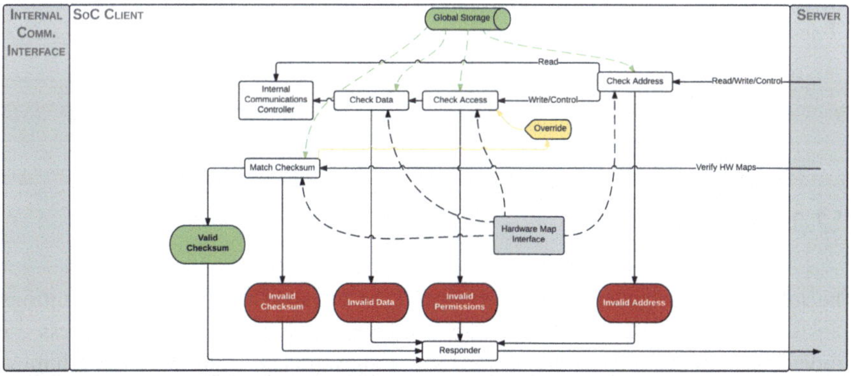

Fig. C.4 Overview of the ironman client

chain for us (known as "deferreds" or "promises"). The server knows the format of the packet entirely and will unpack the data into a Request Packet Object that will be used by the rest of the software downstream. It is at this point a few checks are

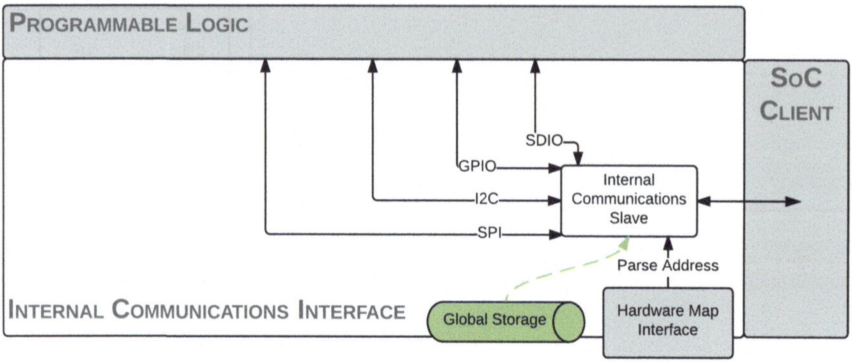

Fig. C.5 Overview of the ironman communications with the hardware

done, such as checking that the data can be unpacked as well as making sure the headers are valid.

If the basic sanity checks look good, then it must decide what to do with the packet. If the request requires communication with the hardware, then it will pass along the packet to the SoC client to dispatch the request. If it simply requires information about the history of packets sent (such as a Resend Packet), then it will return the packets it records in history.

This leads to the other part of the server which is to maintain a log of all inbound/request and outbound/response packets. It is at this point which the board communicates with the outside world and makes it a suitable place to implement the history recording. For example, this can be done with:

```
from ironman.packet import IPBusPacket
from ironman.history import History
from twisted.internet.defer import Deferred

history = History()
Deferred().addCallback(IPBusPacket).addCallback(history.record)
```

C.2.2 Hardware

The job of the hardware interface (Fig. C.3) is to parse the hardware definitions transferred to the board and build up a cached, global mapping of address $\rightarrow$ properties about the address. In Python terminology, this is a giant dictionary. It must assess that a single address is not taken up by two different hardware definitions (no conflicts) and that the hardware map is parseable and valid.[1] It will also provide a

[1] Validity has not been defined yet.

way to compute the checksum of the hardware map files to ensure that the board is running on the same definitions that the monkey has communicated to the board with.

The hardware manager is our primary means of interfacing. For example, this can be done with this short code:

```
from ironman.hardware import HardwareManager, HardwareMap

hw_map = HardwareMap(file('/path/to/my/hardware/map.yml'))
hw = HardwareManager()
hw.add(hw_map, 'clockConfiguration')
```

where we added a hardware map specifying the addresses allowed for clock configuration.

C.2.3 Jarvis, the Client

The job of the client (Fig. C.4) is to analyze the packet more thoroughly. If the client is handling the packet, then it must be a request packet. It will then communicate with the hardware interface to determine whether or not the transaction packet is good: valid address, valid permissions, valid data. If all of these things pass, it then passes the packet along to the internal communications which will build up a response. It should be noted that the client is not allowed to modify the response packet at all. Only the server and internal communications are allowed to do this. In ironman, the client is known as Jarvis.[2] Jarvis is used like so:

```
from ironman.communicator import Jarvis

j = Jarvis()
j.set_hardware_manager(hw)
```

In particular, Jarvis is one of the easiest things to set up since it contains a lot of internal logic to route requests appropriately and execute controllers for you. In this way, Jarvis is a lot like a router.

C.2.4 Internal Communications

Lastly, the internal communications (Fig. C.5) is primarily custom code written by the developers to do exactly that: communicate with the board. Depending on how the board is set up, there may be a virtual filesystem or raw pointers or custom drivers that the code will need to access and interface with. Since this is something that will vary on a board-by-board basis, I leave most of this code up to the user and

[2]Get it? It's his assistant...

only provide a few simple cases for file reading and writing. Continuing on with the code examples, you could use jarvis to register a controller for clock configuration that you added a hardware map for:

```
from ironman.hardware import ComplexIO
@j.register('clockConfiguration')
class ClockConfigurator(ComplexIO):
  # ... insert custom code here for clock configuration
```

And you are done. This will execute your custom code as you have defined in the controller.

C.3 Technical Details

C.3.1 Dependencies

Ironman is written on top of two major pieces of software:

- Twisted: http://twistedmatrix.com/trac/
- Construct: http://construct.readthedocs.io/en/latest/

Twisted is an event-driven networking engine written in Python and licensed under the open source MIT license. Twisted is a platform for developing Internet applications focused on integration and low-level socket interface. It currently supports a large number of existing protocols including HTTP, XMPP, NNTP, IMAP, SSH, IRC, and FTP. Twisted allows for the separation between stream-based protocols and physical transport layers and relies on the event-driven programming paradigm, using callbacks ("deferreds") to trigger and process events.

Construct is a powerful declarative and symmetrical parser and builder for binary data in Python. Instead of writing imperative code to parse binary data, "construct" is used to declaratively define a data structure that describes the binary data. Since the data structure is just a declaration, it can be used to parse binary data into Python objects and to build Python objects back into binary data—also known as packing and unpacking data.

C.4 Code Examples

C.4.1 Parse and Build IPBus Packets

It is common to use ironman to parse and build IPBus packets. As this is most likely a major usage of the software to aid with future debugging, you can easily debug in a few simple lines:

```
from ironman.constructs.ipbus import IPBusConstruct

packet = IPBusConstruct.parse(b'\x20\x00\x00\xf0\x20\x00\x01\x0f\x00
\x00\x00\x03')
print packet

packet.transactions[0].data = b'Hello World'
packet.transactions[0].header.info_code = 'SUCCESS'
response = IPBusConstruct.build(packet)

print response.encode('hex')
```

This will parse a IPBus packet as binary data into a Python object that is easier to interface with. In this example, we have a `READ` request from address `0x3` so we will use the same python object to add `b'Hello World'` data, mark the transaction as a `SUCCESS` (according to IPBus specifications), and then build the binary data for the response packet.

C.4.2 Implementing IPBus

Ironman is currently written with the IPBus protocol as of writing this thesis. The top level construct defines the logical structure of the entire IPBus packet:

```
from construct import *

IPBusConstruct = Struct(
        "pointer" / Pointer(3, Int8ub),
        "bigendian" / Computed(this.pointer == 0xf0),
        "header" / IfThenElse(this.bigendian, PacketHeaderStruct,
        ByteSwapped(PacketHeaderStruct)),
        "transactions" / If(this.header.type_id == "CONTROL",
        GreedyRange(ControlStruct)),
        "status" / If(this.header.type_id == "STATUS", StatusRequestStruct),
        "resend" / If(this.header.type_id == "RESEND", ResendStruct),
        Terminated
)
```

Above is code showing how the top-level IPBus packet is described. As mentioned in the specifications [202], the packet header contains a 4-bit endianness specification. In order to remember this specification while flipping the packet to big-endian (network-default) if needed, a pointer is used to peek at the value of the 3rd byte, and the `bigendian` entry allows us to compute whether the packet is big-endian or little-endian. At this point, we can byte-swap everything as needed to account for the endian-ness of the packet when parsing the binary data, or building a Python object into binary data to send back as response in the same endian-ness as the request packet. The last entry here is `Terminated` which tells the parser/builder that the entire packet should be fully described by this declarative structure. This entire portion is written using the "construct" library which allows for maximum flexibility.

C.4.3 Implementing Jarvis

Jarvis is the friendly router inside ironman. It is otherwise known as a communication slave and its job is to inspect the received packet, identify the address the request wishes to access (control or monitoring), check the hardware map for the board to determine if the address is valid, then dispatch the request to the available route for handling. Jarvis maintains an internal registry of available routes that can handle requests, as well as having access to a hardware manager that provides information about the machinery of the board. One uses Jarvis like so:

```
from ironman.hardware import HardwareManager
from ironman.communicator import Jarvis, SimpleIO

hw = HardwareManager()
hw.add(HardwareMap(file('/path/to/simpleHardwareMap.yml')),
'simpleRegisterController')

j = Jarvis()
j.set_hardware_manager(manager)

@j.register('simpleRegisterController')
class SimpleRegisterFile(SimpleIO)
  __f__ = '/path/to/simpleRegister.txt'
```

Then when Jarvis is called with a packet requesting access to a register inside `/path/to/simpleHardwareMap.yml`, it will dispatch that call to the control assigned to `simpleRegisterController` which is the `SimpleRegisterFile` class, which provides simple read/write file access capabilities:

```
class SimpleIO(object):
    implements(ICommunicationDriver)
    __f__ = None

    def read(self, offset, size):
        with open(self.__f__, 'rb') as f:
            f.seek(offset)
            return f.read(4*size)

    def write(self, offset, data):
        with open(self.__f__, 'r+b') as f:
            f.seek(offset)
            return f.write(data)
```

C.5 Implementing Callback Chain

At the end of the day, one would like to implement the callback chain. This is the declarative structure that tells `ironman` exactly what you would like to execute each time a packet is received. An example is below:

```
from twisted.internet.defer import Deferred
from ironman.constructs.ipbus import IPBusConstruct
from ironman.communicator import Jarvis

# set up jarvis...

function callbackChain():
  return Deferred().addCallback(IPBusConstruct.parse).addCallback(j).
  addCallback(IPBusConstruct.build)

# example of a single callback: callbackChain().callback(<packet>)

# set up a server
from ironman.server import ServerFactory
from twisted.internet import reactor
reactor.listenUDP(8888, ServerFactory('udp', callbackChain))
reactor.listenTCP(8889, ServerFactory('tcp', callbackChain))
reactor.run()
```

where `callbackChain()` represents the full callback chain. Jarvis, as you see in the middle, handles receiving the input packet object, and generating an output packet object. When you call this chain with the packet to handle, it will return a binary packet response associated with that input. In the example code above, a server factory provided by ironman generates the necessary information for receiving and sending packets over UDP or TCP/IP using Twisted, adds extra callbacks to the end of the callback chain provided, and executes the `callback` function on the `callbackChain()`.

Appendix D
N-1 Plots

D.1 0-Lepton

The following plots (Figs. D.1, D.2, and D.3) show N-1 distributions of the variables of interest in the SR for the 0-lepton regions. Each of the variables shows significant discrimination power for the signals shown.

D.2 1-Lepton

The following plots (Figs. D.4, D.5, and D.6) show N-1 distributions of the variables of interest in the SR for the 1-lepton regions. Each of the variables shows significant discrimination power for the signals shown.

© Springer Nature Switzerland AG 2020

G. Stark, *The Search for Supersymmetry in Hadronic Final States Using Boosted Object Reconstruction*, Springer Theses,
https://doi.org/10.1007/978-3-030-34548-8

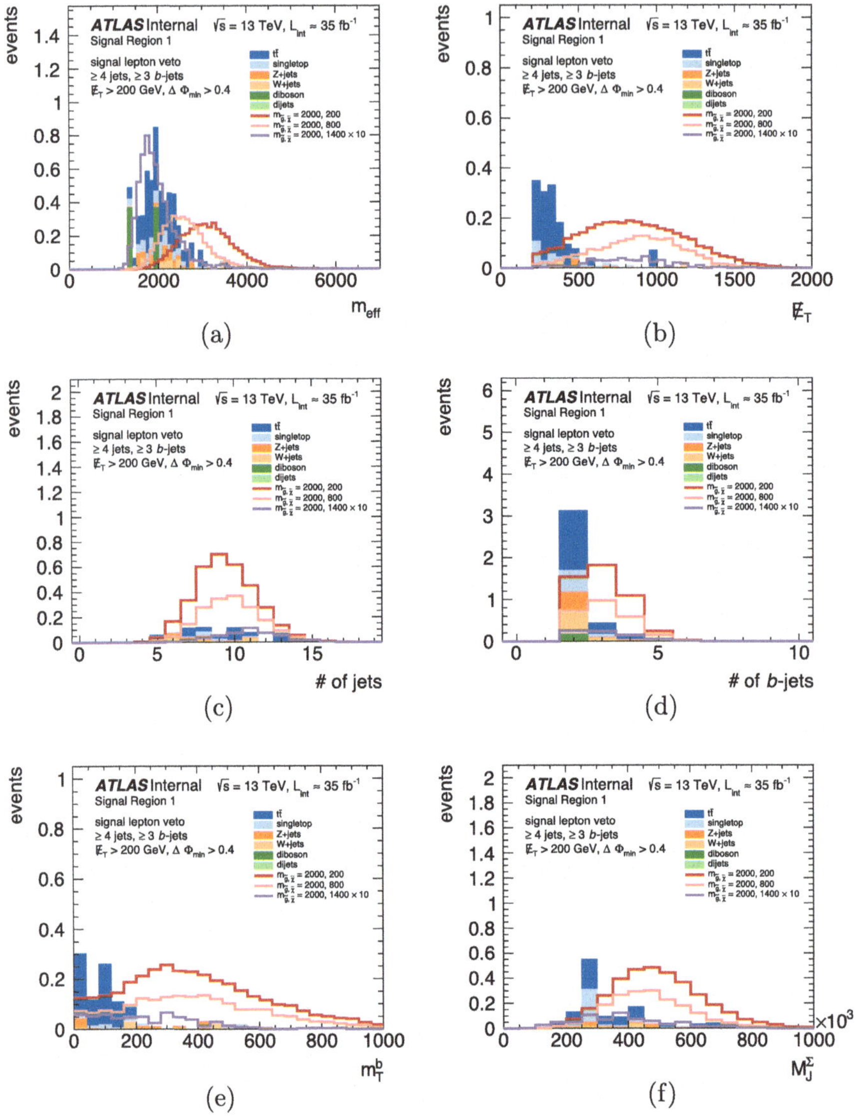

Fig. D.1 N-1 plots for the analysis variables in the boosted signal region for the Gtt 0L analysis. (**a**) m_{eff}. (**b**) $E_{\mathrm{T}}^{\mathrm{miss}}$. (**c**) N_{jet}. (**d**) $N_{b\text{-jets}}$. (**e**) $m_{\mathrm{T,min}}^{b\text{-jets}}$. (**f**) M_J^{Σ}

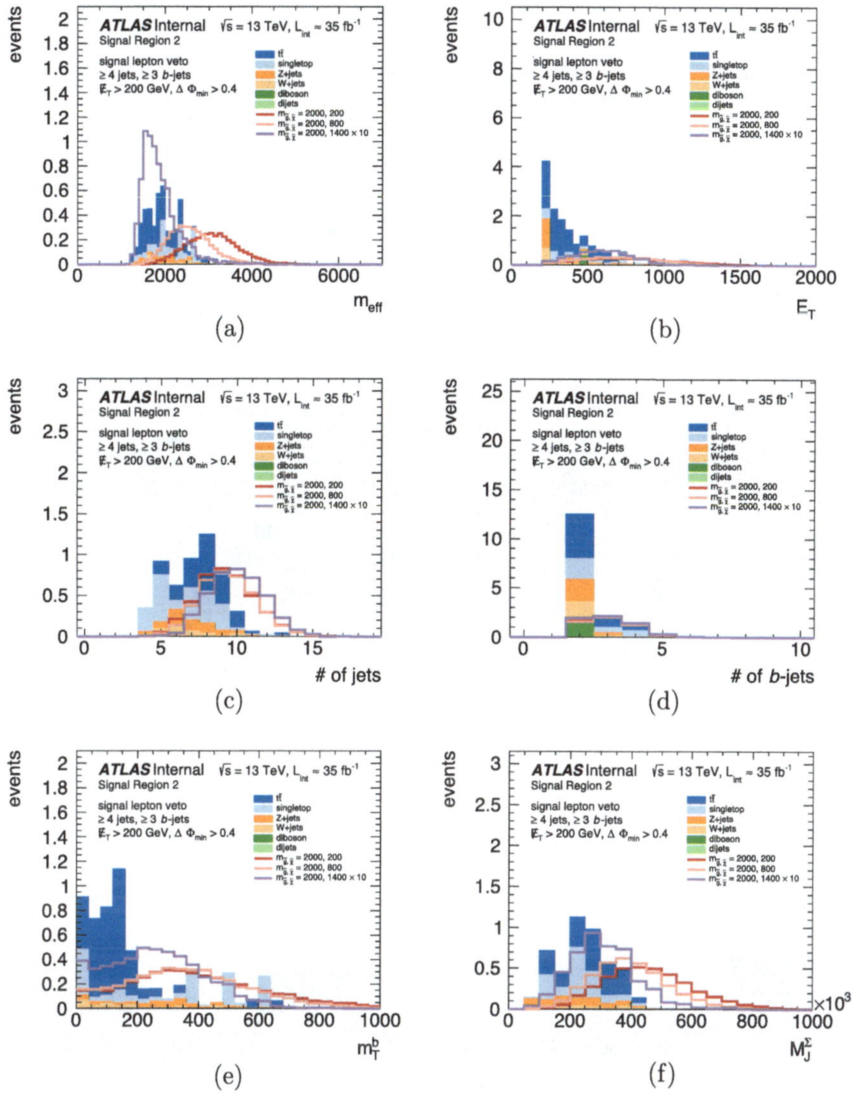

Fig. D.2 N-1 plots for the analysis variables in moderate signal region for the Gtt 0L analysis. (**a**) m_{eff}. (**b**) $E_{\text{T}}^{\text{miss}}$. (**c**) N_{jet}. (**d**) $N_{b\text{-jets}}$. (**e**) $m_{\text{T,min}}^{b\text{-jets}}$. (**f**) M_J^{Σ}

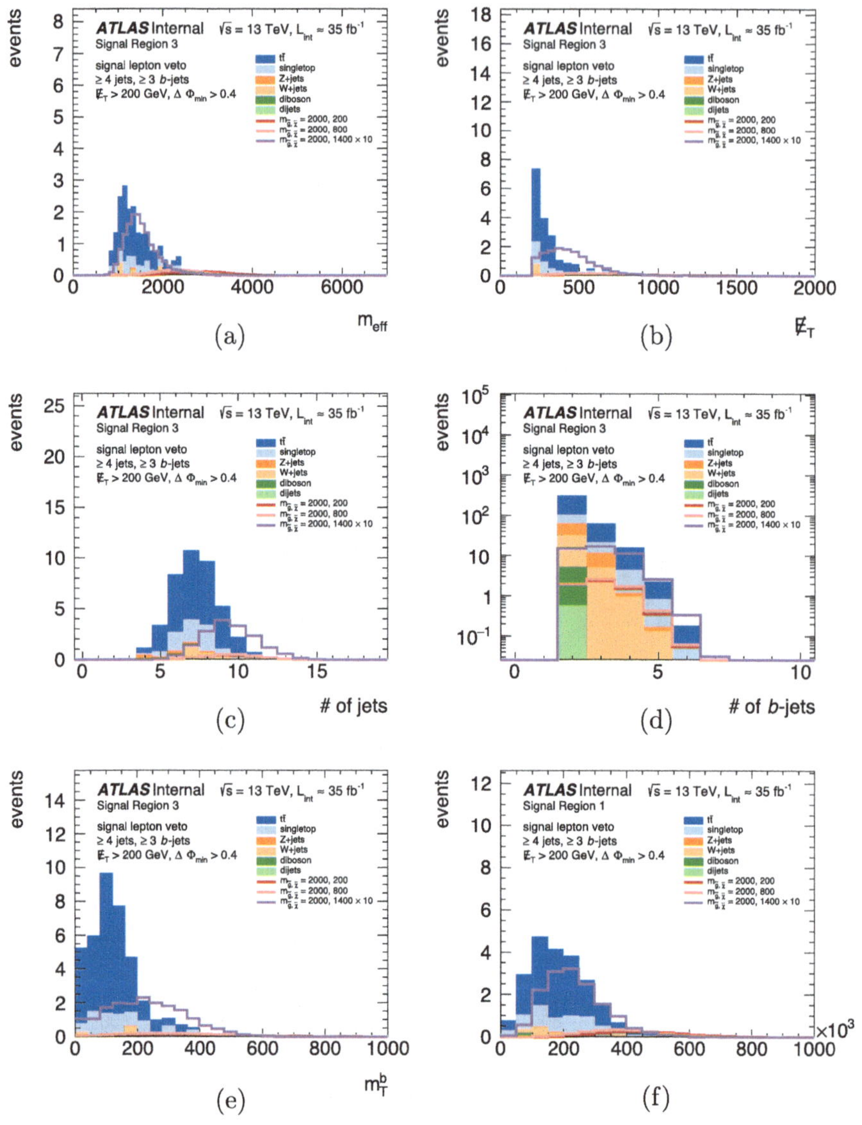

Fig. D.3 N-1 plots for the analysis variables in compressed signal region for the Gtt 0L analysis. (**a**) m_{eff}. (**b**) $E_{\text{T}}^{\text{miss}}$. (**c**) N_{jet}. (**d**) $N_{b\text{-jets}}$. (**e**) $m_{\text{T,min}}^{b\text{-jets}}$. (**f**) M_J^{Σ}

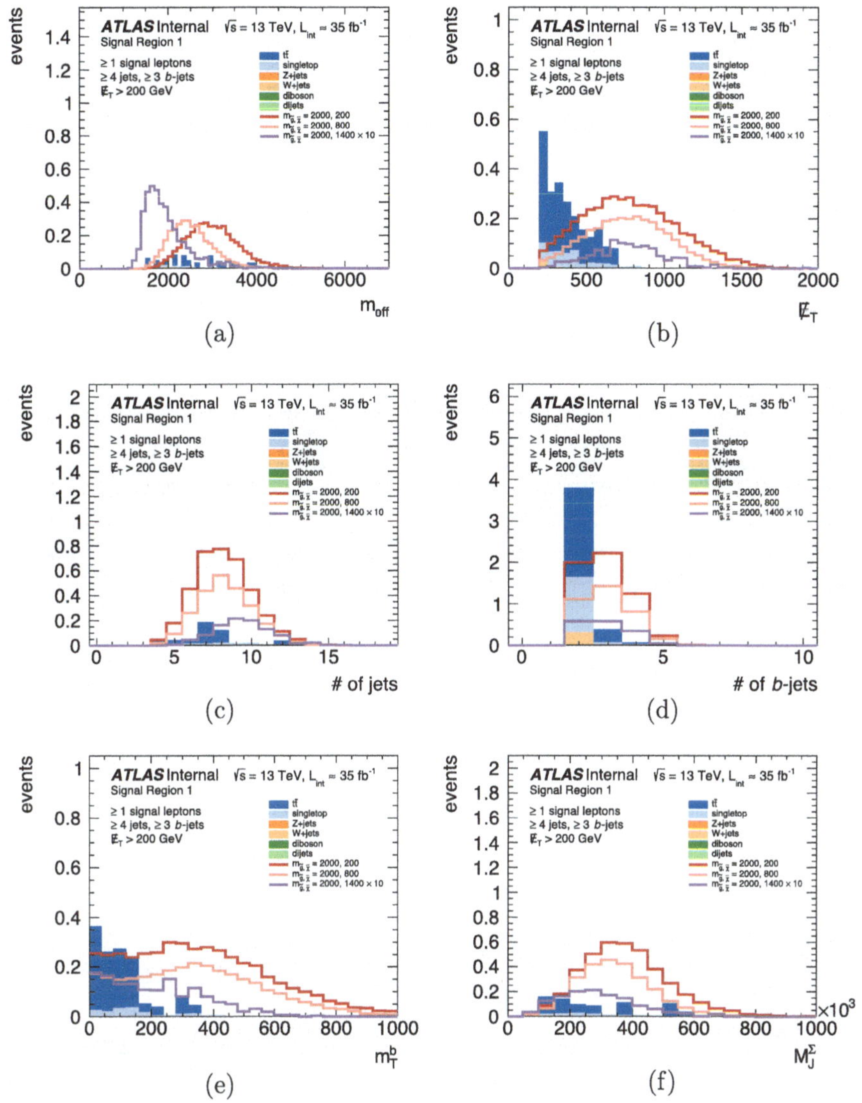

Fig. D.4 *N*-1 plots for the analysis variables in boosted signal region for the Gtt 1L analysis. (**a**) $m_{\text{eff}}^{\text{incl}}$. (**b**) $E_{\text{T}}^{\text{miss}}$. (**c**) N_{jet}. (**d**) $N_{b\text{-jets}}$. (**e**) $m_{\text{T,min}}^{b\text{-jets}}$. (**f**) M_J^{Σ}

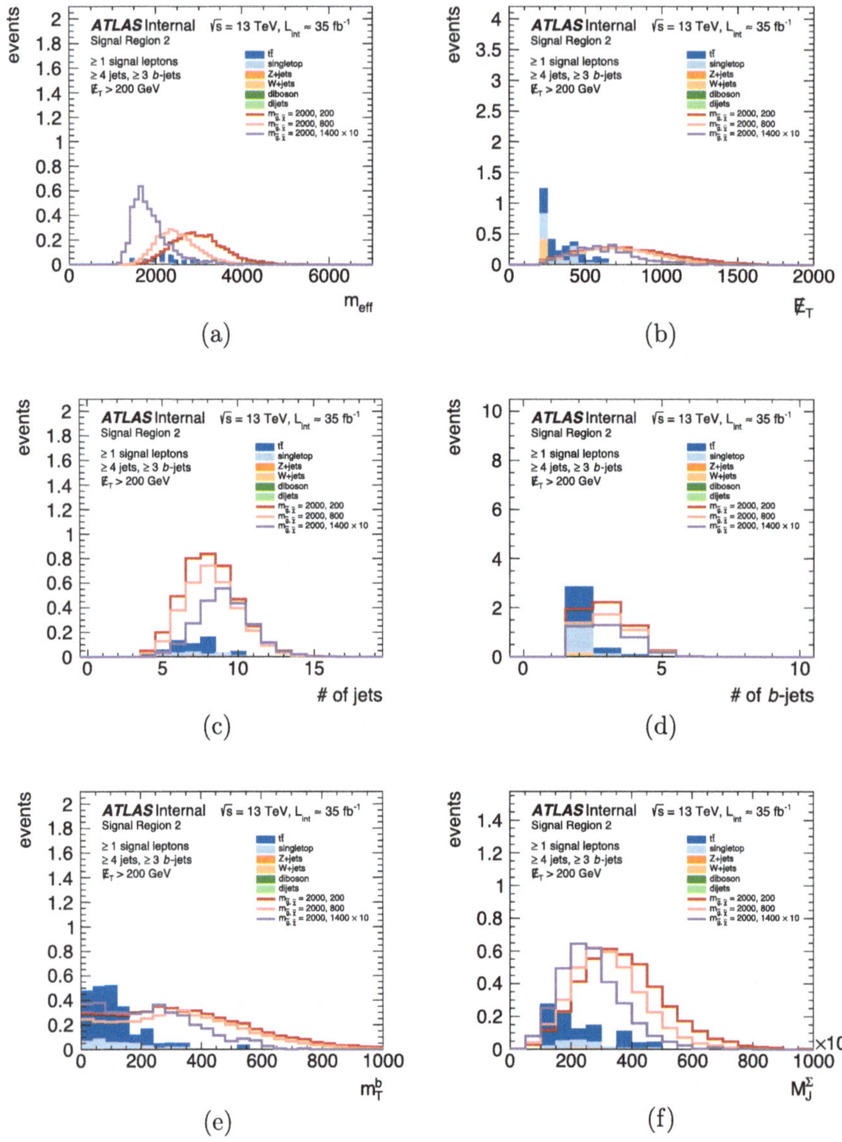

Fig. D.5 *N*-1 plots for the analysis variables in moderate signal region for the Gtt 1L analysis. (**a**) $m_{\mathrm{eff}}^{\mathrm{incl}}$. (**b**) $E_{\mathrm{T}}^{\mathrm{miss}}$. (**c**) N_{jet}. (**d**) $N_{b\text{-jets}}$. (**e**) $m_{\mathrm{T,min}}^{b\text{-jets}}$. (**f**) M_J^{Σ}

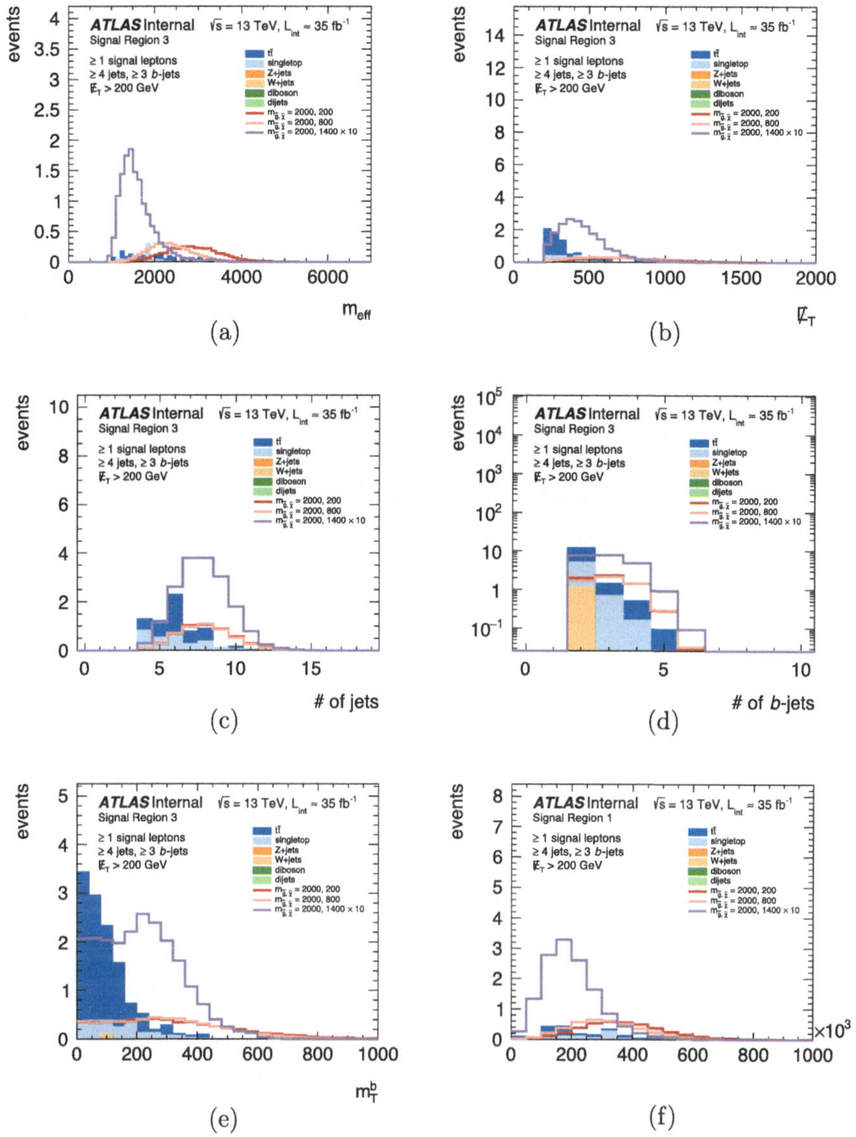

Fig. D.6 N-1 plots for the analysis variables in compressed signal region for the Gtt 1L analysis. **(a)** $m_{\text{eff}}^{\text{incl}}$. **(b)** $E_{\text{T}}^{\text{miss}}$. **(c)** N_{jet}. **(d)** $N_{b\text{-jets}}$. **(e)** $m_{\text{T,min}}^{b\text{-jets}}$. **(f)** M_J^{Σ}

Appendix E
ttbar Heavy-Flavor Classification/Flavor Contamination

E.1 0-Lepton Composition

These plots are appended for Sect. 7.4.3.4. Figure E.1 shows the heavy flavor composition of the $t\bar{t}$ component of the background. Figure E.2 shows the breakdown of the $t\bar{t}$ component of the background into its single-lepton and dilepton part. In Figs. E.3 and E.4 each of the two components is further analyzed. Figure E.5 shows the composition of the single top component of the background in terms of Wt, t-channel, and s-channel. Note that all results here are pre-fit.

E.2 1-Lepton Composition

These plots are appended for Sect. 7.4.4.4. Figure E.6 shows the heavy flavor composition of the $t\bar{t}$ component of the background. Figure E.7 shows the breakdown of the $t\bar{t}$ component of the background into its single-lepton and dilepton part. In Figs. E.8 and E.9 each of the two components is further analyzed. Figure E.10 shows the composition of the single top component of the background in terms of Wt, t-channel, and s-channel. Note that all results here are pre-fit.

© Springer Nature Switzerland AG 2020

G. Stark, *The Search for Supersymmetry in Hadronic Final States Using Boosted Object Reconstruction*, Springer Theses,
https://doi.org/10.1007/978-3-030-34548-8

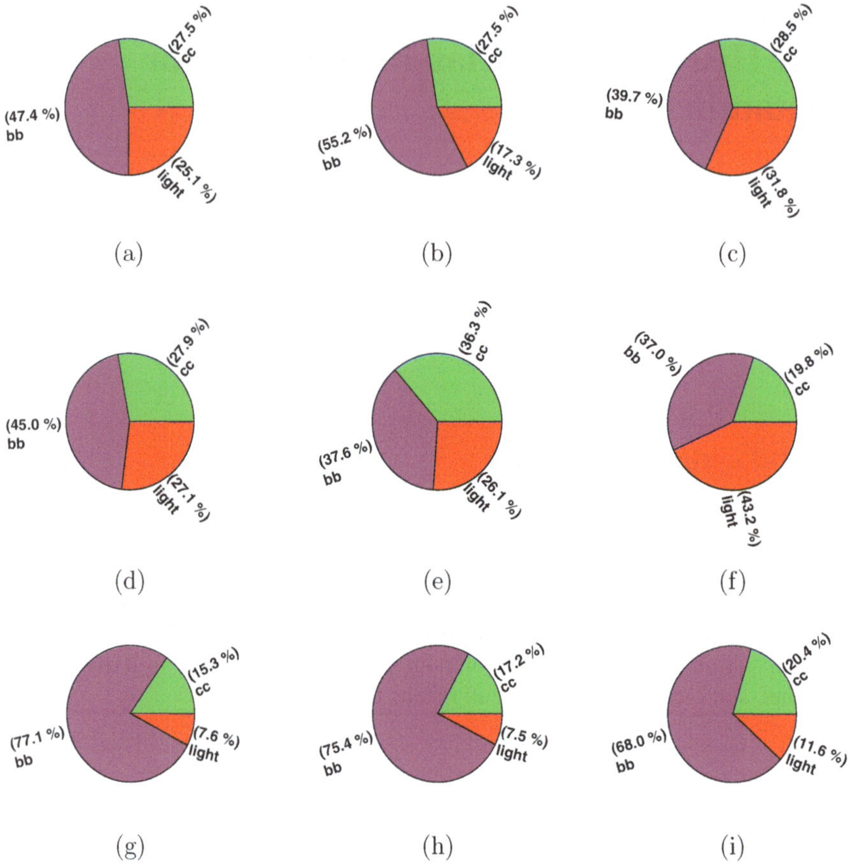

Fig. E.1 Heavy flavor composition of the $t\bar{t}$ component of the background in the optimized Gtt-0L regions [21]. (**a**) CR-0L-B. (**b**) SR-0L-B. (**c**) VR-0L-B. (**d**) CR-0L-M. (**e**) SR-0L-M. (**f**) VR-0L-M. (**g**) CR-0L-C. (**h**) SR-0L-C. (**i**) VR-0L-C

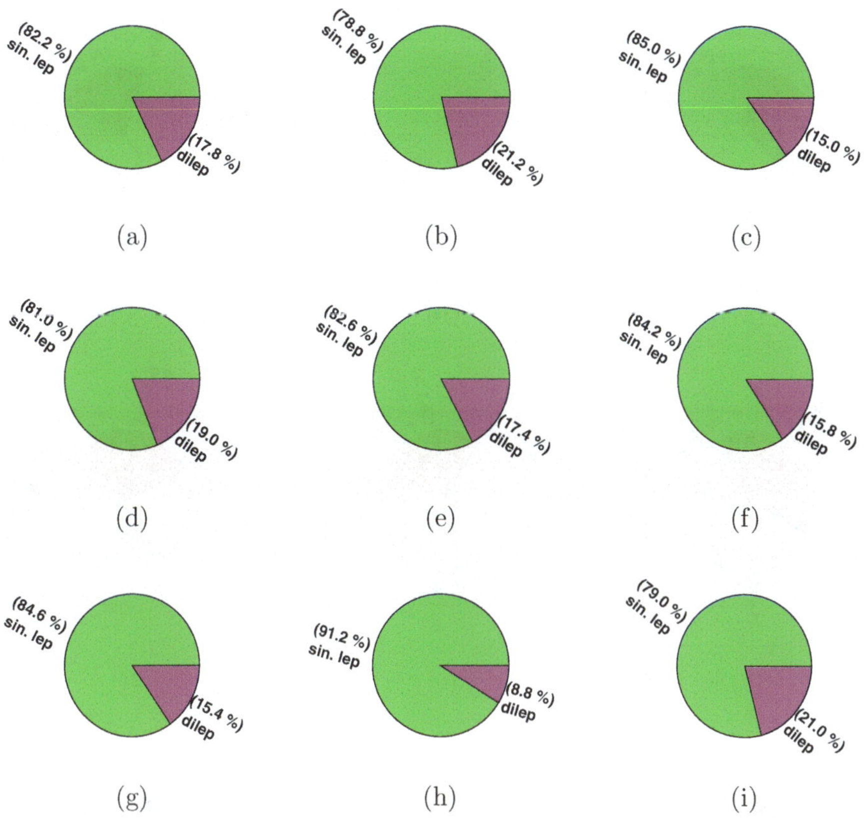

Fig. E.2 Decay type the $t\bar{t}$ component of the background in the optimized Gtt-0L regions [21]. (**a**) CR-0L-B. (**b**) SR-0L-B. (**c**) VR-0L-B. (**d**) CR-0L-M. (**e**) SR-0L-M. (**f**) VR-0L-M. (**g**) CR-0L-C. (**h**) SR-0L-C. (**i**) VR-0L-C

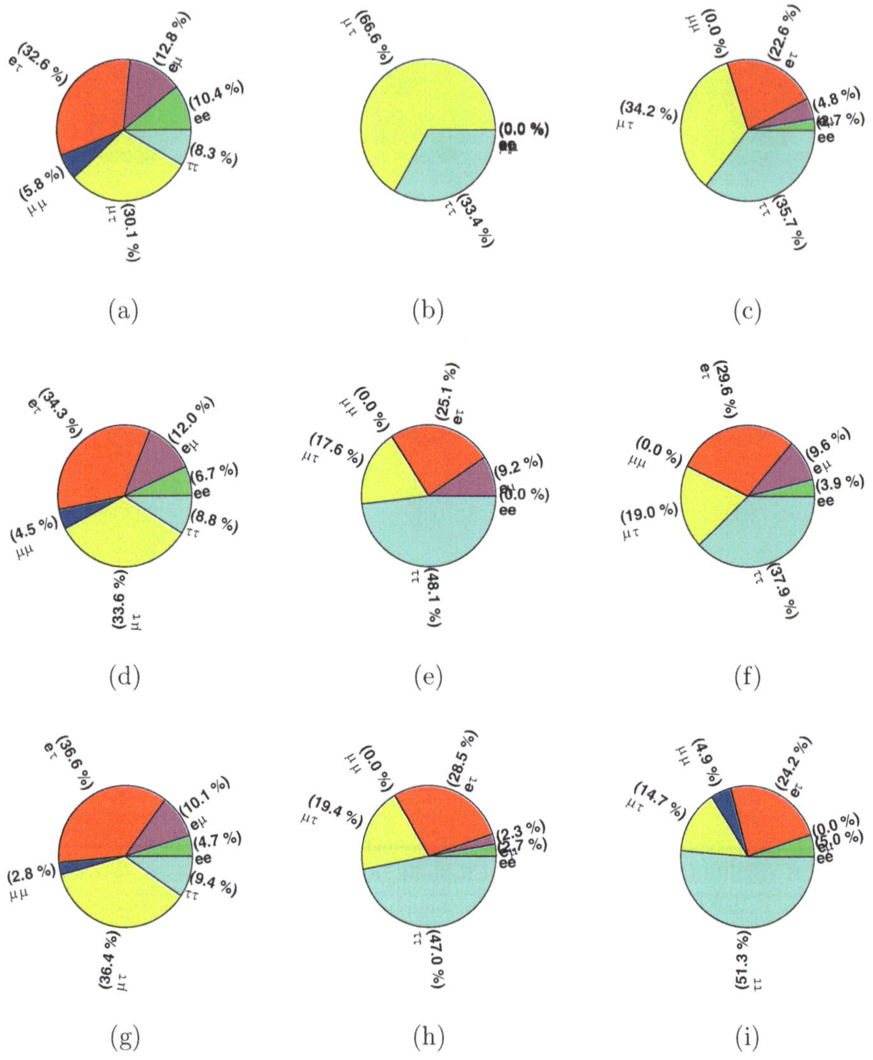

Fig. E.3 Composition of the dileptonic component of the $t\bar{t}$ background in the optimized Gtt-0L regions [21]. (**a**) CR-0L-B. (**b**) SR-0L-B. (**c**) VR-0L-B. (**d**) CR-0L-M. (**e**) SR-0L-M. (**f**) VR-0L-M. (**g**) CR-0L-C. (**h**) SR-0L-C. (**i**) VR-0L-C

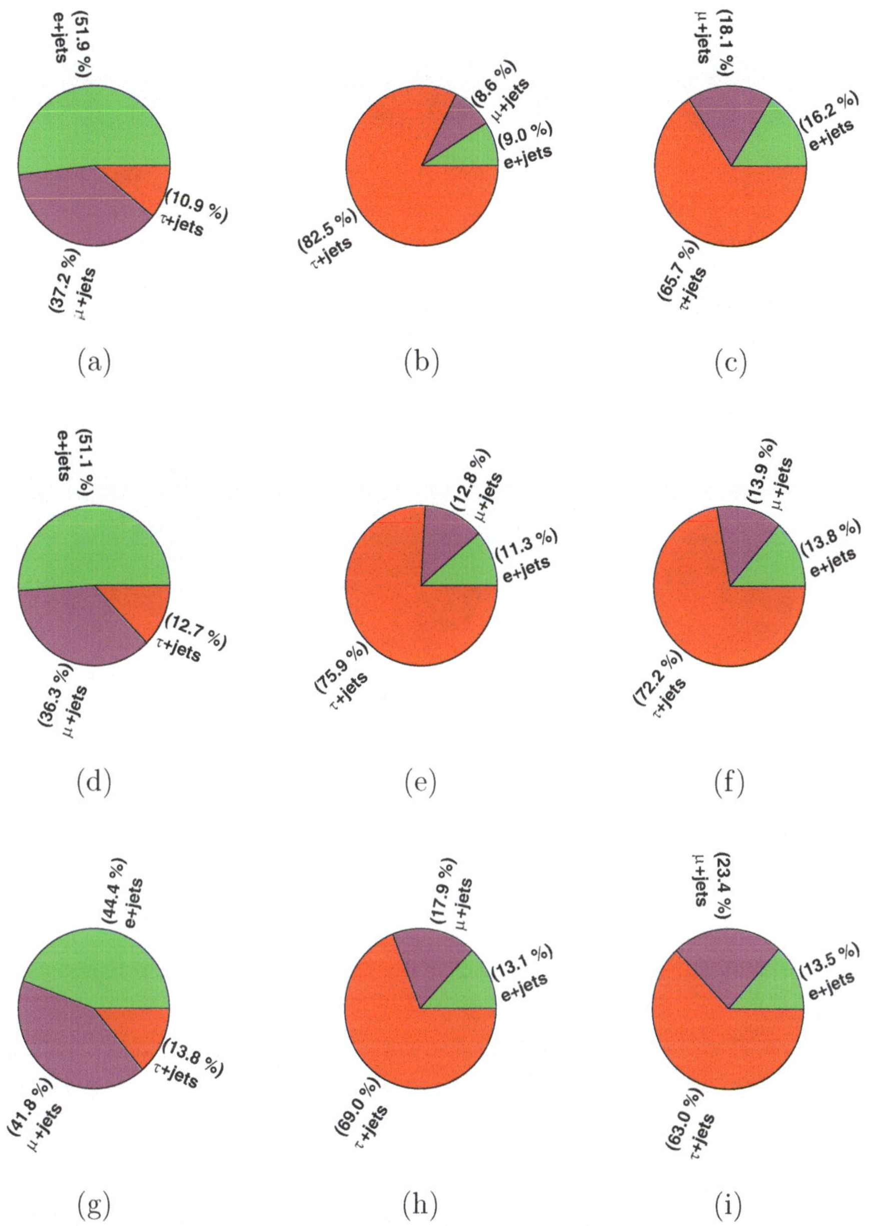

Fig. E.4 Composition of the single-lepton component of the $t\bar{t}$ background in the optimized Gtt-0L regions [21]. (**a**) CR-0L-B. (**b**) SR-0L-B. (**c**) VR-0L-B. (**d**) CR-0L-M. (**e**) SR-0L-M. (**f**) VR-0L-M. (**g**) CR-0L-C. (**h**) SR-0L-C. (**i**) VR-0L-C

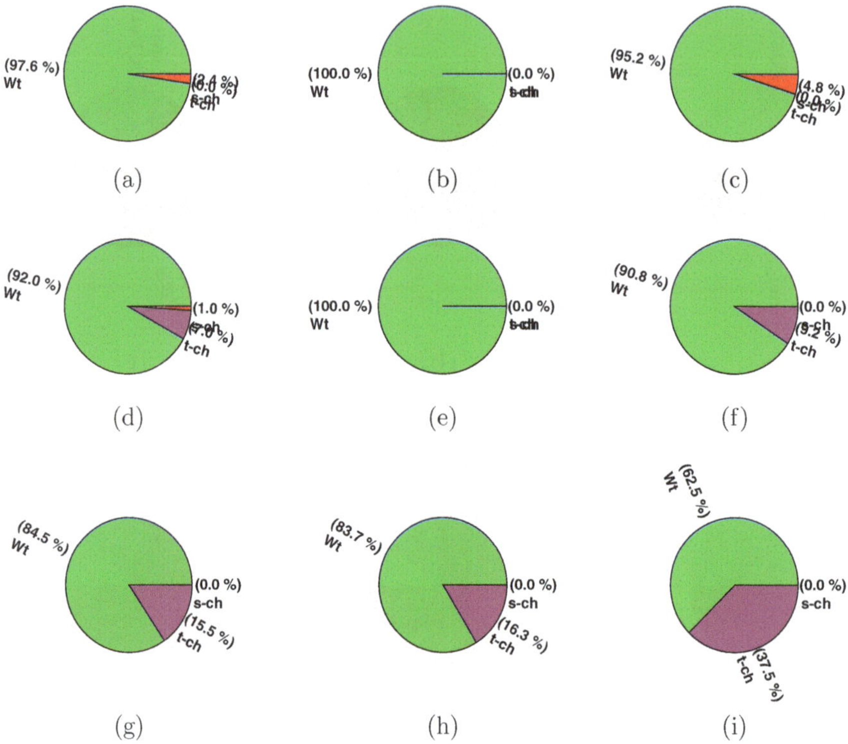

Fig. E.5 Composition of the single top component of the background in terms of Wt, t-channel, and s-channel [21]. (**a**) CR-0L-B. (**b**) SR-0L-B. (**c**) VR-0L-B. (**d**) CR-0L-M. (**e**) SR-0L-M. (**f**) VR-0L-M. (**g**) CR-0L-C. (**h**) SR-0L-C. (**i**) VR-0L-C

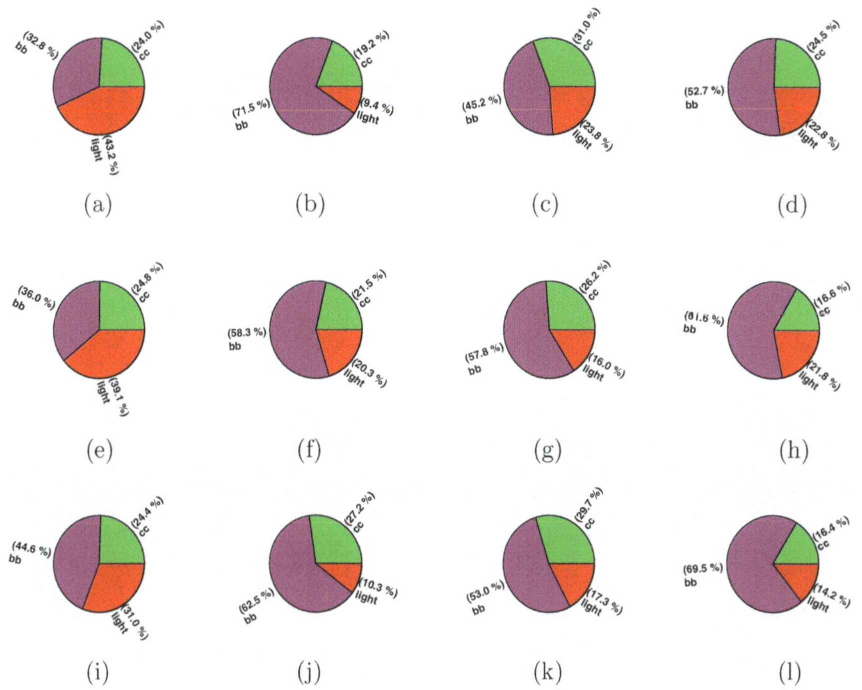

Fig. E.6 Heavy flavor composition of the $t\bar{t}$ component of the background in the optimized Gtt-1L regions [21]. (**a**) CR-1L-B. (**b**) SR-1L-B. (**c**) VR-m_{T}-1L-B. (**d**) VR-$m_{\mathrm{T,min}}^{b\text{-jets}}$-1L-B. (**e**) CR-1L-M. (**f**) SR-1L-M. (**g**) VR-m_{T}-1L-M. (**h**) VR-$m_{\mathrm{T,min}}^{b\text{-jets}}$-1L-M. (**i**) CR-1L-C. (**j**) SR-1L-C. (**k**) VR-m_{T}-1L-C. (**l**) VR-$m_{\mathrm{T,min}}^{b\text{-jets}}$-1L-C

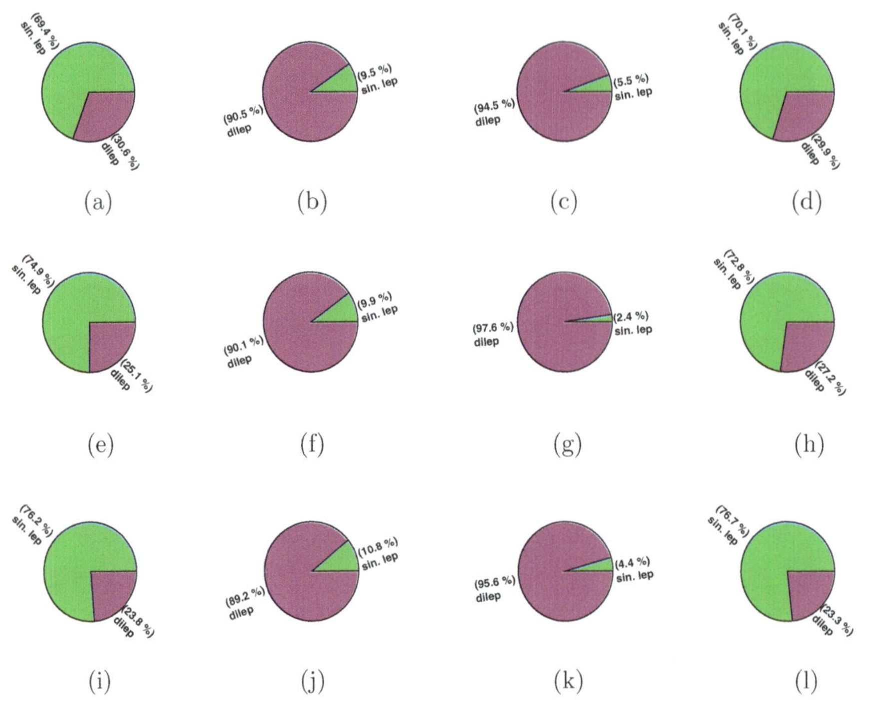

Fig. E.7 Breakdown of the $t\bar{t}$ component of the background in the optimized Gtt-1L regions [21]. (**a**) CR-1L-B. (**b**) SR-1L-B. (**c**) VR-m_{T}-1L-B. (**d**) VR-$m_{\mathrm{T,min}}^{b\text{-jets}}$-1L-B. (**e**) CR-1L-M. (**f**) SR-1L-M. (**g**) VR-m_{T}-1L-M. (**h**) VR-$m_{\mathrm{T,min}}^{b\text{-jets}}$-1L-M. (**i**) CR-1L-C. (**j**) SR-1L-C. (**k**) VR-m_{T}-1L-C. (**l**) VR-$m_{\mathrm{T,min}}^{b\text{-jets}}$-1L-C

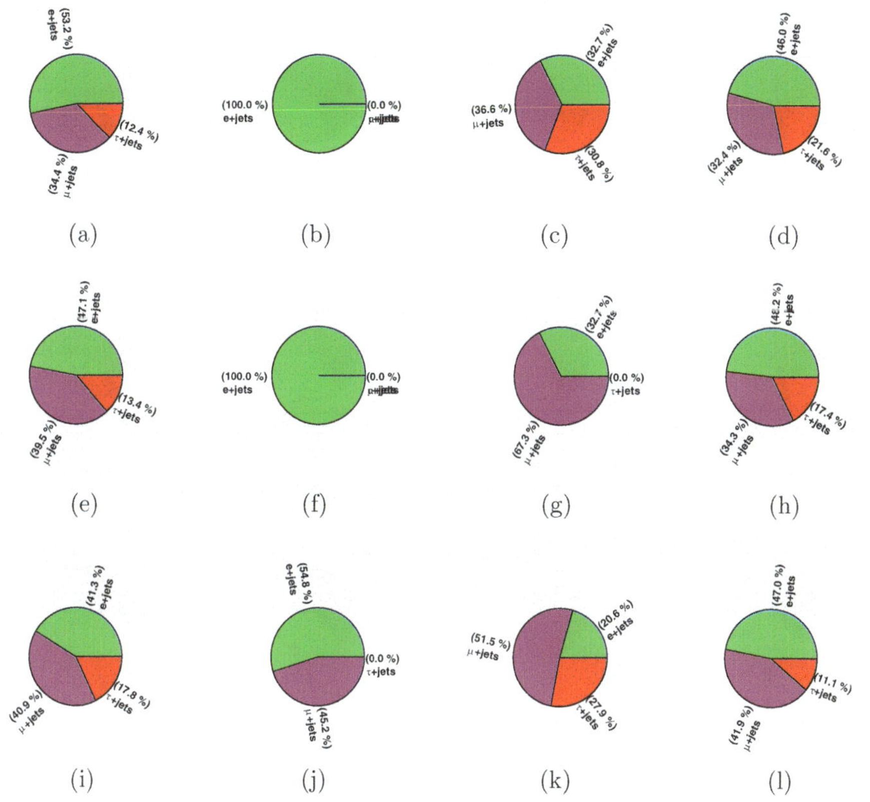

Fig. E.8 Breakdown of the semi-leptonic component of the $t\bar{t}$ background in the optimized Gtt-1L regions [21]. (**a**) CR-1L-B. (**b**) SR-1L-B. (**c**) VR-m_{T}-1L-B. (**d**) VR-$m_{\mathrm{T,min}}^{b\text{-jets}}$-1L-B. (**e**) CR-1L-M. (**f**) SR-1L-M. (**g**) VR-m_{T}-1L-M. (**h**) VR-$m_{\mathrm{T,min}}^{b\text{-jets}}$-1L-M. (**i**) CR-1L-C. (**j**) SR-1L-C. (**k**) VR-m_{T}-1L-C. (**l**) VR-$m_{\mathrm{T,min}}^{b\text{-jets}}$-1L-C

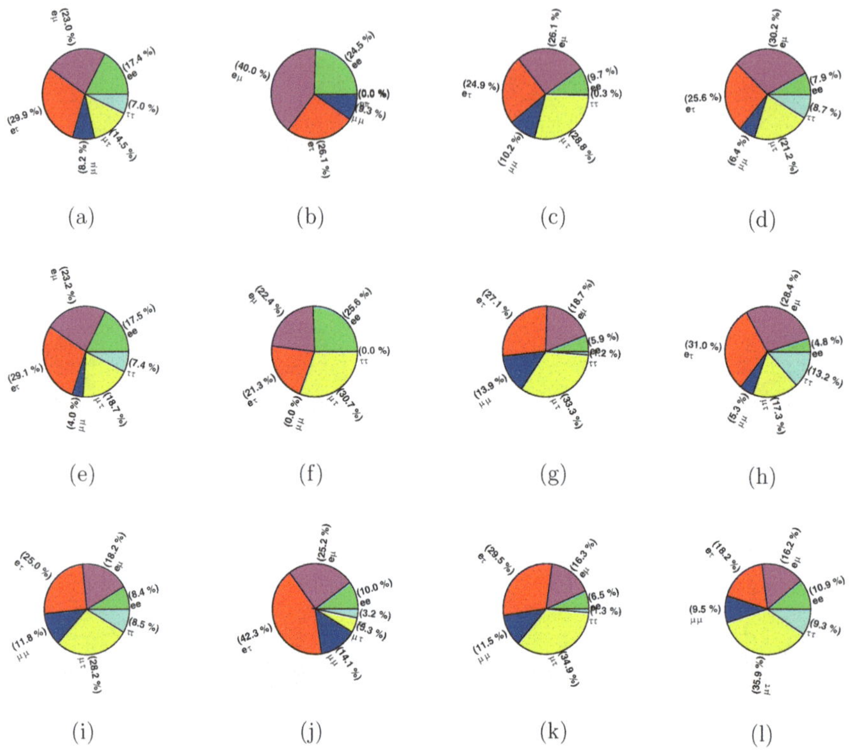

Fig. E.9 Breakdown of the dileptonic component of the $t\bar{t}$ background in the optimized Gtt-1L regions [21]. (**a**) CR-1L-B. (**b**) SR-1L-B. (**c**) VR-m_{T}-1L-B. (**d**) VR-$m_{\mathrm{T,min}}^{b\text{-jets}}$-1L-B. (**e**) CR-1L-M. (**f**) SR-1L-M. (**g**) VR-m_{T}-1L-M. (**h**) VR-$m_{\mathrm{T,min}}^{b\text{-jets}}$-1L-M. (**i**) CR-1L-C. (**j**) SR-1L-C. (**k**) VR-m_{T}-1L-C. (**l**) VR-$m_{\mathrm{T,min}}^{b\text{-jets}}$-1L-C

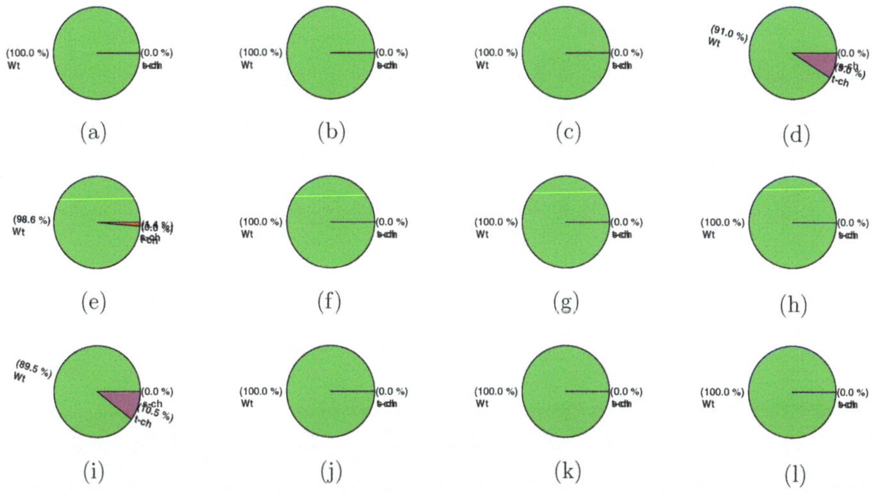

Fig. E.10 Composition of the single top component of the background in terms of Wt, t-channel, and s-channel [21]. (**a**) CR-1L-B. (**b**) SR-1L-B. (**c**) VR-m_{T}-1L-B. (**d**) VR-$m_{\mathrm{T,min}}^{b\text{-jets}}$-1L-B. (**e**) CR-1L-M. (**f**) SR-1L-M. (**g**) VR-m_{T}-1L-M. (**h**) VR-$m_{\mathrm{T,min}}^{b\text{-jets}}$-1L-M. (**i**) CR-1L-C. (**j**) SR-1L-C. (**k**) VR-m_{T}-1L-C. (**l**) VR-$m_{\mathrm{T,min}}^{b\text{-jets}}$-1L-C

Appendix F
Sample List

© Springer Nature Switzerland AG 2020

G. Stark, *The Search for Supersymmetry in Hadronic Final States Using Boosted Object Reconstruction*, Springer Theses,
https://doi.org/10.1007/978-3-030-34548-8

F.1 $t\bar{t}$ + Jets

F.1.1 Nominal

mc15_13TeV.410000.PowhegPythiaEvtGen_P2012_ttbar_hdamp172p5_nonallhad.merge.DAOD_SUSY10.e3698_s2608_s2183_r7326_r6282_p2559/
mc15_13TeV:mc15_13TeV.407009.PowhegPythiaEvtGen_P2012CT10_ttbarHT6c_1k_hdamp172p5_nonAH.merge.DAOD_SUSY10.e4023_s2608_r7725_r7676_p2622
mc15_13TeV:mc15_13TeV.407010.PowhegPythiaEvtGen_P2012CT10_ttbarHT1k_1k5_hdamp172p5_nonAH.merge.DAOD_SUSY10.e4023_s2608_r7725_r7676_p2622
mc15_13TeV:mc15_13TeV.407011.PowhegPythiaEvtGen_P2012CT10_ttbarHT1k5_hdamp172p5_nonAH.merge.DAOD_SUSY10.e4023_s2608_r7725_r7676_p2622
mc15_13TeV:mc15_13TeV.407012.PowhegPythiaEvtGen_P2012CT10_ttbarMET200_hdamp172p5_nonAH.merge.DAOD_SUSY10.e4023_s2608_r7725_r7676_p2622
mc15_13TeV:mc15_13TeV.410000.PowhegPythiaEvtGen_P2012_ttbar_hdamp172p5_nonallhad.merge.DAOD_SUSY10.e3698_a766_a818_r7676_p2622
mc15_13TeV:mc15_13TeV.410000.PowhegPythiaEvtGen_P2012_ttbar_hdamp172p5_nonallhad.merge.DAOD_SUSY10.e3698_s2608_s2183_r7725_r7676_p2622

F.1.2 Systematic Samples

mc15_13TeV:mc15_13TeV.410001.PowhegPythiaEvtGen_P2012radHi_ttbar_hdamp345_down_nonallhad.merge.DAOD_SUSY10.e3783_s2608_r7725_r7676_p2622
mc15_13TeV:mc15_13TeV.410002.PowhegPythiaEvtGen_P2012radLo_ttbar_hdamp172_up_nonallhad.merge.DAOD_SUSY10.e3783_s2608_r7725_r7676_p2622

F.2 Single-Top Samples

```
mc15_13TeV:mc15_13TeV.407018.PowhegPythiaEvtGen_P2012CT10_Wt_inclusive_top_HT500.merge.DAOD_SUSY10.e4024_s2608_r7725_
r7676_p2622/
mc15_13TeV:mc15_13TeV.407019.PowhegPythiaEvtGen_P2012CT10_Wt_inclusive_top_MET200.merge.DAOD_SUSY10.e4024_s2608_r7725_
r7676_p2622/
mc15_13TeV:mc15_13TeV.407020.PowhegPythiaEvtGen_P2012CT10_Wt_inclusive_tbar_HT500.merge.DAOD_SUSY10.e4024_s2608_r7725_
r7676_p2622/
mc15_13TeV:mc15_13TeV.407021.PowhegPythiaEvtGen_P2012CT10_Wt_inclusive_tbar_MET200.merge.DAOD_SUSY10.e4024_s2608_r7725_
r7676_p2622/
mc15_13TeV:mc15_13TeV.410013.PowhegPythiaEvtGen_P2012_Wt_inclusive_top.merge.DAOD_SUSY10.e3753_s2608_s2183_r7725_r7676_p2622/
mc15_13TeV:mc15_13TeV.410014.PowhegPythiaEvtGen_P2012_Wt_inclusive_antitop.merge.DAOD_SUSY10.e3753_s2608_s2183_r7725_
r7676_p2622/
mc15_13TeV:mc15_13TeV.410011.PowhegPythiaEvtGen_P2012_singletop_tchan_lept_top.merge.DAOD_SUSY10.e3824_s2608_s2183_r7725_
r7676_p2622/
mc15_13TeV:mc15_13TeV.410012.PowhegPythiaEvtGen_P2012_singletop_tchan_lept_antitop.merge.DAOD_SUSY10.e3824_s2608_s2183_r7725_
r7676_p2622/
mc15_13TeV:mc15_13TeV.410025.PowhegPythiaEvtGen_P2012_SingleTopSchan_noAllHad_top.merge.DAOD_SUSY10.e3998_s2608_s2183_r7725_
r7676_p2622
mc15_13TeV:mc15_13TeV.410026.PowhegPythiaEvtGen_P2012_SingleTopSchan_noAllHad_antitop.merge.DAOD_SUSY10.e3998_s2608_s2183_
r7725_r7676_p2622
```

F.3 $t\bar{t}$+X ($X = W, Z, WW, H, t\bar{t}$)

```
mc15_13TeV:mc15_13TeV.410066.MadGraphPythia8EvtGen_A14NNPDF23LO_ttW_Np0.merge.DAOD_SUSY10.e4111_s2608_s2183_r7725_r7676_p2622
mc15_13TeV:mc15_13TeV.410067.MadGraphPythia8EvtGen_A14NNPDF23LO_ttW_Np1.merge.DAOD_SUSY10.e4111_s2608_s2183_r7725_r7676_p2622
mc15_13TeV:mc15_13TeV.410068.MadGraphPythia8EvtGen_A14NNPDF23LO_ttW_Np2.merge.DAOD_SUSY10.e4111_s2608_s2183_r7725_r7676_p2622
mc15_13TeV:mc15_13TeV.410073.MadGraphPythia8EvtGen_A14NNPDF23LO_ttZnnqq_Np0.merge.DAOD_SUSY10.e4631_s2726_r7725_r7676_p2622
mc15_13TeV:mc15_13TeV.410074.MadGraphPythia8EvtGen_A14NNPDF23LO_ttZnnqq_Np1.merge.DAOD_SUSY10.e4631_s2726_r7725_r7676_p2622
mc15_13TeV:mc15_13TeV.410075.MadGraphPythia8EvtGen_A14NNPDF23LO_ttZnnqq_Np2.merge.DAOD_SUSY10.e4631_s2726_r7725_r7676_p2622
mc15_13TeV:mc15_13TeV.410111.MadGraphPythia8EvtGen_A14NNPDF23LO_ttee_Np0.merge.DAOD_SUSY10.e4632_s2726_r7725_r7676_p2622
mc15_13TeV:mc15_13TeV.410112.MadGraphPythia8EvtGen_A14NNPDF23LO_ttee_Np1.merge.DAOD_SUSY10.e4632_s2726_r7725_r7676_p2622
mc15_13TeV:mc15_13TeV.410113.MadGraphPythia8EvtGen_A14NNPDF23LO_ttmumu_Np0.merge.DAOD_SUSY10.e4632_s2726_r7725_r7676_p2622
mc15_13TeV:mc15_13TeV.410114.MadGraphPythia8EvtGen_A14NNPDF23LO_ttmumu_Np1.merge.DAOD_SUSY10.e4632_s2726_r7725_r7676_p2622
mc15_13TeV:mc15_13TeV.410115.MadGraphPythia8EvtGen_A14NNPDF23LO_tttautau_Np0.merge.DAOD_SUSY10.e4632_s2726_r7725_r7676_p2622
mc15_13TeV:mc15_13TeV.410116.MadGraphPythia8EvtGen_A14NNPDF23LO_tttautau_Np1.merge.DAOD_SUSY10.e4632_s2726_r7725_r7676_p2622
mc15_13TeV:mc15_13TeV.410080.MadGraphPythia8EvtGen_A14NNPDF23_4topSM.merge.DAOD_SUSY10.*p2622/
```

F.4 $W+$Jets

```
mc15_13TeV:mc15_13TeV.363331.Sherpa_NNPDF30NNLO_Wtaunu_Pt0_70_CVetoBVeto.merge.DAOD_SUSY10.e4709_s2726_r7725_r7676_p2622
mc15_13TeV:mc15_13TeV.363332.Sherpa_NNPDF30NNLO_Wtaunu_Pt0_70_CFilterBVeto.merge.DAOD_SUSY10.e4709_s2726_r7725_r7676_p2622
mc15_13TeV:mc15_13TeV.363333.Sherpa_NNPDF30NNLO_Wtaunu_Pt0_70_BFilter.merge.DAOD_SUSY10.e4709_s2726_r7725_r7676_p2622
mc15_13TeV:mc15_13TeV.363334.Sherpa_NNPDF30NNLO_Wtaunu_Pt70_140_CVetoBVeto.merge.DAOD_SUSY10.e4709_s2726_r7725_r7676_p2622
mc15_13TeV:mc15_13TeV.363335.Sherpa_NNPDF30NNLO_Wtaunu_Pt70_140_CFilterBVeto.merge.DAOD_SUSY10.e4709_s2726_r7725_r7676_p2622
mc15_13TeV:mc15_13TeV.363336.Sherpa_NNPDF30NNLO_Wtaunu_Pt70_140_BFilter.merge.DAOD_SUSY10.e4779_s2726_r7725_r7676_p2622
mc15_13TeV:mc15_13TeV.363337.Sherpa_NNPDF30NNLO_Wtaunu_Pt140_280_CVetoBVeto.merge.DAOD_SUSY10.e4709_s2726_r7725_r7676_p2622
mc15_13TeV:mc15_13TeV.363338.Sherpa_NNPDF30NNLO_Wtaunu_Pt140_280_CFilterBVeto.merge.DAOD_SUSY10.e4709_s2726_r7725_r7676_p2622
mc15_13TeV:mc15_13TeV.363339.Sherpa_NNPDF30NNLO_Wtaunu_Pt140_280_BFilter.merge.DAOD_SUSY10.e4709_s2726_r7725_r7676_p2622
mc15_13TeV:mc15_13TeV.363340.Sherpa_NNPDF30NNLO_Wtaunu_Pt280_500_CVetoBVeto.merge.DAOD_SUSY10.e4709_s2726_r7725_r7676_p2622
mc15_13TeV:mc15_13TeV.363341.Sherpa_NNPDF30NNLO_Wtaunu_Pt280_500_CFilterBVeto.merge.DAOD_SUSY10.e4779_s2726_r7725_r7676_p2622
mc15_13TeV:mc15_13TeV.363342.Sherpa_NNPDF30NNLO_Wtaunu_Pt280_500_BFilter.merge.DAOD_SUSY10.e4779_s2726_r7725_r7676_p2622
mc15_13TeV:mc15_13TeV.363343.Sherpa_NNPDF30NNLO_Wtaunu_Pt500_700_CVetoBVeto.merge.DAOD_SUSY10.e4709_s2726_r7725_r7676_p2622
mc15_13TeV:mc15_13TeV.363344.Sherpa_NNPDF30NNLO_Wtaunu_Pt500_700_CFilterBVeto.merge.DAOD_SUSY10.e4709_s2726_r7725_r7676_p2622
mc15_13TeV:mc15_13TeV.363345.Sherpa_NNPDF30NNLO_Wtaunu_Pt500_700_BFilter.merge.DAOD_SUSY10.e4779_s2726_r7725_r7676_p2622
mc15_13TeV:mc15_13TeV.363346.Sherpa_NNPDF30NNLO_Wtaunu_Pt700_1000_CVetoBVeto.merge.DAOD_SUSY10.e4709_s2726_r7725_r7676_p2622
mc15_13TeV:mc15_13TeV.363347.Sherpa_NNPDF30NNLO_Wtaunu_Pt700_1000_CFilterBVeto.merge.DAOD_SUSY10.e4709_s2726_r7725_r7676_p2622
mc15_13TeV:mc15_13TeV.363348.Sherpa_NNPDF30NNLO_Wtaunu_Pt700_1000_BFilter.merge.DAOD_SUSY10.e4779_s2726_r7725_r7676_p2622
mc15_13TeV:mc15_13TeV.363349.Sherpa_NNPDF30NNLO_Wtaunu_Pt1000_2000_CVetoBVeto.merge.DAOD_SUSY10.e4709_s2726_r7725_r7676_p2622
mc15_13TeV:mc15_13TeV.363350.Sherpa_NNPDF30NNLO_Wtaunu_Pt1000_2000_CFilterBVeto.merge.DAOD_SUSY10.e4709_s2726_r7725_r7676_p2622
mc15_13TeV:mc15_13TeV.363351.Sherpa_NNPDF30NNLO_Wtaunu_Pt1000_2000_BFilter.merge.DAOD_SUSY10.e4779_s2726_r7725_r7676_p2622
mc15_13TeV:mc15_13TeV.363352.Sherpa_NNPDF30NNLO_Wtaunu_Pt2000_E_CMS_CVetoBVeto.merge.DAOD_SUSY10.e4709_s2726_r7725_r7676_p2622
mc15_13TeV:mc15_13TeV.363353.Sherpa_NNPDF30NNLO_Wtaunu_Pt2000_E_CMS_CFilterBVeto.merge.DAOD_SUSY10.e4709_s2726_r7725_r7676_p2622
mc15_13TeV:mc15_13TeV.363354.Sherpa_NNPDF30NNLO_Wtaunu_Pt2000_E_CMS_BFilter.merge.DAOD_SUSY10.e4709_s2726_r7725_r7676_p2622
mc15_13TeV:mc15_13TeV.363436.Sherpa_NNPDF30NNLO_Wmunu_Pt0_70_CVetoBVeto.merge.DAOD_SUSY10.e4715_s2726_r7725_r7676_p2622
mc15_13TeV:mc15_13TeV.363437.Sherpa_NNPDF30NNLO_Wmunu_Pt0_70_CFilterBVeto.merge.DAOD_SUSY10.e4715_s2726_r7725_r7676_p2622
mc15_13TeV:mc15_13TeV.363438.Sherpa_NNPDF30NNLO_Wmunu_Pt0_70_BFilter.merge.DAOD_SUSY10.e4715_s2726_r7725_r7676_p2622
mc15_13TeV:mc15_13TeV.363439.Sherpa_NNPDF30NNLO_Wmunu_Pt70_140_CVetoBVeto.merge.DAOD_SUSY10.e4715_s2726_r7725_r7676_p2622
mc15_13TeV:mc15_13TeV.363440.Sherpa_NNPDF30NNLO_Wmunu_Pt70_140_CFilterBVeto.merge.DAOD_SUSY10.e4715_s2726_r7725_r7676_p2622
mc15_13TeV:mc15_13TeV.363441.Sherpa_NNPDF30NNLO_Wmunu_Pt70_140_BFilter.merge.DAOD_SUSY10.e4771_s2726_r7725_r7676_p2622
mc15_13TeV:mc15_13TeV.363442.Sherpa_NNPDF30NNLO_Wmunu_Pt140_280_CVetoBVeto.merge.DAOD_SUSY10.e4715_s2726_r7725_r7676_p2622
mc15_13TeV:mc15_13TeV.363443.Sherpa_NNPDF30NNLO_Wmunu_Pt140_280_CFilterBVeto.merge.DAOD_SUSY10.e4715_s2726_r7725_r7676_p2622
mc15_13TeV:mc15_13TeV.363444.Sherpa_NNPDF30NNLO_Wmunu_Pt140_280_BFilter.merge.DAOD_SUSY10.e4715_s2726_r7725_r7676_p2622
mc15_13TeV:mc15_13TeV.363445.Sherpa_NNPDF30NNLO_Wmunu_Pt280_500_CVetoBVeto.merge.DAOD_SUSY10.e4715_s2726_r7725_r7676_p2622
mc15_13TeV:mc15_13TeV.363446.Sherpa_NNPDF30NNLO_Wmunu_Pt280_500_CFilterBVeto.merge.DAOD_SUSY10.e4715_s2726_r7725_r7676_p2622
mc15_13TeV:mc15_13TeV.363447.Sherpa_NNPDF30NNLO_Wmunu_Pt280_500_BFilter.merge.DAOD_SUSY10.e4715_s2726_r7725_r7676_p2622
mc15_13TeV:mc15_13TeV.363448.Sherpa_NNPDF30NNLO_Wmunu_Pt500_700_CVetoBVeto.merge.DAOD_SUSY10.e4715_s2726_r7725_r7676_p2622
mc15_13TeV:mc15_13TeV.363449.Sherpa_NNPDF30NNLO_Wmunu_Pt500_700_CFilterBVeto.merge.DAOD_SUSY10.e4715_s2726_r7725_r7676_p2622
```

```
mc15_13TeV:mc15_13TeV.363450.Sherpa_NNPDF30NNLO_Wmunu_Pt500_700_BFilter.merge.DAOD_SUSY10.e4715_s2726_r7725_r7676_p2622
mc15_13TeV:mc15_13TeV.363451.Sherpa_NNPDF30NNLO_Wmunu_Pt700_1000_CVetoBVeto.merge.DAOD_SUSY10.e4715_s2726_r7725_r7676_p2622
mc15_13TeV:mc15_13TeV.363452.Sherpa_NNPDF30NNLO_Wmunu_Pt700_1000_CFilterBVeto.merge.DAOD_SUSY10.e4715_s2726_r7725_r7676_p2622
mc15_13TeV:mc15_13TeV.363453.Sherpa_NNPDF30NNLO_Wmunu_Pt700_1000_BFilter.merge.DAOD_SUSY10.e4715_s2726_r7725_r7676_p2622
mc15_13TeV:mc15_13TeV.363454.Sherpa_NNPDF30NNLO_Wmunu_Pt1000_2000_CVetoBVeto.merge.DAOD_SUSY10.e4715_s2726_r7725_r7676_p2622
mc15_13TeV:mc15_13TeV.363455.Sherpa_NNPDF30NNLO_Wmunu_Pt1000_2000_CFilterBVeto.merge.DAOD_SUSY10.e4715_s2726_r7725_r7676_p2622
mc15_13TeV:mc15_13TeV.363456.Sherpa_NNPDF30NNLO_Wmunu_Pt1000_2000_BFilter.merge.DAOD_SUSY10.e4715_s2726_r7725_r7676_p2622
mc15_13TeV:mc15_13TeV.363457.Sherpa_NNPDF30NNLO_Wmunu_Pt2000_E_CMS_CVetoBVeto.merge.DAOD_SUSY10.e4715_s2726_r7725_r7676_p2622
mc15_13TeV:mc15_13TeV.363458.Sherpa_NNPDF30NNLO_Wmunu_Pt2000_E_CMS_CFilterBVeto.merge.DAOD_SUSY10.e4715_s2726_r7725_r7676_p2622
mc15_13TeV:mc15_13TeV.363459.Sherpa_NNPDF30NNLO_Wmunu_Pt2000_E_CMS_BFilter.merge.DAOD_SUSY10.e4715_s2726_r7725_r7676_p2622
mc15_13TeV:mc15_13TeV.363460.Sherpa_NNPDF30NNLO_Wenu_Pt0_70_CVetoBVeto.merge.DAOD_SUSY10.e4715_s2726_r7725_r7676_p2622
mc15_13TeV:mc15_13TeV.363461.Sherpa_NNPDF30NNLO_Wenu_Pt0_70_CFilterBVeto.merge.DAOD_SUSY10.e4715_s2726_r7725_r7676_p2622
mc15_13TeV:mc15_13TeV.363462.Sherpa_NNPDF30NNLO_Wenu_Pt0_70_BFilter.merge.DAOD_SUSY10.e4715_s2726_r7725_r7676_p2622
mc15_13TeV:mc15_13TeV.363463.Sherpa_NNPDF30NNLO_Wenu_Pt70_140_CVetoBVeto.merge.DAOD_SUSY10.e4715_s2726_r7725_r7676_p2622
mc15_13TeV:mc15_13TeV.363464.Sherpa_NNPDF30NNLO_Wenu_Pt70_140_CFilterBVeto.merge.DAOD_SUSY10.e4715_s2726_r7725_r7676_p2622
mc15_13TeV:mc15_13TeV.363465.Sherpa_NNPDF30NNLO_Wenu_Pt70_140_BFilter.merge.DAOD_SUSY10.e4715_s2726_r7725_r7676_p2622
mc15_13TeV:mc15_13TeV.363466.Sherpa_NNPDF30NNLO_Wenu_Pt140_280_CVetoBVeto.merge.DAOD_SUSY10.e4715_s2726_r7725_r7676_p2622
mc15_13TeV:mc15_13TeV.363467.Sherpa_NNPDF30NNLO_Wenu_Pt140_280_CFilterBVeto.merge.DAOD_SUSY10.e4715_s2726_r7725_r7676_p2622
mc15_13TeV:mc15_13TeV.363468.Sherpa_NNPDF30NNLO_Wenu_Pt140_280_BFilter.merge.DAOD_SUSY10.e4715_s2726_r7725_r7676_p2622
mc15_13TeV:mc15_13TeV.363469.Sherpa_NNPDF30NNLO_Wenu_Pt280_500_CVetoBVeto.merge.DAOD_SUSY10.e4715_s2726_r7725_r7676_p2622
mc15_13TeV:mc15_13TeV.363470.Sherpa_NNPDF30NNLO_Wenu_Pt280_500_CFilterBVeto.merge.DAOD_SUSY10.e4715_s2726_r7725_r7676_p2622
mc15_13TeV:mc15_13TeV.363471.Sherpa_NNPDF30NNLO_Wenu_Pt280_500_BFilter.merge.DAOD_SUSY10.e4715_s2726_r7725_r7676_p2622
mc15_13TeV:mc15_13TeV.363472.Sherpa_NNPDF30NNLO_Wenu_Pt500_700_CVetoBVeto.merge.DAOD_SUSY10.e4715_s2726_r7725_r7676_p2622
mc15_13TeV:mc15_13TeV.363473.Sherpa_NNPDF30NNLO_Wenu_Pt500_700_CFilterBVeto.merge.DAOD_SUSY10.e4715_s2726_r7725_r7676_p2622
mc15_13TeV:mc15_13TeV.363474.Sherpa_NNPDF30NNLO_Wenu_Pt500_700_BFilter.merge.DAOD_SUSY10.e4771_s2726_r7725_r7676_p2622
mc15_13TeV:mc15_13TeV.363475.Sherpa_NNPDF30NNLO_Wenu_Pt700_1000_CVetoBVeto.merge.DAOD_SUSY10.e4715_s2726_r7725_r7676_p2622
mc15_13TeV:mc15_13TeV.363476.Sherpa_NNPDF30NNLO_Wenu_Pt700_1000_CFilterBVeto.merge.DAOD_SUSY10.e4715_s2726_r7725_r7676_p2622
mc15_13TeV:mc15_13TeV.363477.Sherpa_NNPDF30NNLO_Wenu_Pt700_1000_BFilter.merge.DAOD_SUSY10.e4715_s2726_r7725_r7676_p2622
mc15_13TeV:mc15_13TeV.363478.Sherpa_NNPDF30NNLO_Wenu_Pt1000_2000_CVetoBVeto.merge.DAOD_SUSY10.e4715_s2726_r7725_r7676_p2622
mc15_13TeV:mc15_13TeV.363479.Sherpa_NNPDF30NNLO_Wenu_Pt1000_2000_CFilterBVeto.merge.DAOD_SUSY10.e4715_s2726_r7725_r7676_p2622
mc15_13TeV:mc15_13TeV.363480.Sherpa_NNPDF30NNLO_Wenu_Pt1000_2000_BFilter.merge.DAOD_SUSY10.e4715_s2726_r7725_r7676_p2622
mc15_13TeV:mc15_13TeV.363481.Sherpa_NNPDF30NNLO_Wenu_Pt2000_E_CMS_CVetoBVeto.merge.DAOD_SUSY10.e4715_s2726_r7725_r7676_p2622
mc15_13TeV:mc15_13TeV.363483.Sherpa_NNPDF30NNLO_Wenu_Pt2000_E_CMS_BFilter.merge.DAOD_SUSY10.e4715_s2726_r7725_r7676_p2622
```

F.5 $Z + \text{Jets}$

```
mc15_13TeV:mc15_13TeV.363102.Sherpa_NNPDF30NNLO_Ztautau_Pt70_140_CVetoBVeto.merge.DAOD_SUSY10.e4742_s2726_r7725_r7676_p2622
mc15_13TeV:mc15_13TeV.363103.Sherpa_NNPDF30NNLO_Ztautau_Pt70_140_CFilterBVeto.merge.DAOD_SUSY10.e4742_s2726_r7725_r7676_p2622
mc15_13TeV:mc15_13TeV.363104.Sherpa_NNPDF30NNLO_Ztautau_Pt70_140_BFilter.merge.DAOD_SUSY10.e4792_s2726_r7725_r7676_p2622
mc15_13TeV:mc15_13TeV.363105.Sherpa_NNPDF30NNLO_Ztautau_Pt140_280_CVetoBVeto.merge.DAOD_SUSY10.e4666_s2726_r7725_r7676_p2622
mc15_13TeV:mc15_13TeV.363106.Sherpa_NNPDF30NNLO_Ztautau_Pt140_280_CFilterBVeto.merge.DAOD_SUSY10.e4666_s2726_r7725_r7676_p2622
mc15_13TeV:mc15_13TeV.363107.Sherpa_NNPDF30NNLO_Ztautau_Pt140_280_BFilter.merge.DAOD_SUSY10.e4742_s2726_r7725_r7676_p2622
mc15_13TeV:mc15_13TeV.363108.Sherpa_NNPDF30NNLO_Ztautau_Pt280_500_CVetoBVeto.merge.DAOD_SUSY10.e4666_s2726_r7725_r7676_p2622
mc15_13TeV:mc15_13TeV.363109.Sherpa_NNPDF30NNLO_Ztautau_Pt280_500_CFilterBVeto.merge.DAOD_SUSY10.e4792_s2726_r7725_r7676_p2622
mc15_13TeV:mc15_13TeV.363110.Sherpa_NNPDF30NNLO_Ztautau_Pt280_500_BFilter.merge.DAOD_SUSY10.e4792_s2726_r7725_r7676_p2622
mc15_13TeV:mc15_13TeV.363111.Sherpa_NNPDF30NNLO_Ztautau_Pt500_700_CVetoBVeto.merge.DAOD_SUSY10.e4666_s2726_r7725_r7676_p2622
mc15_13TeV:mc15_13TeV.363112.Sherpa_NNPDF30NNLO_Ztautau_Pt500_700_CFilterBVeto.merge.DAOD_SUSY10.e4742_s2726_r7725_r7676_p2622
mc15_13TeV:mc15_13TeV.363113.Sherpa_NNPDF30NNLO_Ztautau_Pt500_700_BFilter.merge.DAOD_SUSY10.e4742_s2726_r7725_r7676_p2622
mc15_13TeV:mc15_13TeV.363114.Sherpa_NNPDF30NNLO_Ztautau_Pt700_1000_CVetoBVeto.merge.DAOD_SUSY10.e4742_s2726_r7725_r7676_p2622
mc15_13TeV:mc15_13TeV.363115.Sherpa_NNPDF30NNLO_Ztautau_Pt700_1000_CFilterBVeto.merge.DAOD_SUSY10.e4792_s2726_r7725_r7676_p2622
mc15_13TeV:mc15_13TeV.363116.Sherpa_NNPDF30NNLO_Ztautau_Pt700_1000_BFilter.merge.DAOD_SUSY10.e4742_s2726_r7725_r7676_p2622
mc15_13TeV:mc15_13TeV.363117.Sherpa_NNPDF30NNLO_Ztautau_Pt1000_2000_CVetoBVeto.merge.DAOD_SUSY10.e4666_s2726_r7725_r7676_p2622
mc15_13TeV:mc15_13TeV.363118.Sherpa_NNPDF30NNLO_Ztautau_Pt1000_2000_CFilterBVeto.merge.DAOD_SUSY10.e4666_s2726_r7725_r7676_p2622
mc15_13TeV:mc15_13TeV.363119.Sherpa_NNPDF30NNLO_Ztautau_Pt1000_2000_BFilter.merge.DAOD_SUSY10.e4666_s2726_r7725_r7676_p2622
mc15_13TeV:mc15_13TeV.363120.Sherpa_NNPDF30NNLO_Ztautau_Pt2000_E_CMS_CVetoBVeto.merge.DAOD_SUSY10.e4690_s2726_r7725_r7676_p2622
mc15_13TeV:mc15_13TeV.363121.Sherpa_NNPDF30NNLO_Ztautau_Pt2000_E_CMS_CFilterBVeto.merge.DAOD_SUSY10.e4690_s2726_r7725_r7676_p2622
mc15_13TeV:mc15_13TeV.363122.Sherpa_NNPDF30NNLO_Ztautau_Pt2000_E_CMS_BFilter.merge.DAOD_SUSY10.e4792_s2726_r7725_r7676_p2622
mc15_13TeV:mc15_13TeV.363361.Sherpa_NNPDF30NNLO_Ztautau_Pt0_70_CVetoBVeto.merge.DAOD_SUSY10.e4689_s2726_r7725_r7676_p2622
mc15_13TeV:mc15_13TeV.363362.Sherpa_NNPDF30NNLO_Ztautau_Pt0_70_CFilterBVeto.merge.DAOD_SUSY10.e4689_s2726_r7725_r7676_p2622
mc15_13TeV:mc15_13TeV.363363.Sherpa_NNPDF30NNLO_Ztautau_Pt0_70_BFilter.merge.DAOD_SUSY10.e4743_s2726_r7725_r7676_p2622
mc15_13TeV:mc15_13TeV.363364.Sherpa_NNPDF30NNLO_Zmumu_Pt0_70_CVetoBVeto.merge.DAOD_SUSY10.e4716_s2726_r7725_r7676_p2622
mc15_13TeV:mc15_13TeV.363365.Sherpa_NNPDF30NNLO_Zmumu_Pt0_70_CFilterBVeto.merge.DAOD_SUSY10.e4716_s2726_r7725_r7676_p2622
mc15_13TeV:mc15_13TeV.363366.Sherpa_NNPDF30NNLO_Zmumu_Pt0_70_BFilter.merge.DAOD_SUSY10.e4716_s2726_r7725_r7676_p2622
mc15_13TeV:mc15_13TeV.363367.Sherpa_NNPDF30NNLO_Zmumu_Pt70_140_CVetoBVeto.merge.DAOD_SUSY10.e4716_s2726_r7725_r7676_p2622
mc15_13TeV:mc15_13TeV.363368.Sherpa_NNPDF30NNLO_Zmumu_Pt70_140_CFilterBVeto.merge.DAOD_SUSY10.e4716_s2726_r7725_r7676_p2622
mc15_13TeV:mc15_13TeV.363369.Sherpa_NNPDF30NNLO_Zmumu_Pt70_140_BFilter.merge.DAOD_SUSY10.e4716_s2726_r7725_r7676_p2622
mc15_13TeV:mc15_13TeV.363370.Sherpa_NNPDF30NNLO_Zmumu_Pt140_280_CVetoBVeto.merge.DAOD_SUSY10.e4716_s2726_r7725_r7676_p2622
mc15_13TeV:mc15_13TeV.363371.Sherpa_NNPDF30NNLO_Zmumu_Pt140_280_CFilterBVeto.merge.DAOD_SUSY10.e4716_s2726_r7725_r7676_p2622
mc15_13TeV:mc15_13TeV.363372.Sherpa_NNPDF30NNLO_Zmumu_Pt140_280_BFilter.merge.DAOD_SUSY10.e4716_s2726_r7725_r7676_p2622
mc15_13TeV:mc15_13TeV.363373.Sherpa_NNPDF30NNLO_Zmumu_Pt280_500_CVetoBVeto.merge.DAOD_SUSY10.e4716_s2726_r7725_r7676_p2622
mc15_13TeV:mc15_13TeV.363374.Sherpa_NNPDF30NNLO_Zmumu_Pt280_500_CFilterBVeto.merge.DAOD_SUSY10.e4716_s2726_r7725_r7676_p2622
mc15_13TeV:mc15_13TeV.363375.Sherpa_NNPDF30NNLO_Zmumu_Pt280_500_BFilter.merge.DAOD_SUSY10.e4772_s2726_r7725_r7676_p2622
mc15_13TeV:mc15_13TeV.363376.Sherpa_NNPDF30NNLO_Zmumu_Pt500_700_CVetoBVeto.merge.DAOD_SUSY10.e4716_s2726_r7725_r7676_p2622
mc15_13TeV:mc15_13TeV.363377.Sherpa_NNPDF30NNLO_Zmumu_Pt500_700_CFilterBVeto.merge.DAOD_SUSY10.e4716_s2726_r7725_r7676_p2622
```

```
mc15_13TeV:mc15_13TeV.363378.Sherpa_NNPDF30NNLO_Zmumu_Pt500_700_BFilter.merge.DAOD_SUSY10.e4772_s2726_r7725_r7676_p2622
mc15_13TeV:mc15_13TeV.363379.Sherpa_NNPDF30NNLO_Zmumu_Pt700_1000_CVetoBVeto.merge.DAOD_SUSY10.e4716_s2726_r7725_r7676_p2622
mc15_13TeV:mc15_13TeV.363380.Sherpa_NNPDF30NNLO_Zmumu_Pt700_1000_CFilterBVeto.merge.DAOD_SUSY10.e4716_s2726_r7725_r7676_p2622
mc15_13TeV:mc15_13TeV.363381.Sherpa_NNPDF30NNLO_Zmumu_Pt700_1000_BFilter.merge.DAOD_SUSY10.e4716_s2726_r7725_r7676_p2622
mc15_13TeV:mc15_13TeV.363382.Sherpa_NNPDF30NNLO_Zmumu_Pt1000_2000_CVetoBVeto.merge.DAOD_SUSY10.e4716_s2726_r7725_r7676_p2622
mc15_13TeV:mc15_13TeV.363383.Sherpa_NNPDF30NNLO_Zmumu_Pt1000_2000_CFilterBVeto.merge.DAOD_SUSY10.e4716_s2726_r7725_r7676_p2622
mc15_13TeV:mc15_13TeV.363384.Sherpa_NNPDF30NNLO_Zmumu_Pt1000_2000_BFilter.merge.DAOD_SUSY10.e4716_s2726_r7725_r7676_p2622
mc15_13TeV:mc15_13TeV.363385.Sherpa_NNPDF30NNLO_Zmumu_Pt2000_E_CMS_CVetoBVeto.merge.DAOD_SUSY10.e4716_s2726_r7725_r7676_p2622
mc15_13TeV:mc15_13TeV.363386.Sherpa_NNPDF30NNLO_Zmumu_Pt2000_E_CMS_CFilterBVeto.merge.DAOD_SUSY10.e4716_s2726_r7725_r7676_p2622
mc15_13TeV:mc15_13TeV.363387.Sherpa_NNPDF30NNLO_Zmumu_Pt2000_E_CMS_BFilter.merge.DAOD_SUSY10.e4716_s2726_r7725_r7676_p2622
mc15_13TeV:mc15_13TeV.363388.Sherpa_NNPDF30NNLO_Zee_Pt0_70_CVetoBVeto.merge.DAOD_SUSY10.e4716_s2726_r7725_r7676_p2622
mc15_13TeV:mc15_13TeV.363389.Sherpa_NNPDF30NNLO_Zee_Pt0_70_CFilterBVeto.merge.DAOD_SUSY10.e4716_s2726_r7725_r7676_p2622
mc15_13TeV:mc15_13TeV.363390.Sherpa_NNPDF30NNLO_Zee_Pt0_70_BFilter.merge.DAOD_SUSY10.e4716_s2726_r7725_r7676_p2622
mc15_13TeV:mc15_13TeV.363391.Sherpa_NNPDF30NNLO_Zee_Pt70_140_CVetoBVeto.merge.DAOD_SUSY10.e4716_s2726_r7725_r7676_p2622
mc15_13TeV:mc15_13TeV.363392.Sherpa_NNPDF30NNLO_Zee_Pt70_140_CFilterBVeto.merge.DAOD_SUSY10.e4772_s2726_r7725_r7676_p2622
mc15_13TeV:mc15_13TeV.363393.Sherpa_NNPDF30NNLO_Zee_Pt70_140_BFilter.merge.DAOD_SUSY10.e4716_s2726_r7725_r7676_p2622
mc15_13TeV:mc15_13TeV.363394.Sherpa_NNPDF30NNLO_Zee_Pt140_280_CVetoBVeto.merge.DAOD_SUSY10.e4716_s2726_r7725_r7676_p2622
mc15_13TeV:mc15_13TeV.363395.Sherpa_NNPDF30NNLO_Zee_Pt140_280_CFilterBVeto.merge.DAOD_SUSY10.e4716_s2726_r7725_r7676_p2622
mc15_13TeV:mc15_13TeV.363396.Sherpa_NNPDF30NNLO_Zee_Pt140_280_BFilter.merge.DAOD_SUSY10.e4772_s2726_r7725_r7676_p2622
mc15_13TeV:mc15_13TeV.363397.Sherpa_NNPDF30NNLO_Zee_Pt280_500_CVetoBVeto.merge.DAOD_SUSY10.e4716_s2726_r7725_r7676_p2622
mc15_13TeV:mc15_13TeV.363398.Sherpa_NNPDF30NNLO_Zee_Pt280_500_CFilterBVeto.merge.DAOD_SUSY10.e4716_s2726_r7725_r7676_p2622
mc15_13TeV:mc15_13TeV.363399.Sherpa_NNPDF30NNLO_Zee_Pt280_500_BFilter.merge.DAOD_SUSY10.e4772_s2726_r7725_r7676_p2622
mc15_13TeV:mc15_13TeV.363400.Sherpa_NNPDF30NNLO_Zee_Pt500_700_CVetoBVeto.merge.DAOD_SUSY10.e4716_s2726_r7725_r7676_p2622
mc15_13TeV:mc15_13TeV.363401.Sherpa_NNPDF30NNLO_Zee_Pt500_700_CFilterBVeto.merge.DAOD_SUSY10.e4716_s2726_r7725_r7676_p2622
mc15_13TeV:mc15_13TeV.363402.Sherpa_NNPDF30NNLO_Zee_Pt500_700_BFilter.merge.DAOD_SUSY10.e4716_s2726_r7725_r7676_p2622
mc15_13TeV:mc15_13TeV.363403.Sherpa_NNPDF30NNLO_Zee_Pt700_1000_CVetoBVeto.merge.DAOD_SUSY10.e4716_s2726_r7725_r7676_p2622
mc15_13TeV:mc15_13TeV.363404.Sherpa_NNPDF30NNLO_Zee_Pt700_1000_CFilterBVeto.merge.DAOD_SUSY10.e4716_s2726_r7725_r7676_p2622
mc15_13TeV:mc15_13TeV.363405.Sherpa_NNPDF30NNLO_Zee_Pt700_1000_BFilter.merge.DAOD_SUSY10.e4716_s2726_r7725_r7676_p2622
mc15_13TeV:mc15_13TeV.363406.Sherpa_NNPDF30NNLO_Zee_Pt1000_2000_CVetoBVeto.merge.DAOD_SUSY10.e4716_s2726_r7725_r7676_p2622
mc15_13TeV:mc15_13TeV.363407.Sherpa_NNPDF30NNLO_Zee_Pt1000_2000_CFilterBVeto.merge.DAOD_SUSY10.e4716_s2726_r7725_r7676_p2622
mc15_13TeV:mc15_13TeV.363408.Sherpa_NNPDF30NNLO_Zee_Pt1000_2000_BFilter.merge.DAOD_SUSY10.e4772_s2726_r7725_r7676_p2622
mc15_13TeV:mc15_13TeV.363409.Sherpa_NNPDF30NNLO_Zee_Pt2000_E_CMS_CVetoBVeto.merge.DAOD_SUSY10.e4716_s2726_r7725_r7676_p2622
mc15_13TeV:mc15_13TeV.363410.Sherpa_NNPDF30NNLO_Zee_Pt2000_E_CMS_CFilterBVeto.merge.DAOD_SUSY10.e4716_s2726_r7725_r7676_p2622
mc15_13TeV:mc15_13TeV.363411.Sherpa_NNPDF30NNLO_Zee_Pt2000_E_CMS_BFilter.merge.DAOD_SUSY10.e4772_s2726_r7725_r7676_p2622
mc15_13TeV:mc15_13TeV.363412.Sherpa_NNPDF30NNLO_Znunu_Pt0_70_CVetoBVeto.merge.DAOD_SUSY10.e4716_s2726_r7772_r7676_p2666
mc15_13TeV:mc15_13TeV.363413.Sherpa_NNPDF30NNLO_Znunu_Pt0_70_CFilterBVeto.merge.DAOD_SUSY10.e4716_s2726_r7772_r7676_p2666
mc15_13TeV:mc15_13TeV.363414.Sherpa_NNPDF30NNLO_Znunu_Pt0_70_BFilter.merge.DAOD_SUSY10.e4716_s2726_r7772_r7676_p2666
mc15_13TeV:mc15_13TeV.363415.Sherpa_NNPDF30NNLO_Znunu_Pt70_140_CVetoBVeto.merge.DAOD_SUSY10.e4716_s2726_r7772_r7676_p2666
mc15_13TeV:mc15_13TeV.363416.Sherpa_NNPDF30NNLO_Znunu_Pt70_140_CFilterBVeto.merge.DAOD_SUSY10.e4716_s2726_r7772_r7676_p2666
mc15_13TeV:mc15_13TeV.363417.Sherpa_NNPDF30NNLO_Znunu_Pt70_140_BFilter.merge.DAOD_SUSY10.e4772_s2726_r7772_r7676_p2666
mc15_13TeV:mc15_13TeV.363418.Sherpa_NNPDF30NNLO_Znunu_Pt140_280_CVetoBVeto.merge.DAOD_SUSY10.e4716_s2726_r7772_r7676_p2666
```

```
mc15_13TeV:mc15_13TeV.363419.Sherpa_NNPDF30NNLO_Znunu_Pt140_280_CFilterBVeto.merge.DAOD_SUSY10.e4716_s2726_r7772_r7676_p2666
mc15_13TeV:mc15_13TeV.363420.Sherpa_NNPDF30NNLO_Znunu_Pt140_280_BFilter.merge.DAOD_SUSY10.e4772_s2726_r7772_r7676_p2666
mc15_13TeV:mc15_13TeV.363421.Sherpa_NNPDF30NNLO_Znunu_Pt280_500_CVetoBVeto.merge.DAOD_SUSY10.e4716_s2726_r7772_r7676_p2666
mc15_13TeV:mc15_13TeV.363422.Sherpa_NNPDF30NNLO_Znunu_Pt280_500_CFilterBVeto.merge.DAOD_SUSY10.e4716_s2726_r7772_r7676_p2666
mc15_13TeV:mc15_13TeV.363423.Sherpa_NNPDF30NNLO_Znunu_Pt280_500_BFilter.merge.DAOD_SUSY10.e4716_s2726_r7772_r7676_p2666
mc15_13TeV:mc15_13TeV.363424.Sherpa_NNPDF30NNLO_Znunu_Pt500_700_CVetoBVeto.merge.DAOD_SUSY10.e4716_s2726_r7772_r7676_p2666
mc15_13TeV:mc15_13TeV.363425.Sherpa_NNPDF30NNLO_Znunu_Pt500_700_CFilterBVeto.merge.DAOD_SUSY10.e4716_s2726_r7772_r7676_p2666
mc15_13TeV:mc15_13TeV.363426.Sherpa_NNPDF30NNLO_Znunu_Pt500_700_BFilter.merge.DAOD_SUSY10.e4716_s2726_r7772_r7676_p2666
mc15_13TeV:mc15_13TeV.363427.Sherpa_NNPDF30NNLO_Znunu_Pt700_1000_CVetoBVeto.merge.DAOD_SUSY10.e4716_s2726_r7772_r7676_p2666
mc15_13TeV:mc15_13TeV.363428.Sherpa_NNPDF30NNLO_Znunu_Pt700_1000_CFilterBVeto.merge.DAOD_SUSY10.e4716_s2726_r7772_r7676_p2666
mc15_13TeV:mc15_13TeV.363429.Sherpa_NNPDF30NNLO_Znunu_Pt700_1000_BFilter.merge.DAOD_SUSY10.e4716_s2726_r7772_r7676_p2666
mc15_13TeV:mc15_13TeV.363430.Sherpa_NNPDF30NNLO_Znunu_Pt1000_2000_CVetoBVeto.merge.DAOD_SUSY10.e4716_s2726_r7772_r7676_p2666
mc15_13TeV:mc15_13TeV.363431.Sherpa_NNPDF30NNLO_Znunu_Pt1000_2000_CFilterBVeto.merge.DAOD_SUSY10.e4716_s2726_r7772_r7676_p2666
mc15_13TeV:mc15_13TeV.363432.Sherpa_NNPDF30NNLO_Znunu_Pt1000_2000_BFilter.merge.DAOD_SUSY10.e4772_s2726_r7772_r7676_p2666
mc15_13TeV:mc15_13TeV.363433.Sherpa_NNPDF30NNLO_Znunu_Pt2000_E_CMS_CVetoBVeto.merge.DAOD_SUSY10.e4716_s2726_r7772_r7676_p2666
mc15_13TeV:mc15_13TeV.363434.Sherpa_NNPDF30NNLO_Znunu_Pt2000_E_CMS_CFilterBVeto.merge.DAOD_SUSY10.e4716_s2726_r7772_r7676_p2666
mc15_13TeV:mc15_13TeV.363435.Sherpa_NNPDF30NNLO_Znunu_Pt2000_E_CMS_BFilter.merge.DAOD_SUSY10.e4772_s2726_r7772_r7676_p2666
```

F.6 Gtt Signal (Off-Shell)

```
mc15_13TeV:mc15_13TeV.370100.MGPy8EG_A14N_GG_ttn1_900_5000_1.merge.DAOD_SUSY10.e4049_s2608_r7772_r7676_p2666
mc15_13TeV:mc15_13TeV.370101.MGPy8EG_A14N_GG_ttn1_900_5000_200.merge.DAOD_SUSY10.e4049_s2608_r7772_r7676_p2666
mc15_13TeV:mc15_13TeV.370102.MGPy8EG_A14N_GG_ttn1_900_5000_400.merge.DAOD_SUSY10.e4049_s2608_r7772_r7676_p2666
mc15_13TeV:mc15_13TeV.370103.MGPy8EG_A14N_GG_ttn1_900_5000_545.merge.DAOD_SUSY10.e4049_s2608_r7772_r7676_p2666
mc15_13TeV:mc15_13TeV.370104.MGPy8EG_A14N_GG_ttn1_1000_5000_1.merge.DAOD_SUSY10.e4049_s2608_r7772_r7676_p2666
mc15_13TeV:mc15_13TeV.370105.MGPy8EG_A14N_GG_ttn1_1000_5000_200.merge.DAOD_SUSY10.e4049_s2608_r7772_r7676_p2666
mc15_13TeV:mc15_13TeV.370106.MGPy8EG_A14N_GG_ttn1_1000_5000_400.merge.DAOD_SUSY10.e4049_s2608_r7772_r7676_p2666
mc15_13TeV:mc15_13TeV.370107.MGPy8EG_A14N_GG_ttn1_1000_5000_600.merge.DAOD_SUSY10.e4049_s2608_r7772_r7676_p2666
mc15_13TeV:mc15_13TeV.370108.MGPy8EG_A14N_GG_ttn1_1000_5000_645.merge.DAOD_SUSY10.e4049_s2608_r7772_r7676_p2666
mc15_13TeV:mc15_13TeV.370109.MGPy8EG_A14N_GG_ttn1_1100_5000_1.merge.DAOD_SUSY10.e4049_s2608_r7772_r7676_p2666
mc15_13TeV:mc15_13TeV.370110.MGPy8EG_A14N_GG_ttn1_1100_5000_200.merge.DAOD_SUSY10.e4049_s2608_r7772_r7676_p2666
mc15_13TeV:mc15_13TeV.370111.MGPy8EG_A14N_GG_ttn1_1100_5000_400.merge.DAOD_SUSY10.e4049_s2608_r7772_r7676_p2666
mc15_13TeV:mc15_13TeV.370112.MGPy8EG_A14N_GG_ttn1_1100_5000_600.merge.DAOD_SUSY10.e4049_s2608_r7772_r7676_p2666
mc15_13TeV:mc15_13TeV.370113.MGPy8EG_A14N_GG_ttn1_1100_5000_745.merge.DAOD_SUSY10.e4049_s2608_r7772_r7676_p2666
mc15_13TeV:mc15_13TeV.370114.MGPy8EG_A14N_GG_ttn1_1200_5000_1.merge.DAOD_SUSY10.e4049_s2608_r7772_r7676_p2666
mc15_13TeV:mc15_13TeV.370115.MGPy8EG_A14N_GG_ttn1_1200_5000_100.merge.DAOD_SUSY10.e4049_s2608_r7772_r7676_p2666
mc15_13TeV:mc15_13TeV.370116.MGPy8EG_A14N_GG_ttn1_1200_5000_200.merge.DAOD_SUSY10.e4049_s2608_r7772_r7676_p2666
mc15_13TeV:mc15_13TeV.370117.MGPy8EG_A14N_GG_ttn1_1200_5000_400.merge.DAOD_SUSY10.e4049_s2608_r7772_r7676_p2666
mc15_13TeV:mc15_13TeV.370118.MGPy8EG_A14N_GG_ttn1_1200_5000_600.merge.DAOD_SUSY10.e4049_s2608_r7772_r7676_p2666
mc15_13TeV:mc15_13TeV.370119.MGPy8EG_A14N_GG_ttn1_1200_5000_700.merge.DAOD_SUSY10.e4049_s2608_r7772_r7676_p2666
```

```
mc15_13TeV:mc15_13TeV.370120.MGPy8EG_A14N_GG_ttn1_1200_5000_800.merge.DAOD_SUSY10.e4049_s2608_r7772_r7676_p2666
mc15_13TeV:mc15_13TeV.370121.MGPy8EG_A14N_GG_ttn1_1200_5000_845.merge.DAOD_SUSY10.e4049_s2608_r7772_r7676_p2666
mc15_13TeV:mc15_13TeV.370122.MGPy8EG_A14N_GG_ttn1_1200_5000_855.merge.DAOD_SUSY10.e4049_s2608_r7772_r7676_p2666
mc15_13TeV:mc15_13TeV.370123.MGPy8EG_A14N_GG_ttn1_1300_5000_1.merge.DAOD_SUSY10.e4049_s2608_r7772_r7676_p2666
mc15_13TeV:mc15_13TeV.370124.MGPy8EG_A14N_GG_ttn1_1300_5000_200.merge.DAOD_SUSY10.e4049_s2608_r7772_r7676_p2666
mc15_13TeV:mc15_13TeV.370125.MGPy8EG_A14N_GG_ttn1_1300_5000_400.merge.DAOD_SUSY10.e4049_s2608_r7772_r7676_p2666
mc15_13TeV:mc15_13TeV.370126.MGPy8EG_A14N_GG_ttn1_1300_5000_600.merge.DAOD_SUSY10.e4049_s2608_r7772_r7676_p2666
mc15_13TeV:mc15_13TeV.370127.MGPy8EG_A14N_GG_ttn1_1300_5000_800.merge.DAOD_SUSY10.e4049_s2608_r7772_r7676_p2666
mc15_13TeV:mc15_13TeV.370128.MGPy8EG_A14N_GG_ttn1_1300_5000_945.merge.DAOD_SUSY10.e4049_s2608_r7772_r7676_p2666
mc15_13TeV:mc15_13TeV.370129.MGPy8EG_A14N_GG_ttn1_1400_5000_1.merge.DAOD_SUSY10.e4049_s2608_r7772_r7676_p2666
mc15_13TeV:mc15_13TeV.370130.MGPy8EG_A14N_GG_ttn1_1400_5000_100.merge.DAOD_SUSY10.e4049_s2608_r7772_r7676_p2666
mc15_13TeV:mc15_13TeV.370131.MGPy8EG_A14N_GG_ttn1_1400_5000_200.merge.DAOD_SUSY10.e4049_s2608_r7772_r7676_p2666
mc15_13TeV:mc15_13TeV.370132.MGPy8EG_A14N_GG_ttn1_1400_5000_400.merge.DAOD_SUSY10.e4049_s2608_r7772_r7676_p2666
mc15_13TeV:mc15_13TeV.370133.MGPy8EG_A14N_GG_ttn1_1400_5000_600.merge.DAOD_SUSY10.e4049_s2608_r7772_r7676_p2666
mc15_13TeV:mc15_13TeV.370134.MGPy8EG_A14N_GG_ttn1_1400_5000_800.merge.DAOD_SUSY10.e4049_s2608_r7772_r7676_p2666
mc15_13TeV:mc15_13TeV.370135.MGPy8EG_A14N_GG_ttn1_1400_5000_1000.merge.DAOD_SUSY10.e4049_s2608_r7772_r7676_p2666
mc15_13TeV:mc15_13TeV.370136.MGPy8EG_A14N_GG_ttn1_1400_5000_1045.merge.DAOD_SUSY10.e4049_s2608_r7772_r7676_p2666
mc15_13TeV:mc15_13TeV.370137.MGPy8EG_A14N_GG_ttn1_1500_5000_1.merge.DAOD_SUSY10.e4049_s2608_r7772_r7676_p2666
mc15_13TeV:mc15_13TeV.370138.MGPy8EG_A14N_GG_ttn1_1500_5000_200.merge.DAOD_SUSY10.e4049_s2608_r7772_r7676_p2666
mc15_13TeV:mc15_13TeV.370139.MGPy8EG_A14N_GG_ttn1_1500_5000_400.merge.DAOD_SUSY10.e4049_s2608_r7772_r7676_p2666
mc15_13TeV:mc15_13TeV.370140.MGPy8EG_A14N_GG_ttn1_1500_5000_600.merge.DAOD_SUSY10.e4049_s2608_r7772_r7676_p2666
mc15_13TeV:mc15_13TeV.370141.MGPy8EG_A14N_GG_ttn1_1500_5000_800.merge.DAOD_SUSY10.e4049_s2608_r7772_r7676_p2666
mc15_13TeV:mc15_13TeV.370142.MGPy8EG_A14N_GG_ttn1_1500_5000_1000.merge.DAOD_SUSY10.e4049_s2608_r7772_r7676_p2666
mc15_13TeV:mc15_13TeV.370143.MGPy8EG_A14N_GG_ttn1_1500_5000_1145.merge.DAOD_SUSY10.e4049_s2608_r7772_r7676_p2666
mc15_13TeV:mc15_13TeV.370144.MGPy8EG_A14N_GG_ttn1_1600_5000_1.merge.DAOD_SUSY10.e4049_s2608_r7772_r7676_p2666
mc15_13TeV:mc15_13TeV.370145.MGPy8EG_A14N_GG_ttn1_1600_5000_200.merge.DAOD_SUSY10.e4049_s2608_r7772_r7676_p2666
mc15_13TeV:mc15_13TeV.370146.MGPy8EG_A14N_GG_ttn1_1600_5000_400.merge.DAOD_SUSY10.e4049_s2608_r7772_r7676_p2666
mc15_13TeV:mc15_13TeV.370147.MGPy8EG_A14N_GG_ttn1_1600_5000_600.merge.DAOD_SUSY10.e4049_s2608_r7772_r7676_p2666
mc15_13TeV:mc15_13TeV.370148.MGPy8EG_A14N_GG_ttn1_1600_5000_800.merge.DAOD_SUSY10.e4049_s2608_r7772_r7676_p2666
mc15_13TeV:mc15_13TeV.370149.MGPy8EG_A14N_GG_ttn1_1600_5000_1000.merge.DAOD_SUSY10.e4049_s2608_r7772_r7676_p2666
mc15_13TeV:mc15_13TeV.370150.MGPy8EG_A14N_GG_ttn1_1600_5000_1200.merge.DAOD_SUSY10.e4049_s2608_r7772_r7676_p2666
mc15_13TeV:mc15_13TeV.370151.MGPy8EG_A14N_GG_ttn1_1600_5000_1245.merge.DAOD_SUSY10.e4049_s2608_r7772_r7676_p2666
mc15_13TeV:mc15_13TeV.370152.MGPy8EG_A14N_GG_ttn1_1700_5000_1.merge.DAOD_SUSY10.e4049_s2608_r7772_r7676_p2666
mc15_13TeV:mc15_13TeV.370153.MGPy8EG_A14N_GG_ttn1_1700_5000_200.merge.DAOD_SUSY10.e4049_s2608_r7772_r7676_p2666
mc15_13TeV:mc15_13TeV.370154.MGPy8EG_A14N_GG_ttn1_1700_5000_400.merge.DAOD_SUSY10.e4049_s2608_r7772_r7676_p2666
mc15_13TeV:mc15_13TeV.370155.MGPy8EG_A14N_GG_ttn1_1700_5000_600.merge.DAOD_SUSY10.e4049_s2608_r7772_r7676_p2666
mc15_13TeV:mc15_13TeV.370156.MGPy8EG_A14N_GG_ttn1_1700_5000_800.merge.DAOD_SUSY10.e4049_s2608_r7772_r7676_p2666
mc15_13TeV:mc15_13TeV.370157.MGPy8EG_A14N_GG_ttn1_1700_5000_1000.merge.DAOD_SUSY10.e4049_s2608_r7772_r7676_p2666
mc15_13TeV:mc15_13TeV.370158.MGPy8EG_A14N_GG_ttn1_1700_5000_1200.merge.DAOD_SUSY10.e4049_s2608_r7772_r7676_p2666
mc15_13TeV:mc15_13TeV.370159.MGPy8EG_A14N_GG_ttn1_1700_5000_1345.merge.DAOD_SUSY10.e4049_s2608_r7772_r7676_p2666
mc15_13TeV:mc15_13TeV.370160.MGPy8EG_A14N_GG_ttn1_1800_5000_1.merge.DAOD_SUSY10.e4049_s2608_r7772_r7676_p2666
```

mc15_13TeV:mc15_13TeV.370161.MGPy8EG_A14N_GG_ttn1_1800_5000_200.merge.DAOD_SUSY10.e4049_s2608_r7772_r7676_p2666
mc15_13TeV:mc15_13TeV.370162.MGPy8EG_A14N_GG_ttn1_1800_5000_400.merge.DAOD_SUSY10.e4049_s2608_r7772_r7676_p2666
mc15_13TeV:mc15_13TeV.370163.MGPy8EG_A14N_GG_ttn1_1800_5000_600.merge.DAOD_SUSY10.e4049_s2608_r7772_r7676_p2666
mc15_13TeV:mc15_13TeV.370164.MGPy8EG_A14N_GG_ttn1_1800_5000_800.merge.DAOD_SUSY10.e4049_s2608_r7772_r7676_p2666
mc15_13TeV:mc15_13TeV.370165.MGPy8EG_A14N_GG_ttn1_1800_5000_1000.merge.DAOD_SUSY10.e4049_s2608_r7772_r7676_p2666
mc15_13TeV:mc15_13TeV.370166.MGPy8EG_A14N_GG_ttn1_1800_5000_1200.merge.DAOD_SUSY10.e4049_s2608_r7772_r7676_p2666
mc15_13TeV:mc15_13TeV.370167.MGPy8EG_A14N_GG_ttn1_1800_5000_1400.merge.DAOD_SUSY10.e4049_s2608_r7772_r7676_p2666
mc15_13TeV:mc15_13TeV.370168.MGPy8EG_A14N_GG_ttn1_1800_5000_1445.merge.DAOD_SUSY10.e4049_s2608_r7772_r7676_p2666
mc15_13TeV:mc15_13TeV.370169.MGPy8EG_A14N_GG_ttn1_1900_5000_1.merge.DAOD_SUSY10.e4049_s2608_r7772_r7676_p2666
mc15_13TeV:mc15_13TeV.370170.MGPy8EG_A14N_GG_ttn1_1900_5000_200.merge.DAOD_SUSY10.e4049_s2608_r7772_r7676_p2666
mc15_13TeV:mc15_13TeV.370171.MGPy8EG_A14N_GG_ttn1_1900_5000_400.merge.DAOD_SUSY10.e4049_s2608_r7772_r7676_p2666
mc15_13TeV:mc15_13TeV.370172.MGPy8EG_A14N_GG_ttn1_1900_5000_600.merge.DAOD_SUSY10.e4049_s2608_r7772_r7676_p2666
mc15_13TeV:mc15_13TeV.370173.MGPy8EG_A14N_GG_ttn1_1900_5000_800.merge.DAOD_SUSY10.e4049_s2608_r7772_r7676_p2666
mc15_13TeV:mc15_13TeV.370174.MGPy8EG_A14N_GG_ttn1_1900_5000_1000.merge.DAOD_SUSY10.e4049_s2608_r7772_r7676_p2666
mc15_13TeV:mc15_13TeV.370175.MGPy8EG_A14N_GG_ttn1_1900_5000_1200.merge.DAOD_SUSY10.e4049_s2608_r7772_r7676_p2666
mc15_13TeV:mc15_13TeV.370176.MGPy8EG_A14N_GG_ttn1_1900_5000_1400.merge.DAOD_SUSY10.e4049_s2608_r7772_r7676_p2666
mc15_13TeV:mc15_13TeV.370177.MGPy8EG_A14N_GG_ttn1_1900_5000_1545.merge.DAOD_SUSY10.e4049_s2608_r7772_r7676_p2666
mc15_13TeV:mc15_13TeV.370178.MGPy8EG_A14N_GG_ttn1_2000_5000_1.merge.DAOD_SUSY10.e4049_s2608_r7772_r7676_p2666
mc15_13TeV:mc15_13TeV.370179.MGPy8EG_A14N_GG_ttn1_2000_5000_200.merge.DAOD_SUSY10.e4049_s2608_r7772_r7676_p2666
mc15_13TeV:mc15_13TeV.370180.MGPy8EG_A14N_GG_ttn1_2000_5000_400.merge.DAOD_SUSY10.e4049_s2608_r7772_r7676_p2666
mc15_13TeV:mc15_13TeV.370181.MGPy8EG_A14N_GG_ttn1_2000_5000_600.merge.DAOD_SUSY10.e4049_s2608_r7772_r7676_p2666
mc15_13TeV:mc15_13TeV.370182.MGPy8EG_A14N_GG_ttn1_2000_5000_800.merge.DAOD_SUSY10.e4049_s2608_r7772_r7676_p2666
mc15_13TeV:mc15_13TeV.370183.MGPy8EG_A14N_GG_ttn1_2000_5000_1000.merge.DAOD_SUSY10.e4049_s2608_r7772_r7676_p2666
mc15_13TeV:mc15_13TeV.370184.MGPy8EG_A14N_GG_ttn1_2000_5000_1200.merge.DAOD_SUSY10.e4049_s2608_r7772_r7676_p2666
mc15_13TeV:mc15_13TeV.370185.MGPy8EG_A14N_GG_ttn1_2000_5000_1400.merge.DAOD_SUSY10.e4049_s2608_r7772_r7676_p2666
mc15_13TeV:mc15_13TeV.370186.MGPy8EG_A14N_GG_ttn1_2000_5000_1600.merge.DAOD_SUSY10.e4049_s2608_r7772_r7676_p2666
mc15_13TeV:mc15_13TeV.370187.MGPy8EG_A14N_GG_ttn1_2000_5000_1645.merge.DAOD_SUSY10.e4049_s2608_r7772_r7676_p2666
mc15_13TeV:mc15_13TeV.370239.MGPy8EG_A14N_GG_ttn1_2100_5000_1.merge.DAOD_SUSY10.e5089_s2726_r7772_r7676_p2666
mc15_13TeV:mc15_13TeV.370240.MGPy8EG_A14N_GG_ttn1_2100_5000_200.merge.DAOD_SUSY10.e5089_s2726_r7772_r7676_p2666
mc15_13TeV:mc15_13TeV.370241.MGPy8EG_A14N_GG_ttn1_2100_5000_400.merge.DAOD_SUSY10.e5089_s2726_r7772_r7676_p2666
mc15_13TeV:mc15_13TeV.370243.MGPy8EG_A14N_GG_ttn1_2100_5000_800.merge.DAOD_SUSY10.e5089_s2726_r7772_r7676_p2666
mc15_13TeV:mc15_13TeV.370244.MGPy8EG_A14N_GG_ttn1_2100_5000_1000.merge.DAOD_SUSY10.e5089_s2726_r7772_r7676_p2666
mc15_13TeV:mc15_13TeV.370245.MGPy8EG_A14N_GG_ttn1_2100_5000_1200.merge.DAOD_SUSY10.e5089_s2726_r7772_r7676_p2666
mc15_13TeV:mc15_13TeV.370246.MGPy8EG_A14N_GG_ttn1_2100_5000_1400.merge.DAOD_SUSY10.e5089_s2726_r7772_r7676_p2666
mc15_13TeV:mc15_13TeV.370247.MGPy8EG_A14N_GG_ttn1_2100_5000_1600.merge.DAOD_SUSY10.e5089_s2726_r7772_r7676_p2666
mc15_13TeV:mc15_13TeV.370248.MGPy8EG_A14N_GG_ttn1_2100_5000_1745.merge.DAOD_SUSY10.e5089_s2726_r7772_r7676_p2666
mc15_13TeV:mc15_13TeV.370249.MGPy8EG_A14N_GG_ttn1_2200_5000_1.merge.DAOD_SUSY10.e5089_s2726_r7772_r7676_p2666
mc15_13TeV:mc15_13TeV.373420.MGPy8EG_A14N_GG_ttn1_2200_5000_200.merge.DAOD_SUSY10.e5089_s2726_r7772_r7676_p2666
mc15_13TeV:mc15_13TeV.373421.MGPy8EG_A14N_GG_ttn1_2200_5000_400.merge.DAOD_SUSY10.e5089_s2726_r7772_r7676_p2666
mc15_13TeV:mc15_13TeV.373422.MGPy8EG_A14N_GG_ttn1_2200_5000_600.merge.DAOD_SUSY10.e5089_s2726_r7772_r7676_p2666
mc15_13TeV:mc15_13TeV.373423.MGPy8EG_A14N_GG_ttn1_2200_5000_800.merge.DAOD_SUSY10.e5089_s2726_r7772_r7676_p2666

```
mc15_13TeV:mc15_13TeV.373424.MGPy8EG_A14N_GG_ttn1_2200_5000_1000.merge.DAOD_SUSY10.e5089_s2726_r7772_r7676_p2666
mc15_13TeV:mc15_13TeV.373425.MGPy8EG_A14N_GG_ttn1_2200_5000_1200.merge.DAOD_SUSY10.e5089_s2726_r7772_r7676_p2666
mc15_13TeV:mc15_13TeV.373426.MGPy8EG_A14N_GG_ttn1_2200_5000_1400.merge.DAOD_SUSY10.e5089_s2726_r7772_r7676_p2666
mc15_13TeV:mc15_13TeV.373427.MGPy8EG_A14N_GG_ttn1_2200_5000_1600.merge.DAOD_SUSY10.e5089_s2726_r7772_r7676_p2666
mc15_13TeV:mc15_13TeV.373428.MGPy8EG_A14N_GG_ttn1_2200_5000_1800.merge.DAOD_SUSY10.e5089_s2726_r7772_r7676_p2666
mc15_13TeV:mc15_13TeV.373429.MGPy8EG_A14N_GG_ttn1_2300_5000_1.merge.DAOD_SUSY10.e5089_s2726_r7772_r7676_p2666
mc15_13TeV:mc15_13TeV.373430.MGPy8EG_A14N_GG_ttn1_2300_5000_200.merge.DAOD_SUSY10.e5089_s2726_r7772_r7676_p2666
mc15_13TeV:mc15_13TeV.373431.MGPy8EG_A14N_GG_ttn1_2300_5000_400.merge.DAOD_SUSY10.e5089_s2726_r7772_r7676_p2666
mc15_13TeV:mc15_13TeV.373432.MGPy8EG_A14N_GG_ttn1_2300_5000_600.merge.DAOD_SUSY10.e5089_s2726_r7772_r7676_p2666
mc15_13TeV:mc15_13TeV.373433.MGPy8EG_A14N_GG_ttn1_2300_5000_800.merge.DAOD_SUSY10.e5089_s2726_r7772_r7676_p2666
mc15_13TeV:mc15_13TeV.373434.MGPy8EG_A14N_GG_ttn1_2300_5000_1000.merge.DAOD_SUSY10.e5089_s2726_r7772_r7676_p2666
mc15_13TeV:mc15_13TeV.373435.MGPy8EG_A14N_GG_ttn1_2300_5000_1200.merge.DAOD_SUSY10.e5089_s2726_r7772_r7676_p2666
mc15_13TeV:mc15_13TeV.373436.MGPy8EG_A14N_GG_ttn1_2300_5000_1400.merge.DAOD_SUSY10.e5089_s2726_r7772_r7676_p2666
mc15_13TeV:mc15_13TeV.373437.MGPy8EG_A14N_GG_ttn1_2300_5000_1600.merge.DAOD_SUSY10.e5089_s2726_r7772_r7676_p2666
mc15_13TeV:mc15_13TeV.373438.MGPy8EG_A14N_GG_ttn1_2300_5000_1800.merge.DAOD_SUSY10.e5089_s2726_r7772_r7676_p2666
mc15_13TeV:mc15_13TeV.373439.MGPy8EG_A14N_GG_ttn1_2400_5000_1.merge.DAOD_SUSY10.e5089_s2726_r7772_r7676_p2666
mc15_13TeV:mc15_13TeV.373440.MGPy8EG_A14N_GG_ttn1_2400_5000_200.merge.DAOD_SUSY10.e5089_s2726_r7772_r7676_p2666
mc15_13TeV:mc15_13TeV.373441.MGPy8EG_A14N_GG_ttn1_2400_5000_400.merge.DAOD_SUSY10.e5089_s2726_r7772_r7676_p2666
mc15_13TeV:mc15_13TeV.373442.MGPy8EG_A14N_GG_ttn1_2400_5000_600.merge.DAOD_SUSY10.e5089_s2726_r7772_r7676_p2666
mc15_13TeV:mc15_13TeV.373443.MGPy8EG_A14N_GG_ttn1_2400_5000_800.merge.DAOD_SUSY10.e5089_s2726_r7772_r7676_p2666
mc15_13TeV:mc15_13TeV.373444.MGPy8EG_A14N_GG_ttn1_2400_5000_1000.merge.DAOD_SUSY10.e5089_s2726_r7772_r7676_p2666
mc15_13TeV:mc15_13TeV.373445.MGPy8EG_A14N_GG_ttn1_2400_5000_1200.merge.DAOD_SUSY10.e5089_s2726_r7772_r7676_p2666
mc15_13TeV:mc15_13TeV.373446.MGPy8EG_A14N_GG_ttn1_2400_5000_1400.merge.DAOD_SUSY10.e5089_s2726_r7772_r7676_p2666
mc15_13TeV:mc15_13TeV.373447.MGPy8EG_A14N_GG_ttn1_2400_5000_1600.merge.DAOD_SUSY10.e5089_s2726_r7772_r7676_p2666
mc15_13TeV:mc15_13TeV.373448.MGPy8EG_A14N_GG_ttn1_2400_5000_2000.merge.DAOD_SUSY10.e5089_s2726_r7772_r7676_p2666
```

F.7 Gtt Signal (On-Shell)

```
mc15_13TeV:mc15_13TeV.370400.MGPy8EG_A14N_GG_ttn1OnShellt1_900_450_60.merge.DAOD_SUSY10.e4051_s2608_r7772_r7676_p2666
mc15_13TeV:mc15_13TeV.370401.MGPy8EG_A14N_GG_ttn1OnShellt1_900_650_60.merge.DAOD_SUSY10.e4051_s2608_r7772_r7676_p2666
mc15_13TeV:mc15_13TeV.370402.MGPy8EG_A14N_GG_ttn1OnShellt1_900_720_60.merge.DAOD_SUSY10.e4051_s2608_r7772_r7676_p2666
mc15_13TeV:mc15_13TeV.370403.MGPy8EG_A14N_GG_ttn1OnShellt1_1000_450_60.merge.DAOD_SUSY10.e4051_s2608_r7772_r7676_p2666
mc15_13TeV:mc15_13TeV.370404.MGPy8EG_A14N_GG_ttn1OnShellt1_1000_650_60.merge.DAOD_SUSY10.e4051_s2608_r7772_r7676_p2666
mc15_13TeV:mc15_13TeV.370405.MGPy8EG_A14N_GG_ttn1OnShellt1_1000_820_60.merge.DAOD_SUSY10.e4051_s2608_r7772_r7676_p2666
mc15_13TeV:mc15_13TeV.370406.MGPy8EG_A14N_GG_ttn1OnShellt1_1100_450_60.merge.DAOD_SUSY10.e4051_s2608_r7772_r7676_p2666
mc15_13TeV:mc15_13TeV.370407.MGPy8EG_A14N_GG_ttn1OnShellt1_1100_650_60.merge.DAOD_SUSY10.e4051_s2608_r7772_r7676_p2666
mc15_13TeV:mc15_13TeV.370408.MGPy8EG_A14N_GG_ttn1OnShellt1_1100_850_60.merge.DAOD_SUSY10.e4051_s2608_r7772_r7676_p2666
```

```
mc15_13TeV:mc15_13TeV.370409.MGPy8EG_A14N_GG_ttn1OnShellt1_1100_920_60.merge.DAOD_SUSY10.e4051_s2608_r7772_r7676_p2666
mc15_13TeV:mc15_13TeV.370410.MGPy8EG_A14N_GG_ttn1OnShellt1_1200_450_60.merge.DAOD_SUSY10.e4051_s2608_r7772_r7676_p2666
mc15_13TeV:mc15_13TeV.370411.MGPy8EG_A14N_GG_ttn1OnShellt1_1200_650_60.merge.DAOD_SUSY10.e4051_s2608_r7772_r7676_p2666
mc15_13TeV:mc15_13TeV.370412.MGPy8EG_A14N_GG_ttn1OnShellt1_1200_850_60.merge.DAOD_SUSY10.e4051_s2608_r7772_r7676_p2666
mc15_13TeV:mc15_13TeV.370413.MGPy8EG_A14N_GG_ttn1OnShellt1_1200_1020_60.merge.DAOD_SUSY10.e4051_s2608_r7772_r7676_p2666
mc15_13TeV:mc15_13TeV.370414.MGPy8EG_A14N_GG_ttn1OnShellt1_1300_450_60.merge.DAOD_SUSY10.e4051_s2608_r7772_r7676_p2666
mc15_13TeV:mc15_13TeV.370416.MGPy8EG_A14N_GG_ttn1OnShellt1_1300_850_60.merge.DAOD_SUSY10.e4051_s2608_r7772_r7676_p2666
mc15_13TeV:mc15_13TeV.370417.MGPy8EG_A14N_GG_ttn1OnShellt1_1300_1050_60.merge.DAOD_SUSY10.e4051_s2608_r7772_r7676_p2666
mc15_13TeV:mc15_13TeV.370418.MGPy8EG_A14N_GG_ttn1OnShellt1_1300_1120_60.merge.DAOD_SUSY10.e4051_s2608_r7772_r7676_p2666
mc15_13TeV:mc15_13TeV.370419.MGPy8EG_A14N_GG_ttn1OnShellt1_1400_450_60.merge.DAOD_SUSY10.e4051_s2608_r7772_r7676_p2666
mc15_13TeV:mc15_13TeV.370420.MGPy8EG_A14N_GG_ttn1OnShellt1_1400_650_60.merge.DAOD_SUSY10.e4051_s2608_r7772_r7676_p2666
mc15_13TeV:mc15_13TeV.370421.MGPy8EG_A14N_GG_ttn1OnShellt1_1400_850_60.merge.DAOD_SUSY10.e4051_s2608_r7772_r7676_p2666
mc15_13TeV:mc15_13TeV.370422.MGPy8EG_A14N_GG_ttn1OnShellt1_1400_1050_60.merge.DAOD_SUSY10.e4051_s2608_r7772_r7676_p2666
mc15_13TeV:mc15_13TeV.370423.MGPy8EG_A14N_GG_ttn1OnShellt1_1400_1220_60.merge.DAOD_SUSY10.e4051_s2608_r7772_r7676_p2666
mc15_13TeV:mc15_13TeV.370424.MGPy8EG_A14N_GG_ttn1OnShellt1_1500_450_60.merge.DAOD_SUSY10.e4051_s2608_r7772_r7676_p2666
mc15_13TeV:mc15_13TeV.370425.MGPy8EG_A14N_GG_ttn1OnShellt1_1500_650_60.merge.DAOD_SUSY10.e4051_s2608_r7772_r7676_p2666
mc15_13TeV:mc15_13TeV.370426.MGPy8EG_A14N_GG_ttn1OnShellt1_1500_850_60.merge.DAOD_SUSY10.e4051_s2608_r7772_r7676_p2666
mc15_13TeV:mc15_13TeV.370427.MGPy8EG_A14N_GG_ttn1OnShellt1_1500_1050_60.merge.DAOD_SUSY10.e4051_s2608_r7772_r7676_p2666
mc15_13TeV:mc15_13TeV.370428.MGPy8EG_A14N_GG_ttn1OnShellt1_1500_1250_60.merge.DAOD_SUSY10.e4051_s2608_r7772_r7676_p2666
mc15_13TeV:mc15_13TeV.370429.MGPy8EG_A14N_GG_ttn1OnShellt1_1500_1320_60.merge.DAOD_SUSY10.e4051_s2608_r7772_r7676_p2666
mc15_13TeV:mc15_13TeV.370430.MGPy8EG_A14N_GG_ttn1OnShellt1_1600_450_60.merge.DAOD_SUSY10.e4051_s2608_r7772_r7676_p2666
mc15_13TeV:mc15_13TeV.370431.MGPy8EG_A14N_GG_ttn1OnShellt1_1600_650_60.merge.DAOD_SUSY10.e4051_s2608_r7772_r7676_p2666
mc15_13TeV:mc15_13TeV.370432.MGPy8EG_A14N_GG_ttn1OnShellt1_1600_850_60.merge.DAOD_SUSY10.e4051_s2608_r7772_r7676_p2666
mc15_13TeV:mc15_13TeV.370433.MGPy8EG_A14N_GG_ttn1OnShellt1_1600_1050_60.merge.DAOD_SUSY10.e4051_s2608_r7772_r7676_p2666
mc15_13TeV:mc15_13TeV.370434.MGPy8EG_A14N_GG_ttn1OnShellt1_1600_1250_60.merge.DAOD_SUSY10.e4051_s2608_r7772_r7676_p2666
mc15_13TeV:mc15_13TeV.370435.MGPy8EG_A14N_GG_ttn1OnShellt1_1600_1420_60.merge.DAOD_SUSY10.e4051_s2608_r7772_r7676_p2666
mc15_13TeV:mc15_13TeV.370436.MGPy8EG_A14N_GG_ttn1OnShellt1_1700_450_60.merge.DAOD_SUSY10.e4051_s2608_r7772_r7676_p2666
mc15_13TeV:mc15_13TeV.370437.MGPy8EG_A14N_GG_ttn1OnShellt1_1700_650_60.merge.DAOD_SUSY10.e4051_s2608_r7772_r7676_p2666
mc15_13TeV:mc15_13TeV.370438.MGPy8EG_A14N_GG_ttn1OnShellt1_1700_850_60.merge.DAOD_SUSY10.e4051_s2608_r7772_r7676_p2666
mc15_13TeV:mc15_13TeV.370439.MGPy8EG_A14N_GG_ttn1OnShellt1_1700_1050_60.merge.DAOD_SUSY10.e4051_s2608_r7772_r7676_p2666
mc15_13TeV:mc15_13TeV.370440.MGPy8EG_A14N_GG_ttn1OnShellt1_1700_1250_60.merge.DAOD_SUSY10.e4051_s2608_r7772_r7676_p2666
mc15_13TeV:mc15_13TeV.370441.MGPy8EG_A14N_GG_ttn1OnShellt1_1700_1450_60.merge.DAOD_SUSY10.e4051_s2608_r7772_r7676_p2666
mc15_13TeV:mc15_13TeV.370442.MGPy8EG_A14N_GG_ttn1OnShellt1_1700_1520_60.merge.DAOD_SUSY10.e4051_s2608_r7772_r7676_p2666
mc15_13TeV:mc15_13TeV.370443.MGPy8EG_A14N_GG_ttn1OnShellt1_1800_450_60.merge.DAOD_SUSY10.e4051_s2608_r7772_r7676_p2666
mc15_13TeV:mc15_13TeV.370444.MGPy8EG_A14N_GG_ttn1OnShellt1_1800_650_60.merge.DAOD_SUSY10.e4051_s2608_r7772_r7676_p2666
mc15_13TeV:mc15_13TeV.370445.MGPy8EG_A14N_GG_ttn1OnShellt1_1800_850_60.merge.DAOD_SUSY10.e4051_s2608_r7772_r7676_p2666
mc15_13TeV:mc15_13TeV.370447.MGPy8EG_A14N_GG_ttn1OnShellt1_1800_1250_60.merge.DAOD_SUSY10.e4051_s2608_r7772_r7676_p2666
mc15_13TeV:mc15_13TeV.370448.MGPy8EG_A14N_GG_ttn1OnShellt1_1800_1450_60.merge.DAOD_SUSY10.e4051_s2608_r7772_r7676_p2666
mc15_13TeV:mc15_13TeV.370449.MGPy8EG_A14N_GG_ttn1OnShellt1_1800_1620_60.merge.DAOD_SUSY10.e4051_s2608_r7772_r7676_p2666
mc15_13TeV:mc15_13TeV.370450.MGPy8EG_A14N_GG_ttn1OnShellt1_1900_450_60.merge.DAOD_SUSY10.e4051_s2608_r7772_r7676_p2666
mc15_13TeV:mc15_13TeV.370451.MGPy8EG_A14N_GG_ttn1OnShellt1_1900_650_60.merge.DAOD_SUSY10.e4051_s2608_r7772_r7676_p2666
```

```
mc15_13TeV:mc15_13TeV.370452.MGPy8EG_A14N_GG_ttn1OnShellt1_1900_850_60.merge.DAOD_SUSY10.e4051_s2608_r7772_r7676_p2666
mc15_13TeV:mc15_13TeV.370453.MGPy8EG_A14N_GG_ttn1OnShellt1_1900_1050_60.merge.DAOD_SUSY10.e4051_s2608_r7772_r7676_p2666
mc15_13TeV:mc15_13TeV.370454.MGPy8EG_A14N_GG_ttn1OnShellt1_1900_1250_60.merge.DAOD_SUSY10.e4051_s2608_r7772_r7676_p2666
mc15_13TeV:mc15_13TeV.370455.MGPy8EG_A14N_GG_ttn1OnShellt1_1900_1450_60.merge.DAOD_SUSY10.e4051_s2608_r7772_r7676_p2666
mc15_13TeV:mc15_13TeV.370456.MGPy8EG_A14N_GG_ttn1OnShellt1_1900_1650_60.merge.DAOD_SUSY10.e4051_s2608_r7772_r7676_p2666
mc15_13TeV:mc15_13TeV.370457.MGPy8EG_A14N_GG_ttn1OnShellt1_1900_1720_60.merge.DAOD_SUSY10.e4051_s2608_r7772_r7676_p2666
mc15_13TeV:mc15_13TeV.370458.MGPy8EG_A14N_GG_ttn1OnShellt1_2000_450_60.merge.DAOD_SUSY10.e4051_s2608_r7772_r7676_p2666
mc15_13TeV:mc15_13TeV.370459.MGPy8EG_A14N_GG_ttn1OnShellt1_2000_650_60.merge.DAOD_SUSY10.e4051_s2608_r7772_r7676_p2666
mc15_13TeV:mc15_13TeV.370460.MGPy8EG_A14N_GG_ttn1OnShellt1_2000_850_60.merge.DAOD_SUSY10.e4051_s2608_r7772_r7676_p2666
mc15_13TeV:mc15_13TeV.370461.MGPy8EG_A14N_GG_ttn1OnShellt1_2000_1050_60.merge.DAOD_SUSY10.e4051_s2608_r7772_r7676_p2666
mc15_13TeV:mc15_13TeV.370462.MGPy8EG_A14N_GG_ttn1OnShellt1_2000_1250_60.merge.DAOD_SUSY10.e4051_s2608_r7772_r7676_p2666
mc15_13TeV:mc15_13TeV.370463.MGPy8EG_A14N_GG_ttn1OnShellt1_2000_1450_60.merge.DAOD_SUSY10.e4051_s2608_r7772_r7676_p2666
mc15_13TeV:mc15_13TeV.370464.MGPy8EG_A14N_GG_ttn1OnShellt1_2000_1650_60.merge.DAOD_SUSY10.e4051_s2608_r7772_r7676_p2666
mc15_13TeV:mc15_13TeV.370465.MGPy8EG_A14N_GG_ttn1OnShellt1_2000_1820_60.merge.DAOD_SUSY10.e4051_s2608_r7772_r7676_p2666
mc15_13TeV:mc15_13TeV.370500.MGPy8EG_A14N_GG_ttn1OnShellt1_900_450_250.merge.DAOD_SUSY10.e4051_s2608_r7772_r7676_p2666
mc15_13TeV:mc15_13TeV.370501.MGPy8EG_A14N_GG_ttn1OnShellt1_900_650_250.merge.DAOD_SUSY10.e4051_s2608_r7772_r7676_p2666
mc15_13TeV:mc15_13TeV.370502.MGPy8EG_A14N_GG_ttn1OnShellt1_900_720_250.merge.DAOD_SUSY10.e4051_s2608_r7772_r7676_p2666
mc15_13TeV:mc15_13TeV.370503.MGPy8EG_A14N_GG_ttn1OnShellt1_1000_450_250.merge.DAOD_SUSY10.e4051_s2608_r7772_r7676_p2666
mc15_13TeV:mc15_13TeV.370504.MGPy8EG_A14N_GG_ttn1OnShellt1_1000_650_250.merge.DAOD_SUSY10.e4051_s2608_r7772_r7676_p2666
mc15_13TeV:mc15_13TeV.370505.MGPy8EG_A14N_GG_ttn1OnShellt1_1000_820_250.merge.DAOD_SUSY10.e4051_s2608_r7772_r7676_p2666
mc15_13TeV:mc15_13TeV.370506.MGPy8EG_A14N_GG_ttn1OnShellt1_1100_450_250.merge.DAOD_SUSY10.e4051_s2608_r7772_r7676_p2666
mc15_13TeV:mc15_13TeV.370507.MGPy8EG_A14N_GG_ttn1OnShellt1_1100_650_250.merge.DAOD_SUSY10.e4051_s2608_r7772_r7676_p2666
mc15_13TeV:mc15_13TeV.370508.MGPy8EG_A14N_GG_ttn1OnShellt1_1100_850_250.merge.DAOD_SUSY10.e4051_s2608_r7772_r7676_p2666
mc15_13TeV:mc15_13TeV.370509.MGPy8EG_A14N_GG_ttn1OnShellt1_1100_920_250.merge.DAOD_SUSY10.e4051_s2608_r7772_r7676_p2666
mc15_13TeV:mc15_13TeV.370510.MGPy8EG_A14N_GG_ttn1OnShellt1_1200_450_250.merge.DAOD_SUSY10.e4051_s2608_r7772_r7676_p2666
mc15_13TeV:mc15_13TeV.370511.MGPy8EG_A14N_GG_ttn1OnShellt1_1200_650_250.merge.DAOD_SUSY10.e4051_s2608_r7772_r7676_p2666
mc15_13TeV:mc15_13TeV.370512.MGPy8EG_A14N_GG_ttn1OnShellt1_1200_850_250.merge.DAOD_SUSY10.e4051_s2608_r7772_r7676_p2666
mc15_13TeV:mc15_13TeV.370513.MGPy8EG_A14N_GG_ttn1OnShellt1_1200_1020_250.merge.DAOD_SUSY10.e4051_s2608_r7772_r7676_p2666
mc15_13TeV:mc15_13TeV.370514.MGPy8EG_A14N_GG_ttn1OnShellt1_1300_450_250.merge.DAOD_SUSY10.e4051_s2608_r7772_r7676_p2666
mc15_13TeV:mc15_13TeV.370515.MGPy8EG_A14N_GG_ttn1OnShellt1_1300_650_250.merge.DAOD_SUSY10.e4051_s2608_r7772_r7676_p2666
mc15_13TeV:mc15_13TeV.370516.MGPy8EG_A14N_GG_ttn1OnShellt1_1300_850_250.merge.DAOD_SUSY10.e4051_s2608_r7772_r7676_p2666
mc15_13TeV:mc15_13TeV.370517.MGPy8EG_A14N_GG_ttn1OnShellt1_1300_1050_250.merge.DAOD_SUSY10.e4051_s2608_r7772_r7676_p2666
mc15_13TeV:mc15_13TeV.370518.MGPy8EG_A14N_GG_ttn1OnShellt1_1300_1120_250.merge.DAOD_SUSY10.e4051_s2608_r7772_r7676_p2666
mc15_13TeV:mc15_13TeV.370519.MGPy8EG_A14N_GG_ttn1OnShellt1_1400_450_250.merge.DAOD_SUSY10.e4051_s2608_r7772_r7676_p2666
mc15_13TeV:mc15_13TeV.370520.MGPy8EG_A14N_GG_ttn1OnShellt1_1400_650_250.merge.DAOD_SUSY10.e4051_s2608_r7772_r7676_p2666
mc15_13TeV:mc15_13TeV.370521.MGPy8EG_A14N_GG_ttn1OnShellt1_1400_850_250.merge.DAOD_SUSY10.e4051_s2608_r7772_r7676_p2666
mc15_13TeV:mc15_13TeV.370522.MGPy8EG_A14N_GG_ttn1OnShellt1_1400_1050_250.merge.DAOD_SUSY10.e4051_s2608_r7772_r7676_p2666
mc15_13TeV:mc15_13TeV.370523.MGPy8EG_A14N_GG_ttn1OnShellt1_1400_1220_250.merge.DAOD_SUSY10.e4051_s2608_r7772_r7676_p2666
mc15_13TeV:mc15_13TeV.370524.MGPy8EG_A14N_GG_ttn1OnShellt1_1500_450_250.merge.DAOD_SUSY10.e4051_s2608_r7772_r7676_p2666
mc15_13TeV:mc15_13TeV.370525.MGPy8EG_A14N_GG_ttn1OnShellt1_1500_650_250.merge.DAOD_SUSY10.e4051_s2608_r7772_r7676_p2666
mc15_13TeV:mc15_13TeV.370526.MGPy8EG_A14N_GG_ttn1OnShellt1_1500_850_250.merge.DAOD_SUSY10.e4051_s2608_r7772_r7676_p2666
```

```
mc15_13TeV:mc15_13TeV.370527.MGPy8EG_A14N_GG_ttn1OnShellt1_1500_1050_250.merge.DAOD_SUSY10.e4051_s2608_r7772_r7676_p2666
mc15_13TeV:mc15_13TeV.370528.MGPy8EG_A14N_GG_ttn1OnShellt1_1500_1250_250.merge.DAOD_SUSY10.e4051_s2608_r7772_r7676_p2666
mc15_13TeV:mc15_13TeV.370529.MGPy8EG_A14N_GG_ttn1OnShellt1_1500_1320_250.merge.DAOD_SUSY10.e4051_s2608_r7772_r7676_p2666
mc15_13TeV:mc15_13TeV.370530.MGPy8EG_A14N_GG_ttn1OnShellt1_1600_450_250.merge.DAOD_SUSY10.e4051_s2608_r7772_r7676_p2666
mc15_13TeV:mc15_13TeV.370531.MGPy8EG_A14N_GG_ttn1OnShellt1_1600_650_250.merge.DAOD_SUSY10.e4051_s2608_r7772_r7676_p2666
mc15_13TeV:mc15_13TeV.370532.MGPy8EG_A14N_GG_ttn1OnShellt1_1600_850_250.merge.DAOD_SUSY10.e4051_s2608_r7772_r7676_p2666
mc15_13TeV:mc15_13TeV.370533.MGPy8EG_A14N_GG_ttn1OnShellt1_1600_1050_250.merge.DAOD_SUSY10.e4051_s2608_r7772_r7676_p2666
mc15_13TeV:mc15_13TeV.370534.MGPy8EG_A14N_GG_ttn1OnShellt1_1600_1250_250.merge.DAOD_SUSY10.e4051_s2608_r7772_r7676_p2666
mc15_13TeV:mc15_13TeV.370535.MGPy8EG_A14N_GG_ttn1OnShellt1_1600_1420_250.merge.DAOD_SUSY10.e4051_s2608_r7772_r7676_p2666
mc15_13TeV:mc15_13TeV.370536.MGPy8EG_A14N_GG_ttn1OnShellt1_1700_450_250.merge.DAOD_SUSY10.e4051_s2608_r7772_r7676_p2666
mc15_13TeV:mc15_13TeV.370537.MGPy8EG_A14N_GG_ttn1OnShellt1_1700_650_250.merge.DAOD_SUSY10.e4051_s2608_r7772_r7676_p2666
mc15_13TeV:mc15_13TeV.370538.MGPy8EG_A14N_GG_ttn1OnShellt1_1700_850_250.merge.DAOD_SUSY10.e4051_s2608_r7772_r7676_p2666
mc15_13TeV:mc15_13TeV.370539.MGPy8EG_A14N_GG_ttn1OnShellt1_1700_1050_250.merge.DAOD_SUSY10.e4051_s2608_r7772_r7676_p2666
mc15_13TeV:mc15_13TeV.370540.MGPy8EG_A14N_GG_ttn1OnShellt1_1700_1250_250.merge.DAOD_SUSY10.e4051_s2608_r7772_r7676_p2666
mc15_13TeV:mc15_13TeV.370541.MGPy8EG_A14N_GG_ttn1OnShellt1_1700_1450_250.merge.DAOD_SUSY10.e4051_s2608_r7772_r7676_p2666
mc15_13TeV:mc15_13TeV.370542.MGPy8EG_A14N_GG_ttn1OnShellt1_1700_1520_250.merge.DAOD_SUSY10.e4051_s2608_r7772_r7676_p2666
mc15_13TeV:mc15_13TeV.370543.MGPy8EG_A14N_GG_ttn1OnShellt1_1800_450_250.merge.DAOD_SUSY10.e4051_s2608_r7772_r7676_p2666
mc15_13TeV:mc15_13TeV.370544.MGPy8EG_A14N_GG_ttn1OnShellt1_1800_650_250.merge.DAOD_SUSY10.e4051_s2608_r7772_r7676_p2666
mc15_13TeV:mc15_13TeV.370545.MGPy8EG_A14N_GG_ttn1OnShellt1_1800_850_250.merge.DAOD_SUSY10.e4051_s2608_r7772_r7676_p2666
mc15_13TeV:mc15_13TeV.370546.MGPy8EG_A14N_GG_ttn1OnShellt1_1800_1050_250.merge.DAOD_SUSY10.e4051_s2608_r7772_r7676_p2666
mc15_13TeV:mc15_13TeV.370547.MGPy8EG_A14N_GG_ttn1OnShellt1_1800_1250_250.merge.DAOD_SUSY10.e4051_s2608_r7772_r7676_p2666
mc15_13TeV:mc15_13TeV.370548.MGPy8EG_A14N_GG_ttn1OnShellt1_1800_1450_250.merge.DAOD_SUSY10.e4051_s2608_r7772_r7676_p2666
mc15_13TeV:mc15_13TeV.370549.MGPy8EG_A14N_GG_ttn1OnShellt1_1800_1620_250.merge.DAOD_SUSY10.e4051_s2608_r7772_r7676_p2666
mc15_13TeV:mc15_13TeV.370550.MGPy8EG_A14N_GG_ttn1OnShellt1_1900_450_250.merge.DAOD_SUSY10.e4051_s2608_r7772_r7676_p2666
mc15_13TeV:mc15_13TeV.370551.MGPy8EG_A14N_GG_ttn1OnShellt1_1900_650_250.merge.DAOD_SUSY10.e4051_s2608_r7772_r7676_p2666
mc15_13TeV:mc15_13TeV.370552.MGPy8EG_A14N_GG_ttn1OnShellt1_1900_850_250.merge.DAOD_SUSY10.e4051_s2608_r7772_r7676_p2666
mc15_13TeV:mc15_13TeV.370553.MGPy8EG_A14N_GG_ttn1OnShellt1_1900_1050_250.merge.DAOD_SUSY10.e4051_s2608_r7772_r7676_p2666
mc15_13TeV:mc15_13TeV.370554.MGPy8EG_A14N_GG_ttn1OnShellt1_1900_1250_250.merge.DAOD_SUSY10.e4051_s2608_r7772_r7676_p2666
mc15_13TeV:mc15_13TeV.370555.MGPy8EG_A14N_GG_ttn1OnShellt1_1900_1450_250.merge.DAOD_SUSY10.e4051_s2608_r7772_r7676_p2666
mc15_13TeV:mc15_13TeV.370556.MGPy8EG_A14N_GG_ttn1OnShellt1_1900_1650_250.merge.DAOD_SUSY10.e4051_s2608_r7772_r7676_p2666
mc15_13TeV:mc15_13TeV.370557.MGPy8EG_A14N_GG_ttn1OnShellt1_1900_1720_250.merge.DAOD_SUSY10.e4051_s2608_r7772_r7676_p2666
mc15_13TeV:mc15_13TeV.370558.MGPy8EG_A14N_GG_ttn1OnShellt1_2000_450_250.merge.DAOD_SUSY10.e4051_s2608_r7772_r7676_p2666
mc15_13TeV:mc15_13TeV.370559.MGPy8EG_A14N_GG_ttn1OnShellt1_2000_650_250.merge.DAOD_SUSY10.e4051_s2608_r7772_r7676_p2666
mc15_13TeV:mc15_13TeV.370560.MGPy8EG_A14N_GG_ttn1OnShellt1_2000_850_250.merge.DAOD_SUSY10.e4051_s2608_r7772_r7676_p2666
mc15_13TeV:mc15_13TeV.370561.MGPy8EG_A14N_GG_ttn1OnShellt1_2000_1050_250.merge.DAOD_SUSY10.e4051_s2608_r7772_r7676_p2666
mc15_13TeV:mc15_13TeV.370562.MGPy8EG_A14N_GG_ttn1OnShellt1_2000_1250_250.merge.DAOD_SUSY10.e4051_s2608_r7772_r7676_p2666
mc15_13TeV:mc15_13TeV.370563.MGPy8EG_A14N_GG_ttn1OnShellt1_2000_1450_250.merge.DAOD_SUSY10.e4051_s2608_r7772_r7676_p2666
mc15_13TeV:mc15_13TeV.370564.MGPy8EG_A14N_GG_ttn1OnShellt1_2000_1650_250.merge.DAOD_SUSY10.e4051_s2608_r7772_r7676_p2666
mc15_13TeV:mc15_13TeV.370565.MGPy8EG_A14N_GG_ttn1OnShellt1_2000_1820_250.merge.DAOD_SUSY10.e4051_s2608_r7772_r7676_p2666
```

F.8 Gbb Signal

```
mc15_13TeV:mc15_13TeV.370250.MGPy8EG_A14N_GG_bbn1_900_5000_1.merge.DAOD_SUSY10.e4050_a766_a821_r7676_p2666
mc15_13TeV:mc15_13TeV.370251.MGPy8EG_A14N_GG_bbn1_900_5000_200.merge.DAOD_SUSY10.e4050_a766_a821_r7676_p2666
mc15_13TeV:mc15_13TeV.370252.MGPy8EG_A14N_GG_bbn1_900_5000_400.merge.DAOD_SUSY10.e4050_a766_a821_r7676_p2666
mc15_13TeV:mc15_13TeV.370253.MGPy8EG_A14N_GG_bbn1_900_5000_600.merge.DAOD_SUSY10.e4050_a766_a821_r7676_p2666
mc15_13TeV:mc15_13TeV.370254.MGPy8EG_A14N_GG_bbn1_900_5000_800.merge.DAOD_SUSY10.e4050_a766_a821_r7676_p2666
mc15_13TeV:mc15_13TeV.370255.MGPy8EG_A14N_GG_bbn1_900_5000_880.merge.DAOD_SUSY10.e4050_a766_a821_r7676_p2666
mc15_13TeV:mc15_13TeV.370256.MGPy8EG_A14N_GG_bbn1_1000_5000_1.merge.DAOD_SUSY10.e4050_a766_a821_r7676_p2666
mc15_13TeV:mc15_13TeV.370257.MGPy8EG_A14N_GG_bbn1_1000_5000_200.merge.DAOD_SUSY10.e4050_a766_a821_r7676_p2666
mc15_13TeV:mc15_13TeV.370258.MGPy8EG_A14N_GG_bbn1_1000_5000_400.merge.DAOD_SUSY10.e4050_a766_a821_r7676_p2666
mc15_13TeV:mc15_13TeV.370259.MGPy8EG_A14N_GG_bbn1_1000_5000_600.merge.DAOD_SUSY10.e4050_a766_a821_r7676_p2666
mc15_13TeV:mc15_13TeV.370260.MGPy8EG_A14N_GG_bbn1_1000_5000_800.merge.DAOD_SUSY10.e4050_a766_a821_r7676_p2666
mc15_13TeV:mc15_13TeV.370261.MGPy8EG_A14N_GG_bbn1_1000_5000_980.merge.DAOD_SUSY10.e4050_a766_a821_r7676_p2666
mc15_13TeV:mc15_13TeV.370262.MGPy8EG_A14N_GG_bbn1_1100_5000_1.merge.DAOD_SUSY10.e4050_a766_a821_r7676_p2666
mc15_13TeV:mc15_13TeV.370263.MGPy8EG_A14N_GG_bbn1_1100_5000_200.merge.DAOD_SUSY10.e4050_a766_a821_r7676_p2666
mc15_13TeV:mc15_13TeV.370264.MGPy8EG_A14N_GG_bbn1_1100_5000_400.merge.DAOD_SUSY10.e4050_a766_a821_r7676_p2666
mc15_13TeV:mc15_13TeV.370265.MGPy8EG_A14N_GG_bbn1_1100_5000_600.merge.DAOD_SUSY10.e4050_a766_a821_r7676_p2666
mc15_13TeV:mc15_13TeV.370266.MGPy8EG_A14N_GG_bbn1_1100_5000_800.merge.DAOD_SUSY10.e4050_a766_a821_r7676_p2666
mc15_13TeV:mc15_13TeV.370267.MGPy8EG_A14N_GG_bbn1_1100_5000_1000.merge.DAOD_SUSY10.e4050_a766_a821_r7676_p2666
mc15_13TeV:mc15_13TeV.370268.MGPy8EG_A14N_GG_bbn1_1100_5000_1080.merge.DAOD_SUSY10.e4050_a766_a821_r7676_p2666
mc15_13TeV:mc15_13TeV.370269.MGPy8EG_A14N_GG_bbn1_1200_5000_1.merge.DAOD_SUSY10.e4050_a766_a821_r7676_p2666
mc15_13TeV:mc15_13TeV.370270.MGPy8EG_A14N_GG_bbn1_1200_5000_200.merge.DAOD_SUSY10.e4050_a766_a821_r7676_p2666
mc15_13TeV:mc15_13TeV.370271.MGPy8EG_A14N_GG_bbn1_1200_5000_400.merge.DAOD_SUSY10.e4050_a766_a821_r7676_p2666
mc15_13TeV:mc15_13TeV.370272.MGPy8EG_A14N_GG_bbn1_1200_5000_600.merge.DAOD_SUSY10.e4050_a766_a821_r7676_p2666
mc15_13TeV:mc15_13TeV.370273.MGPy8EG_A14N_GG_bbn1_1200_5000_800.merge.DAOD_SUSY10.e4050_a766_a821_r7676_p2666
mc15_13TeV:mc15_13TeV.370274.MGPy8EG_A14N_GG_bbn1_1200_5000_1000.merge.DAOD_SUSY10.e4050_a766_a821_r7676_p2666
mc15_13TeV:mc15_13TeV.370275.MGPy8EG_A14N_GG_bbn1_1200_5000_1180.merge.DAOD_SUSY10.e4050_a766_a821_r7676_p2666
mc15_13TeV:mc15_13TeV.370276.MGPy8EG_A14N_GG_bbn1_1300_5000_1.merge.DAOD_SUSY10.e4050_a766_a821_r7676_p2666
mc15_13TeV:mc15_13TeV.370277.MGPy8EG_A14N_GG_bbn1_1300_5000_200.merge.DAOD_SUSY10.e4050_a766_a821_r7676_p2666
mc15_13TeV:mc15_13TeV.370278.MGPy8EG_A14N_GG_bbn1_1300_5000_400.merge.DAOD_SUSY10.e4050_a766_a821_r7676_p2666
mc15_13TeV:mc15_13TeV.370279.MGPy8EG_A14N_GG_bbn1_1300_5000_600.merge.DAOD_SUSY10.e4050_a766_a821_r7676_p2666
mc15_13TeV:mc15_13TeV.370280.MGPy8EG_A14N_GG_bbn1_1300_5000_800.merge.DAOD_SUSY10.e4050_a766_a821_r7676_p2666
mc15_13TeV:mc15_13TeV.370281.MGPy8EG_A14N_GG_bbn1_1300_5000_1000.merge.DAOD_SUSY10.e4050_a766_a821_r7676_p2666
mc15_13TeV:mc15_13TeV.370282.MGPy8EG_A14N_GG_bbn1_1300_5000_1200.merge.DAOD_SUSY10.e4050_a766_a821_r7676_p2666
mc15_13TeV:mc15_13TeV.370283.MGPy8EG_A14N_GG_bbn1_1300_5000_1280.merge.DAOD_SUSY10.e4050_a766_a821_r7676_p2666
mc15_13TeV:mc15_13TeV.370284.MGPy8EG_A14N_GG_bbn1_1400_5000_1.merge.DAOD_SUSY10.e4050_a766_a821_r7676_p2666
mc15_13TeV:mc15_13TeV.370285.MGPy8EG_A14N_GG_bbn1_1400_5000_200.merge.DAOD_SUSY10.e4050_a766_a821_r7676_p2666
mc15_13TeV:mc15_13TeV.370286.MGPy8EG_A14N_GG_bbn1_1400_5000_400.merge.DAOD_SUSY10.e4050_a766_a821_r7676_p2666
mc15_13TeV:mc15_13TeV.370287.MGPy8EG_A14N_GG_bbn1_1400_5000_600.merge.DAOD_SUSY10.e4050_a766_a821_r7676_p2666
```

```
mc15_13TeV:mc15_13TeV.370288.MGPy8EG_A14N_GG_bbn1_1400_5000_800.merge.DAOD_SUSY10.e4050_a766_a821_r7676_p2666
mc15_13TeV:mc15_13TeV.370289.MGPy8EG_A14N_GG_bbn1_1400_5000_1000.merge.DAOD_SUSY10.e4050_a766_a821_r7676_p2666
mc15_13TeV:mc15_13TeV.370290.MGPy8EG_A14N_GG_bbn1_1400_5000_1200.merge.DAOD_SUSY10.e4050_a766_a821_r7676_p2666
mc15_13TeV:mc15_13TeV.370291.MGPy8EG_A14N_GG_bbn1_1400_5000_1380.merge.DAOD_SUSY10.e4050_a766_a821_r7676_p2666
mc15_13TeV:mc15_13TeV.370292.MGPy8EG_A14N_GG_bbn1_1500_5000_1.merge.DAOD_SUSY10.e4050_a766_a821_r7676_p2666
mc15_13TeV:mc15_13TeV.370293.MGPy8EG_A14N_GG_bbn1_1500_5000_200.merge.DAOD_SUSY10.e4050_a766_a821_r7676_p2666
mc15_13TeV:mc15_13TeV.370294.MGPy8EG_A14N_GG_bbn1_1500_5000_400.merge.DAOD_SUSY10.e4050_a766_a821_r7676_p2666
mc15_13TeV:mc15_13TeV.370295.MGPy8EG_A14N_GG_bbn1_1500_5000_600.merge.DAOD_SUSY10.e4050_a766_a821_r7676_p2666
mc15_13TeV:mc15_13TeV.370296.MGPy8EG_A14N_GG_bbn1_1500_5000_800.merge.DAOD_SUSY10.e4050_a766_a821_r7676_p2666
mc15_13TeV:mc15_13TeV.370297.MGPy8EG_A14N_GG_bbn1_1500_5000_1000.merge.DAOD_SUSY10.e4050_a766_a821_r7676_p2666
mc15_13TeV:mc15_13TeV.370298.MGPy8EG_A14N_GG_bbn1_1500_5000_1200.merge.DAOD_SUSY10.e4050_a766_a821_r7676_p2666
mc15_13TeV:mc15_13TeV.370299.MGPy8EG_A14N_GG_bbn1_1500_5000_1400.merge.DAOD_SUSY10.e4050_a766_a821_r7676_p2666
mc15_13TeV:mc15_13TeV.370300.MGPy8EG_A14N_GG_bbn1_1500_5000_1480.merge.DAOD_SUSY10.e4050_a766_a821_r7676_p2666
mc15_13TeV:mc15_13TeV.370301.MGPy8EG_A14N_GG_bbn1_1600_5000_1.merge.DAOD_SUSY10.e4050_a766_a821_r7676_p2666
mc15_13TeV:mc15_13TeV.370302.MGPy8EG_A14N_GG_bbn1_1600_5000_200.merge.DAOD_SUSY10.e4050_a766_a821_r7676_p2666
mc15_13TeV:mc15_13TeV.370303.MGPy8EG_A14N_GG_bbn1_1600_5000_400.merge.DAOD_SUSY10.e4050_a766_a821_r7676_p2666
mc15_13TeV:mc15_13TeV.370304.MGPy8EG_A14N_GG_bbn1_1600_5000_600.merge.DAOD_SUSY10.e4050_a766_a821_r7676_p2666
mc15_13TeV:mc15_13TeV.370305.MGPy8EG_A14N_GG_bbn1_1600_5000_800.merge.DAOD_SUSY10.e4050_a766_a821_r7676_p2666
mc15_13TeV:mc15_13TeV.370306.MGPy8EG_A14N_GG_bbn1_1600_5000_1000.merge.DAOD_SUSY10.e4050_a766_a821_r7676_p2666
mc15_13TeV:mc15_13TeV.370307.MGPy8EG_A14N_GG_bbn1_1600_5000_1200.merge.DAOD_SUSY10.e4050_a766_a821_r7676_p2666
mc15_13TeV:mc15_13TeV.370308.MGPy8EG_A14N_GG_bbn1_1600_5000_1400.merge.DAOD_SUSY10.e4050_a766_a821_r7676_p2666
mc15_13TeV:mc15_13TeV.370309.MGPy8EG_A14N_GG_bbn1_1600_5000_1580.merge.DAOD_SUSY10.e4050_a766_a821_r7676_p2666
mc15_13TeV:mc15_13TeV.370310.MGPy8EG_A14N_GG_bbn1_1700_5000_1.merge.DAOD_SUSY10.e4050_a766_a821_r7676_p2666
mc15_13TeV:mc15_13TeV.370311.MGPy8EG_A14N_GG_bbn1_1700_5000_200.merge.DAOD_SUSY10.e4050_a766_a821_r7676_p2666
mc15_13TeV:mc15_13TeV.370312.MGPy8EG_A14N_GG_bbn1_1700_5000_400.merge.DAOD_SUSY10.e4050_a766_a821_r7676_p2666
mc15_13TeV:mc15_13TeV.370313.MGPy8EG_A14N_GG_bbn1_1700_5000_600.merge.DAOD_SUSY10.e4050_a766_a821_r7676_p2666
mc15_13TeV:mc15_13TeV.370314.MGPy8EG_A14N_GG_bbn1_1700_5000_800.merge.DAOD_SUSY10.e4050_a766_a821_r7676_p2666
mc15_13TeV:mc15_13TeV.370315.MGPy8EG_A14N_GG_bbn1_1700_5000_1000.merge.DAOD_SUSY10.e4050_a766_a821_r7676_p2666
mc15_13TeV:mc15_13TeV.370316.MGPy8EG_A14N_GG_bbn1_1700_5000_1200.merge.DAOD_SUSY10.e4050_a766_a821_r7676_p2666
mc15_13TeV:mc15_13TeV.370317.MGPy8EG_A14N_GG_bbn1_1700_5000_1400.merge.DAOD_SUSY10.e4050_a766_a821_r7676_p2666
mc15_13TeV:mc15_13TeV.370318.MGPy8EG_A14N_GG_bbn1_1700_5000_1600.merge.DAOD_SUSY10.e4050_a766_a821_r7676_p2666
mc15_13TeV:mc15_13TeV.370319.MGPy8EG_A14N_GG_bbn1_1700_5000_1680.merge.DAOD_SUSY10.e4050_a766_a821_r7676_p2666
mc15_13TeV:mc15_13TeV.370320.MGPy8EG_A14N_GG_bbn1_1800_5000_1.merge.DAOD_SUSY10.e4050_a766_a821_r7676_p2666
mc15_13TeV:mc15_13TeV.370321.MGPy8EG_A14N_GG_bbn1_1800_5000_200.merge.DAOD_SUSY10.e4050_a766_a821_r7676_p2666
mc15_13TeV:mc15_13TeV.370322.MGPy8EG_A14N_GG_bbn1_1800_5000_400.merge.DAOD_SUSY10.e4050_a766_a821_r7676_p2666
mc15_13TeV:mc15_13TeV.370323.MGPy8EG_A14N_GG_bbn1_1800_5000_600.merge.DAOD_SUSY10.e4050_a766_a821_r7676_p2666
mc15_13TeV:mc15_13TeV.370324.MGPy8EG_A14N_GG_bbn1_1800_5000_800.merge.DAOD_SUSY10.e4050_a766_a821_r7676_p2666
mc15_13TeV:mc15_13TeV.370325.MGPy8EG_A14N_GG_bbn1_1800_5000_1000.merge.DAOD_SUSY10.e4050_a766_a821_r7676_p2666
mc15_13TeV:mc15_13TeV.370326.MGPy8EG_A14N_GG_bbn1_1800_5000_1200.merge.DAOD_SUSY10.e4050_a766_a821_r7676_p2666
mc15_13TeV:mc15_13TeV.370327.MGPy8EG_A14N_GG_bbn1_1800_5000_1400.merge.DAOD_SUSY10.e4050_a766_a821_r7676_p2666
mc15_13TeV:mc15_13TeV.370328.MGPy8EG_A14N_GG_bbn1_1800_5000_1600.merge.DAOD_SUSY10.e4050_a766_a821_r7676_p2666
```

```
mc15_13TeV:mc15_13TeV.370329.MGPy8EG_A14N_GG_bbn1_1800_5000_1780.merge.DAOD_SUSY10.e4050_a766_a821_r7676_p2666
mc15_13TeV:mc15_13TeV.370330.MGPy8EG_A14N_GG_bbn1_1900_5000_1.merge.DAOD_SUSY10.e4050_a766_a821_r7676_p2666
mc15_13TeV:mc15_13TeV.370331.MGPy8EG_A14N_GG_bbn1_1900_5000_200.merge.DAOD_SUSY10.e4050_a766_a821_r7676_p2666
mc15_13TeV:mc15_13TeV.370332.MGPy8EG_A14N_GG_bbn1_1900_5000_400.merge.DAOD_SUSY10.e4050_a766_a821_r7676_p2666
mc15_13TeV:mc15_13TeV.370333.MGPy8EG_A14N_GG_bbn1_1900_5000_600.merge.DAOD_SUSY10.e4050_a766_a821_r7676_p2666
mc15_13TeV:mc15_13TeV.370334.MGPy8EG_A14N_GG_bbn1_1900_5000_800.merge.DAOD_SUSY10.e4050_a766_a821_r7676_p2666
mc15_13TeV:mc15_13TeV.370335.MGPy8EG_A14N_GG_bbn1_1900_5000_1000.merge.DAOD_SUSY10.e4050_a766_a821_r7676_p2666
mc15_13TeV:mc15_13TeV.370336.MGPy8EG_A14N_GG_bbn1_1900_5000_1200.merge.DAOD_SUSY10.e4050_a766_a821_r7676_p2666
mc15_13TeV:mc15_13TeV.370337.MGPy8EG_A14N_GG_bbn1_1900_5000_1400.merge.DAOD_SUSY10.e4050_a766_a821_r7676_p2666
mc15_13TeV:mc15_13TeV.370338.MGPy8EG_A14N_GG_bbn1_1900_5000_1600.merge.DAOD_SUSY10.e4050_a766_a821_r7676_p2666
mc15_13TeV:mc15_13TeV.370339.MGPy8EG_A14N_GG_bbn1_1900_5000_1800.merge.DAOD_SUSY10.e4050_a766_a821_r7676_p2666
mc15_13TeV:mc15_13TeV.370340.MGPy8EG_A14N_GG_bbn1_1900_5000_1880.merge.DAOD_SUSY10.e4050_a766_a821_r7676_p2666
mc15_13TeV:mc15_13TeV.370341.MGPy8EG_A14N_GG_bbn1_2000_5000_1.merge.DAOD_SUSY10.e4050_a766_a821_r7676_p2666
mc15_13TeV:mc15_13TeV.370342.MGPy8EG_A14N_GG_bbn1_2000_5000_200.merge.DAOD_SUSY10.e4050_a766_a821_r7676_p2666
mc15_13TeV:mc15_13TeV.370343.MGPy8EG_A14N_GG_bbn1_2000_5000_400.merge.DAOD_SUSY10.e4050_a766_a821_r7676_p2666
mc15_13TeV:mc15_13TeV.370344.MGPy8EG_A14N_GG_bbn1_2000_5000_600.merge.DAOD_SUSY10.e4050_a766_a821_r7676_p2666
mc15_13TeV:mc15_13TeV.370345.MGPy8EG_A14N_GG_bbn1_2000_5000_800.merge.DAOD_SUSY10.e4050_a766_a821_r7676_p2666
mc15_13TeV:mc15_13TeV.370346.MGPy8EG_A14N_GG_bbn1_2000_5000_1000.merge.DAOD_SUSY10.e4050_a766_a821_r7676_p2666
mc15_13TeV:mc15_13TeV.370347.MGPy8EG_A14N_GG_bbn1_2000_5000_1200.merge.DAOD_SUSY10.e4050_a766_a821_r7676_p2666
mc15_13TeV:mc15_13TeV.370348.MGPy8EG_A14N_GG_bbn1_2000_5000_1400.merge.DAOD_SUSY10.e4050_a766_a821_r7676_p2666
mc15_13TeV:mc15_13TeV.370349.MGPy8EG_A14N_GG_bbn1_2000_5000_1600.merge.DAOD_SUSY10.e4050_a766_a821_r7676_p2666
mc15_13TeV:mc15_13TeV.370350.MGPy8EG_A14N_GG_bbn1_2000_5000_1800.merge.DAOD_SUSY10.e4050_a766_a821_r7676_p2666
mc15_13TeV:mc15_13TeV.370351.MGPy8EG_A14N_GG_bbn1_2000_5000_1980.merge.DAOD_SUSY10.e4050_a766_a821_r7676_p2666
mc15_13TeV:mc15_13TeV.370352.MGPy8EG_A14N_GG_bbn1_2100_5000_1.merge.DAOD_SUSY10.e5087_a766_a813_r7648_p2666
mc15_13TeV:mc15_13TeV.370353.MGPy8EG_A14N_GG_bbn1_2100_5000_200.merge.DAOD_SUSY10.e5087_a766_a813_r7648_p2666
mc15_13TeV:mc15_13TeV.370354.MGPy8EG_A14N_GG_bbn1_2100_5000_400.merge.DAOD_SUSY10.e5087_a766_a813_r7648_p2666
mc15_13TeV:mc15_13TeV.370355.MGPy8EG_A14N_GG_bbn1_2100_5000_600.merge.DAOD_SUSY10.e5087_a766_a813_r7648_p2666
mc15_13TeV:mc15_13TeV.370356.MGPy8EG_A14N_GG_bbn1_2100_5000_800.merge.DAOD_SUSY10.e5087_a766_a813_r7648_p2666
mc15_13TeV:mc15_13TeV.370357.MGPy8EG_A14N_GG_bbn1_2100_5000_1000.merge.DAOD_SUSY10.e5087_a766_a813_r7648_p2666
mc15_13TeV:mc15_13TeV.370358.MGPy8EG_A14N_GG_bbn1_2100_5000_1200.merge.DAOD_SUSY10.e5087_a766_a813_r7648_p2666
mc15_13TeV:mc15_13TeV.370359.MGPy8EG_A14N_GG_bbn1_2100_5000_1400.merge.DAOD_SUSY10.e5087_a766_a813_r7648_p2666
mc15_13TeV:mc15_13TeV.370360.MGPy8EG_A14N_GG_bbn1_2100_5000_1600.merge.DAOD_SUSY10.e5087_a766_a813_r7648_p2666
mc15_13TeV:mc15_13TeV.370361.MGPy8EG_A14N_GG_bbn1_2100_5000_2000.merge.DAOD_SUSY10.e5087_a766_a813_r7648_p2666
mc15_13TeV:mc15_13TeV.370362.MGPy8EG_A14N_GG_bbn1_2200_5000_1.merge.DAOD_SUSY10.e5087_a766_a813_r7648_p2666
mc15_13TeV:mc15_13TeV.370364.MGPy8EG_A14N_GG_bbn1_2200_5000_400.merge.DAOD_SUSY10.e5087_a766_a813_r7648_p2666
mc15_13TeV:mc15_13TeV.370365.MGPy8EG_A14N_GG_bbn1_2200_5000_600.merge.DAOD_SUSY10.e5087_a766_a813_r7648_p2666
mc15_13TeV:mc15_13TeV.370366.MGPy8EG_A14N_GG_bbn1_2200_5000_800.merge.DAOD_SUSY10.e5087_a766_a813_r7648_p2666
mc15_13TeV:mc15_13TeV.370368.MGPy8EG_A14N_GG_bbn1_2200_5000_1200.merge.DAOD_SUSY10.e5087_a766_a813_r7648_p2666
mc15_13TeV:mc15_13TeV.370370.MGPy8EG_A14N_GG_bbn1_2200_5000_1600.merge.DAOD_SUSY10.e5087_a766_a813_r7648_p2666
mc15_13TeV:mc15_13TeV.370371.MGPy8EG_A14N_GG_bbn1_2200_5000_2000.merge.DAOD_SUSY10.e5087_a766_a813_r7648_p2666
mc15_13TeV:mc15_13TeV.370372.MGPy8EG_A14N_GG_bbn1_2300_5000_1.merge.DAOD_SUSY10.e5087_a766_a813_r7648_p2666
```

```
mc15_13TeV:mc15_13TeV.370373.MGPy8EG_A14N_GG_bbn1_2300_5000_200.merge.DAOD_SUSY10.e5087_a766_a813_r7648_p2666
mc15_13TeV:mc15_13TeV.370374.MGPy8EG_A14N_GG_bbn1_2300_5000_400.merge.DAOD_SUSY10.e5087_a766_a813_r7648_p2666
mc15_13TeV:mc15_13TeV.370375.MGPy8EG_A14N_GG_bbn1_2300_5000_600.merge.DAOD_SUSY10.e5087_a766_a813_r7648_p2666
mc15_13TeV:mc15_13TeV.370376.MGPy8EG_A14N_GG_bbn1_2300_5000_800.merge.DAOD_SUSY10.e5087_a766_a813_r7648_p2666
mc15_13TeV:mc15_13TeV.370377.MGPy8EG_A14N_GG_bbn1_2300_5000_1000.merge.DAOD_SUSY10.e5087_a766_a813_r7648_p2666
mc15_13TeV:mc15_13TeV.370378.MGPy8EG_A14N_GG_bbn1_2300_5000_1200.merge.DAOD_SUSY10.e5087_a766_a813_r7648_p2666
mc15_13TeV:mc15_13TeV.370379.MGPy8EG_A14N_GG_bbn1_2300_5000_1400.merge.DAOD_SUSY10.e5087_a766_a813_r7648_p2666
mc15_13TeV:mc15_13TeV.370380.MGPy8EG_A14N_GG_bbn1_2300_5000_1600.merge.DAOD_SUSY10.e5087_a766_a813_r7648_p2666
mc15_13TeV:mc15_13TeV.370381.MGPy8EG_A14N_GG_bbn1_2300_5000_2200.merge.DAOD_SUSY10.e5087_a766_a813_r7648_p2666
mc15_13TeV:mc15_13TeV.370382.MGPy8EG_A14N_GG_bbn1_2400_5000_1.merge.DAOD_SUSY10.e5087_a766_a813_r7648_p2666
mc15_13TeV:mc15_13TeV.370383.MGPy8EG_A14N_GG_bbn1_2400_5000_200.merge.DAOD_SUSY10.e5087_a766_a813_r7648_p2666
mc15_13TeV:mc15_13TeV.370384.MGPy8EG_A14N_GG_bbn1_2400_5000_400.merge.DAOD_SUSY10.e5087_a766_a813_r7648_p2666
mc15_13TeV:mc15_13TeV.370385.MGPy8EG_A14N_GG_bbn1_2400_5000_600.merge.DAOD_SUSY10.e5087_a766_a813_r7648_p2666
mc15_13TeV:mc15_13TeV.370386.MGPy8EG_A14N_GG_bbn1_2400_5000_800.merge.DAOD_SUSY10.e5087_a766_a813_r7648_p2666
mc15_13TeV:mc15_13TeV.370387.MGPy8EG_A14N_GG_bbn1_2400_5000_1000.merge.DAOD_SUSY10.e5087_a766_a813_r7648_p2666
mc15_13TeV:mc15_13TeV.370388.MGPy8EG_A14N_GG_bbn1_2400_5000_1200.merge.DAOD_SUSY10.e5087_a766_a813_r7648_p2666
mc15_13TeV:mc15_13TeV.370389.MGPy8EG_A14N_GG_bbn1_2400_5000_1400.merge.DAOD_SUSY10.e5087_a766_a813_r7648_p2666
mc15_13TeV:mc15_13TeV.370390.MGPy8EG_A14N_GG_bbn1_2400_5000_1600.merge.DAOD_SUSY10.e5087_a766_a813_r7648_p2666
mc15_13TeV:mc15_13TeV.370391.MGPy8EG_A14N_GG_bbn1_2400_5000_2200.merge.DAOD_SUSY10.e5087_a766_a813_r7648_p2666
```

Appendix G
Model-Dependent Limits by Region

This appendix presents the separate model-dependent limits as described in Sect. 8.3 for the six Gtt regions (Fig. G.1).

© Springer Nature Switzerland AG 2020

G. Stark, *The Search for Supersymmetry in Hadronic Final States Using Boosted Object Reconstruction*, Springer Theses,
https://doi.org/10.1007/978-3-030-34548-8

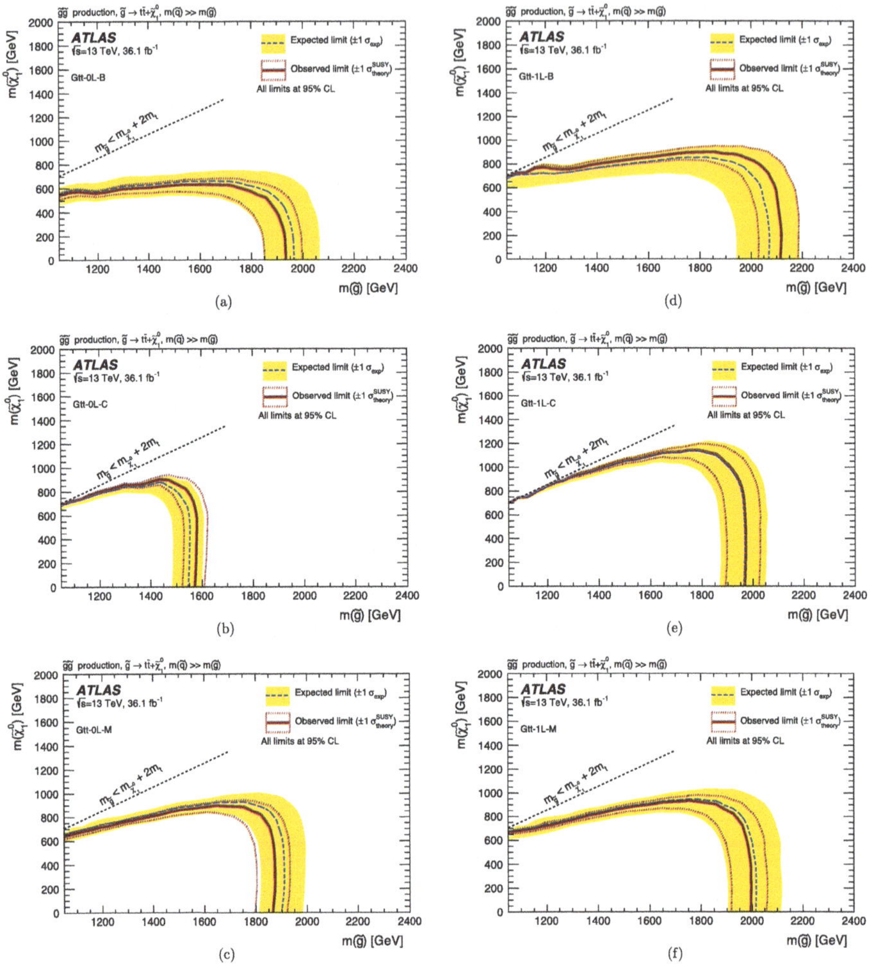

Fig. G.1 Exclusion limits in the $\tilde{\chi}_1^0$ and $\tilde{g}$ mass plane for the Gtt model for each signal region obtained in the context of the cut-and-count analysis. The dashed and solid bold lines show the 95% CL expected and observed limits, respectively. The shaded bands around the expected limits show the impact of the experimental and background uncertainties. The dotted lines show the impact on the observed limit of the variation of the nominal signal cross-section by $\pm 1\sigma$ of its theoretical uncertainty. The 95% CL expected and observed limits from the ATLAS search based on 2015 data [70] are also shown. (**a**) Gtt 0L Boosted. (**b**) Gtt 0L Moderate. (**c**) Gtt 0L Compressed. (**d**) Gtt 1L Boosted. (**e**) Gtt 1L Moderate. (**f**) Gtt 1L Compressed

Appendix H
HEPData Plots

This appendix serves to show all the plots for HEPData [205] which are provided as a reference for theorists and analyzers who want to reinterpret this thesis analysis. See Sect. 8.4 for more details (Figs. H.1, H.2 and H.3).

© Springer Nature Switzerland AG 2020

G. Stark, *The Search for Supersymmetry in Hadronic Final States Using Boosted Object Reconstruction*, Springer Theses,
https://doi.org/10.1007/978-3-030-34548-8

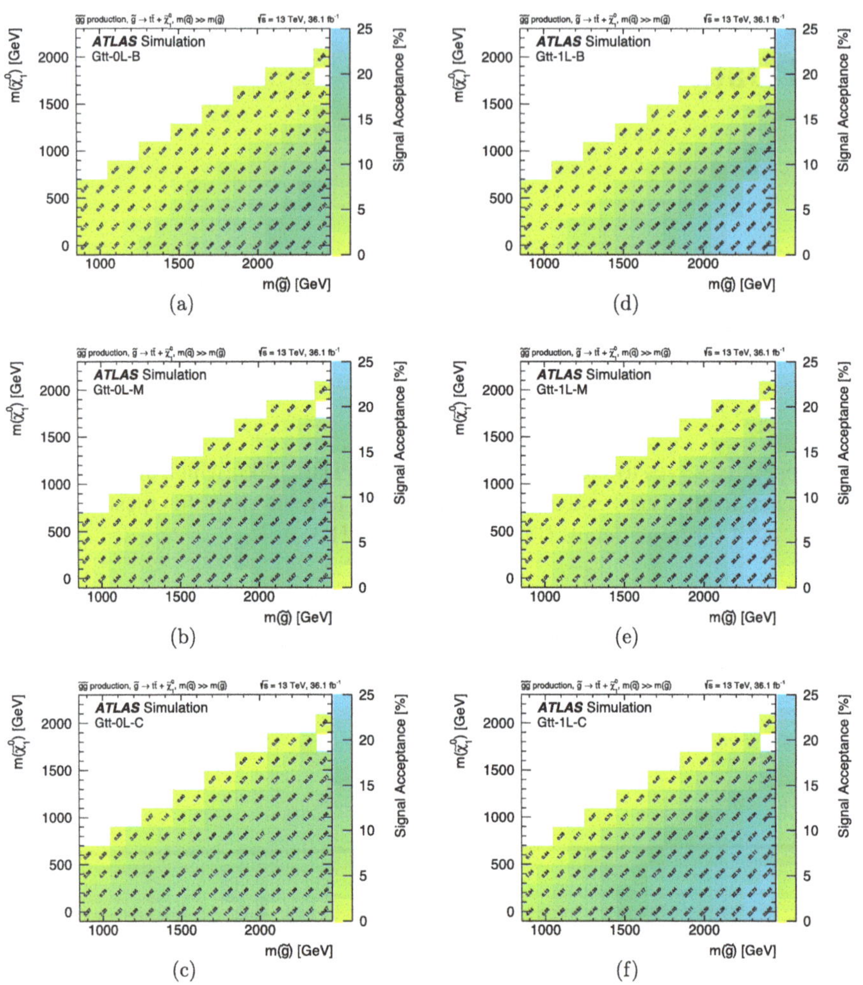

Fig. H.1 Signal acceptances (truth level) for the 0-lepton and 1-lepton signal regions. (**a**) 0-lepton Boosted. (**b**) 0-lepton Moderate. (**c**) 0-lepton Compressed. (**d**) 1-lepton Boosted. (**e**) 1-lepton Moderate. (**f**) 1-lepton Compressed

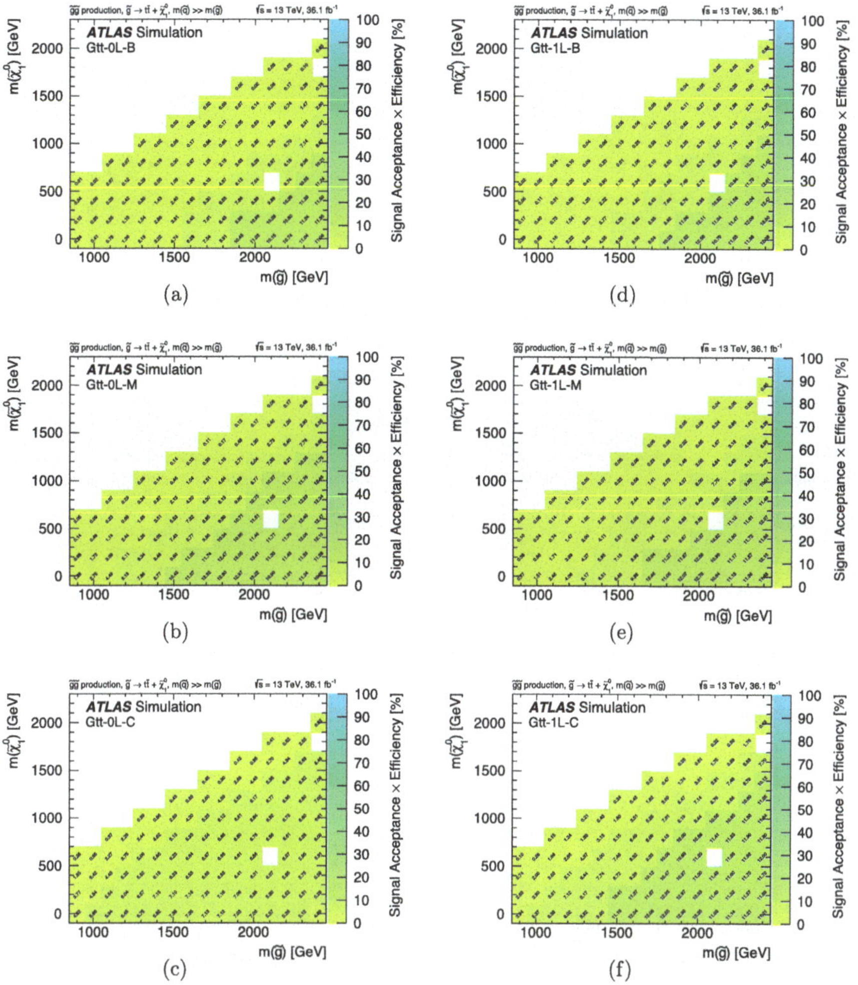

Fig. H.2 Signal acceptances ⊗ efficiencies (reconstruction level) for the 0-lepton and 1-lepton signal regions. (**a**) 0-lepton Boosted. (**b**) 0-lepton Moderate. (**c**) 0-lepton Compressed. (**d**) 1-lepton Boosted. (**e**) 1-lepton Moderate. (**f**) 1-lepton Compressed

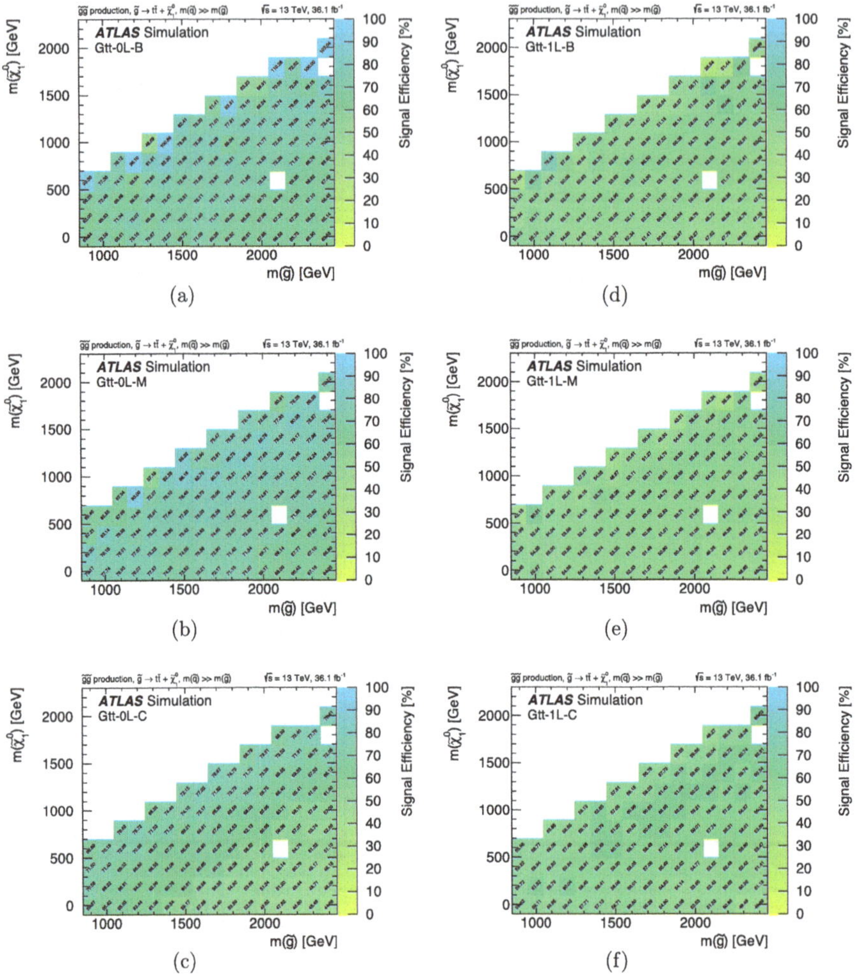

Fig. H.3 Signal efficiencies for the 0-lepton and 1-lepton signal regions. (**a**) 0-lepton Boosted. (**b**) 0-lepton Moderate. (**c**) 0-lepton Compressed. (**d**) 1-lepton Boosted. (**e**) 1-lepton Moderate. (**f**) 1-lepton Compressed

Glossary

topo-cluster topological energy cluster. 100, 101, 104

ATLAS a general-purpose detector at the LHC. xv, xviii, xx, xxi, 3, 4, 15, 27, 35, 36, 45, 63, 65, 67, 68, 71–73, 75, 77, 84, 85, 87, 90, 91, 109, 112, 113, 120, 140, 158, 188

BC Bunch Crossing. 60, 63, 68, 114
BSM Beyond the Standard Model. 20, 67, 93
bunch train a group of bunches. 49

CERN European Organization for Nuclear Research. 27, 49
CL Confidence Level. xxix, xxxviii, 149, 152–154, 236
CMX Common Merger Extended Module. 51
CP Cluster Processor. 49
CPM Cluster Processor Module. 51, 57
CR control region. xxv, xxvi, xxxix, xl, 119, 120, 123–126, 128, 129, 132–138, 140, 142
CSC Cathode Strip Chambers. 45, 46, 90
CTP Central Trigger Processor. 48, 49, 51

eFEX electron Feature EXtractor. xx, 57
EMB LAr electromagnetic barrel. 41, 43, 44
EMCal electromagnetic calorimeter. xxi, 43, 45, 58, 68, 72, 79, 102
EMEC LAr Elecromagnetic End-Cap Calorimeter. 41, 43, 44

© Springer Nature Switzerland AG 2020
G. Stark, *The Search for Supersymmetry in Hadronic Final States Using Boosted Object Reconstruction*, Springer Theses,
https://doi.org/10.1007/978-3-030-34548-8

FCal forward calorimeter. xxi, 43, 44, 58, 72, 73
FELIX Front-End Link EXchange. xx, 57, 58
FEX L1Calo Feature EXtractor. xx, 57, 58, 63
FPGA Field Programmable Array. 49, 52, 57, 58, 63
FSR final-state radiation. 142
FTK Fast TracKer. xix, 48, 52

gBlock Group of contiguous gTowers. Most have a size of 0.6×0.6 in $\Delta\phi \times \Delta\eta$. xx, 60–62, 165
gCaloTower Calorimeter tower transmitted to the gFEX. Most have a size of 0.2×0.2 in $\Delta\phi \times \Delta\eta$. 60
gFEX global Feature EXtractor. x, 57–63, 65, 155, 157–160, 165, 167–173, 177, 178, 181, 182
GRL Good Runs List. 112
GSC Global Sequential Calibration. xxii, 78–80
gTower Tower, formed by summing electromagnetic & hadronic gCaloTowers, as used on the gFEX. Most have a size of 0.2×0.2 in $\Delta\phi \times \Delta\eta$. xx, xxx–xxxv, 60–62, 65, 160, 161, 163–170, 173, 175–178

HCal hadronic calorimeter. 58, 68
HEC LAr Hadronic End-Cap Calorimeter. xxi, 43, 44, 72
HL-LHC High Luminosity LHC. xx, 29, 62
HLT High-Level Trigger. xix, 47, 48, 51–53, 168
Hub Common readout infrastructure for L1Calo. The ROD is a daughter card. xx, 57

I^2C I2C. 63
IBL Insertable B-Layer. 112
ID Inner Detector. xviii, xix, xxi, 34, 38–40, 52, 69, 72, 83, 85, 90–94, 101–105, 114
IP Internet Protocol. See TCP/IP. 187
IPBus IP-based protocol implementing register-level access over Ethernet for module control & monitoring. 187, 188, 192, 193
ironman Transport-neutral, single-threaded Python framework to connect external users to board-specific hardware. 63, 187, 192, 194, 195
ISR initial-state radiation. 142

JEM Jet/Energy Module. 51, 57
JEP Jet/Energy Processor. 49
JER jet energy resolution. 140
JES jet energy scale. xxii, 76, 78, 79, 81, 82, 105, 140
jFEX jet Feature EXtractor. xx, 57
JVT jet-vertex-tagger. 101

L1 Level-1. xix, xxiv, xxix, 47–49, 51–53, 60, 113, 114, 157, 158, 165–167
L1A L1 accept. 48, 49, 52, 57
L1Calo L1 calorimeter trigger. xix, xxxiii, 48–51, 55, 57, 63, 157, 170
L1Muon L1 muon trigger. 48, 49, 51
L1Topo L1 topological processor. xx, 51, 58, 60
LAr Liquid Argon Calorimeter. xix, xxi, 41–43, 49, 55, 57, 69, 72, 73, 92, 112, 113
large-R large-radius. 104, 105, 140, 158, 159
LATOME LAr Trigger prOcessing MEzzanine. Mezzanine card for LAr Carrier Board. Together these form the LDPB. 57, 58
LDPS LAr Digital Processing System. 57
LHC Large Hadron Collider. xvii, xviii, 3, 16, 17, 24, 27, 28, 32, 36, 44, 67, 167, 168, 181, 182, 255
LO leading order. xvi, 14, 54
LSP lightest supersymmetric particle. 21
LTDB LAr Trigger Digitizer Board. 57

MC monte-carlo. xxiv, 54, 107, 112–114, 120, 138, 140
MDT Monitored Drift Tubes. 45, 46, 90
monte-carlo simulated event using random numbers. 54
MPSoC Multi-Processor SoC. 63
MS muon spectrometer. 45, 52, 90, 91
MUCTPI Muon-to-CTP interface. 51

NLL next-to-leading-logarithm. 55, 144
NLO next-to-leading-order. xv, 4, 54, 55, 144
NNLO next-to-next-to-leading-order. xv, 4
NP Nuisance Parameter. 82

PDF parton distribution function. 17, 18, 54, 55
pFPGA Processor FPGA. xx, 58–61
PPM L1Calo PreProcessor Module. xx, 57
PS Proton Synchotron. 28
PSB Proton Synchotron Booster. 28
punch-through For jets at very high transverse momentum it is possible that part of the energy is not deposited in the calorimeter, but leaks out to the detector components beyond the calorimeter. This leads to a systematic reduction in the measured jet energy. Jets that deposit energy beyond the hadronic Tile calorimeter and in the muon system are called punch-through jets. [41]. 41, 75, 101, 115

QCD Quantum Chromodynamics. A theory describing the strong interactions of SM particles. xvi, xxiv, 4, 12–15, 20, 67, 77, 94, 182

QED Quantum Electrodynamics. A theory describing the electromagnetic and weak interactions of SM particles. 4, 8, 9, 12–14, 20

ROD ReadOut Driver. xx, 57

ROI region-of-interest. 48, 51, 58

RPC Resistive Plate Chambers. 45, 46

SCT Semiconductor Tracker. xix, 39, 40, 112, 113

SM Standard Model. 3–5, 8, 9, 15, 20–22, 25, 67, 93, 107, 120, 138, 181

small-R small-radius. 100, 105, 111, 123, 126

SoC System-on-Chip. 58, 63, 188, 190

SPS Super Proton Synchotron. xviii, 28, 36

SR signal region. xxv, xxvi, xxxix, xl, 116, 119, 120, 123–129, 131–138, 140, 197

SSB Spontaneous Symmetry Breaking. 7, 8, 21

super-cell LAr calorimeter region formed by summing transverse energy from cells that are adjacent in η and ϕ. 57

SUSY Supersymmetry. 21, 22, 24, 26, 54, 68, 109, 110, 120, 137, 146, 181, 182

TCP/IP Transmission Control Protocol/Internet Protocol. A conceptual model and set of communications protocols used on the Internet and similar computer networks. 187, 195, 255

TDAQ Trigger and Data Acquisition. xix, 47, 48, 53, 57

TGC Thin Gap Chambers. 45, 46

Tile Tile calorimeter. xix, xxi, 43, 44, 69, 72, 79, 83, 112, 113

TOB Trigger OBject. 51, 58, 60, 61, 65

TRT Transition Radiation Tracker. xix, 39, 40

UDP User Datagram Protocol. Part of the Internet Protocol Suite. It is a very basic transport layer with no handshaking. 187, 195

VR validation region. xxv–xxvii, xxxix, xl, 119, 120, 123–125, 127, 128, 131, 132, 134–138, 140

zFPGA Zynq FPGA. 58, 60

Zynq+® A Xilinx MPSoC composed of an FPGA, ARM processor, real-time processor, and a MALI-400 GPU. xx, 58, 59, 61, 63, 65

Bibliography

1. M. Aaboud et al., Search for squarks and gluinos in events with hadronically decaying tau leptons, jets and missing transverse momentum in proton-proton collisions at $\sqrt{s} = 13\,\text{TeV}$ recorded with the ATLAS detector. Eur. Phys. J. **C76**(12), 683 (2016). https://doi.org/10.1140/epjc/s10052-016-4481-2. arXiv: 1607.05979 [hep-ex]
2. M. Aaboud et al., Search for squarks and gluinos in final states with jets and missing transverse momentum using 36 fb^{-1} of $\sqrt{s} = 13\,\text{TeV}$ pp collision data with the ATLAS detector (2017). arXiv: 1712.02332 [hep-ex]
3. M. Aaboud et al., Search for supersymmetry in final states with two same-sign or three leptons and jets using 36 fb^{-1} of $\sqrt{s} = 13\,\text{TeV}$ pp collision data with the ATLAS detector. J. High Energy Phys. **9**, 84 (2017). https://doi.org/10.1007/JHEP09(2017)084. arXiv: 1706.03731 [hep-ex]
4. M. Aaboud et al., Search for squarks and gluinos in events with an isolated lepton, jets, and missing transverse momentum at $\sqrt{s} = 13\,\text{TeV}$ with the ATLAS detector. Phys. Rev. **D96**(11), 112010 (2017). https://doi.org/10.1103/PhysRevD.96.112010. arXiv: 1708.08232 [hep-ex]
5. M. Aaboud et al., Search for new phenomena in events containing a same-flavour opposite-sign dilepton pair, jets, and large missing transverse momentum in $\sqrt{s} = 13$ pp collisions with the ATLAS detector. Eur. Phys. J. **C77**(3), 144 (2017). https://doi.org/10.1140/epjc/s10052-017-4700-5. arXiv: 1611.05791 [hep-ex]
6. M. Aaboud et al., Search for photonic signatures of gauge-mediated supersymmetry in 13 TeV pp collisions with the ATLAS detector (2018). arXiv: 1802.03158 [hep-ex]
7. G. Aad et al., Observation of a new particle in the search for the Standard Model Higgs boson with the ATLAS detector at the LHC. Phys. Lett. **B716**, 1–29 (2012). https://doi.org/10.1016/j.physletb.2012.08.020. arXiv: 1207.7214 [hep-ex]
8. S. Abachi et al., Observation of the top quark. Phys. Rev. Lett. **74**, 2632–2637 (1995). https://doi.org/10.1103/PhysRevLett.74.2632
9. F. Abe et al., Observation of top quark production in $\bar{p}p$ collisions with the collider detector at Fermilab. Phys. Rev. Lett. **74**, 2626–2631 (1995). https://doi.org/10.1103/PhysRevLett.74.2626
10. P.A.R. Ade et al., Planck 2015 results. XIII. Cosmological parameters. Astron. Astrophys. **594**, A13 (2016). https://doi.org/10.1051/0004-6361/201525830. arXiv: 1502.01589 [astro-ph.CO]
11. S. Agostinelli et al., GEANT4: a simulation toolkit. Nucl. Instrum. Meth. A **506**, 250–303 (2003). https://doi.org/10.1016/S0168-9002(03)01368-8

© Springer Nature Switzerland AG 2020

G. Stark, *The Search for Supersymmetry in Hadronic Final States Using Boosted Object Reconstruction*, Springer Theses,
https://doi.org/10.1007/978-3-030-34548-8

12. M. Aleksa et al., (CERN). ATLAS liquid argon calorimeter phase-I upgrade technical design report. Technical Report CERN-LHCC-2013-017. ATLAS-TDR-022. Final version presented to December 2013 LHCC (2013). https://cds.cern.ch/record/1602230
13. G. Aielli et al., The RPC first level muon trigger in the barrel of the ATLAS experiment. Nucl. Phys. B Proc. Suppl. **158**, 11–15 (2006). ISSN: 0920-5632. https://doi.org/10.1016/j.nuclphysbps.2006.07.031. Proceedings of the 8th International Workshop on Resistive Plate Chambers and Related Detectors
14. S. Alioli et al., A general framework for implementing NLO calculations in shower Monte Carlo programs: the POWHEG BOX. J. High Energy Phys. **6**, 43 (2010). https://doi.org/10.1007/JHEP06(2010)043. arXiv: 1002.2581 [hep-ph]
15. D. Alves, Simplified models for LHC new physics searches. J. Phys. **G39**, 105005 (2012). Ed. by Nima Arkani-Hamed et al. https://doi.org/10.1088/0954-3899/39/10/105005. arXiv: 1105.2838 [hep-ph]
16. J. Alwall, P. Schuster, N. Toro, Simplified models for a first characterization of new physics at the LHC. Phys. Rev. **D79**, 075020 (2009). https://doi.org/10.1103/PhysRevD.79.075020. arXiv: 0810.3921 [hep-ph]
17. J. Alwall et al., The automated computation of tree-level and next-to-leading order differential cross sections, and their matching to parton shower simulations. J. High Energy Phys. **7**, 79 (2014). https://doi.org/10.1007/JHEP07(2014)079. arXiv: 1405.0301 [hep-ph]
18. J.R. Andersen et al., Handbook of LHC higgs cross sections: 3. Higgs properties (2013). https://doi.org/10.5170/CERN-2013-004. arXiv: 1307.1347 [hep-ph]
19. J.T. Anderson et al., FELIX: a high-throughput network approach for interfacing to front end electronics for ATLAS upgrades (2015). Technical Report ATL-DAQ-PROC-2015-014. 8. Geneva: CERN. https://cds.cern.ch/record/2016626
20. J.-F. Arguin et al., Search for gluino-mediated stop and sbottom pair production in events with b-jets and large missing transverse momentum (2015). Technical Report. ATL-COM-PHYS-2015-319. Geneva: CERN. https://cds.cern.ch/record/2011623
21. J.-F. Arguin et al., Search for gluino-mediated stop and sbottom pair production in events with b-jets and large missing transverse momentum (2016). Technical Report. ATL-COM-PHYS-2016-1592. Geneva: CERN. https://cds.cern.ch/record/2231120
22. T. Argyropoulos et al., Cathode strip chambers in ATLAS: installation, commissioning and in situ performance, in *2008 IEEE Nuclear Science Symposium Conference Record* (2008), pp. 2819–2824. https://doi.org/10.1109/NSSMIC.2008.4774958
23. N. Arkani-Hamed, S. Dimopoulos, G.R. Dvali, The hierarchy problem and new dimensions at a millimeter. Phys. Lett. **B429**, 263–272 (1998). https://doi.org/10.1016/S0370-2693(98)00466-3. arXiv: hep-ph/9803315 [hep-ph]
24. P. Artoisenet et al., Automatic spin-entangled decays of heavy resonances in Monte Carlo simulations. J. High Energy Phys. **3**, 15 (2013). https://doi.org/10.1007/JHEP03(2013)015. arXiv: 1212.3460 [hep-ph]
25. A. Arvanitaki et al., The last vestiges of naturalness. J. High Energy Phys. **3**, 22 (2014). https://doi.org/10.1007/JHEP03(2014)022. arXiv: 1309.3568 [hep-ph]
26. ATLAS Collaboration, Optimisation of the ATLAS b-tagging performance for the 2016 LHC run. ATL-PHYS-PUB-2016-012. https://cds.cern.ch/record/2160731
27. ATLAS Collaboration, The silicon microstrip sensors of the ATLAS semiconductor tracker. Nucl. Instrum. Methods Phys. Res., Sect. A **578**(1), 98–118 (2007). ISSN: 0168-9002. https://doi.org/10.1016/j.nima.2007.04.157
28. ATLAS Collaboration, The ATLAS experiment at the CERN large hadron collider. J. Instrum. **3**, S08003 (2008). https://doi.org/10.1088/1748-0221/3/08/S08003
29. ATLAS Collaboration, Drift time measurement in the ATLAS liquid argon electromagnetic calorimeter using cosmic muons. Eur. Phys. J. C **70** (2010), 755. 10.1140/epjc/s10052-010-1403-6. arXiv: 1002.4189 [hep-ex]
30. ATLAS Collaboration, Commissioning of the ATLAS high performance b-tagging algorithms in the 7 TeV collision data (2011). ATLAS-CONF-2011-102. https://cds.cern.ch/record/1369219

31. ATLAS Collaboration, Performance of the ATLAS trigger system in 2010. Eur. Phys. J. C **72**, 1849 (2012). https://doi.org/10.1140/epjc/s10052-011-1849-1. arXiv: 1110.1530 [hep-ex]
32. ATLAS Collaboration, *Performance of the ATLAS Electron and Photon Trigger in pp Collisions at* $\sqrt{s} = 7$ *TeV in 2011* (2012). ATLAS-CONF-2012-048. https://cds.cern.ch/record/1450089
33. ATLAS Collaboration, Summary of ATLAS Pythia 8 tunes (2012). ATL-PHYS-PUB-2012-003. https://cds.cern.ch/record/1474107
34. ATLAS Collaboration, Measurement of the b-tag efficiency in a sample of jets containing muons with 5 fb^{-1} of data from the ATLAS detector (2012). ATLAS-CONF-2012-043. https://cds.cern.ch/record/1435197
35. ATLAS Collaboration, Measurement of the Mistag Rate of b-tagging algorithms with 5 fb^{-1} of Data Collected by the ATLAS Detector (2012). ATLAS-CONF-2012-040. https://cds.cern.ch/record/1435194
36. ATLAS Collaboration, Measurements of $W\gamma$ and $Z\gamma$ production in *pp* collisions at $\sqrt{s} = 7$ TeV with the ATLAS detector at the LHC. Phys. Rev. D **87**, 112003 (2013). https://doi.org/10.1103/PhysRevD.87.112003
37. ATLAS Collaboration, Pile-up subtraction and suppression for jets in ATLAS (2013). ATLAS-CONF-2013-083. https://cds.cern.ch/record/1570994
38. ATLAS Collaboration, Search for new phenomena in final states with large jet multiplicities and missing transverse momentum at $\sqrt{s} = 8$ TeV proton-proton collisions using the ATLAS experiment. J. High Energy Phys. **10**, 130 (2013). https://doi.org/10.1007/JHEP10(2013)130. arXiv: 1308.1841 [hep-ex]. Erratum in: J. High Energy Phys. **1**, 109 (2014). https://doi.org/10.1007/JHEP01(2014)109
39. ATLAS Collaboration, Search for squarks and gluinos with the ATLAS detector in final states with jets and missing transverse momentum using 47fb^{-1} of $\sqrt{s} = 7$ TeV proton-proton collision data. Phys. Rev. D **87**, 012008 (2013). https://doi.org/10.1103/PhysRevD.87.012008. arXiv: 1208.0949 [hep-ex]
40. ATLAS Collaboration, Performance of boosted top quark identification in 2012 ATLAS data (2013). ATLAS-CONF-2013-084. https://cds.cern.ch/record/1571040
41. ATLAS Collaboration, Jet energy measurement with the ATLAS detector in proton-proton collisions at $\sqrt{s} = 7$ TeV. Eur. Phys. J. C **73**, 2304 (2013). https://doi.org/10.1140/epjc/s10052-013-2304-2. arXiv: 1112.6426 [hep-ex]
42. ATLAS Collaboration, Jet energy resolution in proton–proton collisions at $\sqrt{s} = 7$ TeV recorded in 2010 with the ATLAS detector. Eur. Phys. J. C **73**, 2306 (2013). https://doi.org/10.1140/epjc/s10052-013-2306-0. arXiv: 1210.6210 [hep-ex]
43. ATLAS Collaboration, Search for squarks and gluinos with the ATLAS detector in final states with jets and missing transverse momentum using 4.7 fb^{-1} of $\sqrt{s} = 7$ TeV proton-proton collision data. Phys. Rev. D **87**(1), 12008 (2013). https://doi.org/10.1103/PhysRevD.87.012008. arXiv: 1208.0949 [hep-ex]
44. ATLAS Collaboration, Performance of jet substructure techniques for large-R jets in proton–proton collisions at $\sqrt{s} = 7$ TeV using the ATLAS detector. J. High Energy Phys. 9, 76 (2013). https://doi.org/10.1007/JHEP09(2013)076. arXiv: 1306.4945 [hep-ex]
45. ATLAS Collaboration, Performance of shower deconstruction in ATLAS (2014). ATLAS-CONF-2014-003. https://cds.cern.ch/record/1648661
46. ATLAS Collaboration, Search for strong production of supersymmetric particles in final states with missing transverse momentum and at least three b-jets at $\sqrt{s} = 8$ TeV proton-proton collisions with the ATLAS detector. J. High Energy Phys. **10**, 24 (2014). https://doi.org/10.1007/JHEP10(2014)024. arXiv: 1407.0600 [hep-ex]
47. ATLAS Collaboration, ATLAS Pythia 8 tunes to 7 TeV data (2014). ATL-PHYS-PUB-2014-021. https://cds.cern.ch/record/1966419
48. ATLAS Collaboration, Calibration of b-tagging using dileptonic top pair events in a combinatorial likelihood approach with the ATLAS experiment (2014). ATLAS-CONF-2014-004. https://cds.cern.ch/record/1664335

49. ATLAS Collaboration, Calibration of the performance of b-tagging for c and light-flavour jets in the 2012 ATLAS data (2014). ATLAS-CONF-2014-046. https://cds.cern.ch/record/1741020
50. ATLAS Collaboration, Search for high-mass dilepton resonances in *pp* collisions at $\sqrt{s}$ = 8 TeV with the ATLAS detector. Phys. Rev. D **90**, 52005 (2014). https://doi.org/10.1103/PhysRevD.90.052005. arXiv: 1405.4123 [hep-ex]
51. ATLAS Collaboration, Search for new particles in events with one lepton and missing transverse momentum in *pp* collisions at $\sqrt{s}$ = 8 TeV with the ATLAS detector. J. High Energy Phys. **9**, 37 (2014). https://doi.org/10.1007/JHEP09(2014)037. arXiv: 1407.7494 [hep-ex]
52. ATLAS Collaboration, Tagging and suppression of pileup jets with the ATLAS detector (2014). ATLAS-CONF-2014-018. https://cds.cern.ch/record/1700870
53. ATLAS Collaboration, Electron and photon energy calibration with the ATLAS detector using LHC Run 1 data. Eur. Phys. J. C **74**, 3071 (2014). https://doi.org/10.1140/epjc/s10052-014-3071-4. arXiv: 1407.5063 [hep-ex]
54. ATLAS Collaboration, Summary plots from the ATLAS Standard Model physics group (2015). https://atlas.web.cern.ch/Atlas/GROUPS/PHYSICS/CombinedSummaryPlots/SM/. Accessed 24 March 2018
55. ATLAS Collaboration, Summary plots from the ATLAS Supersymmetry physics group (2015). https://twiki.cern.ch/twiki/bin/view/AtlasPublic/SupersymmetryPublicResults. Accessed 08 April 2018
56. ATLAS Collaboration, Global feature extraction (gFEX) performance plots ATL-COM-DAQ-2014-087 (2015). Accessed 10 April 2018. https://twiki.cern.ch/twiki/bin/view/AtlasPublic/JetTriggerPublicResults
57. ATLAS Collaboration, *ATLAS Luminosity Public Results Run 2* (2015). https://twiki.cern.ch/twiki/bin/view/AtlasPublic/LuminosityPublicResultsRun2. Accessed 11 March 2018
58. ATLAS Collaboration, Performance of the ATLAS muon trigger in *pp* collisions at $\sqrt{s}$ = 8 TeV. Eur. Phys. J. C **75**, 120 (2015). https://doi.org/10.1140/epjc/s10052-015-3325-9. arXiv: 1408.3179 [hep-ex]
59. ATLAS Collaboration, Search for massive supersymmetric particles decaying to many jets using the ATLAS detector in *pp* collisions at $\sqrt{s}$ = 8 TeV. Phys. Rev. D **91**, 112016 (2015). https://doi.org/10.1103/PhysRevD.91.112016. arXiv: 1502.05686 [hep-ex]. Erratum in: Phys. Rev. D **93**, 39901 (2016). https://doi.org/10.1103/PhysRevD.93.039901
60. ATLAS Collaboration, Identification and energy calibration of hadronically decaying tau leptons with the ATLAS experiment in *pp* collisions at $\sqrt{s}$ = 8 TeV. Eur. Phys. J. C **75**, 303 (2015). https://doi.org/10.1140/epjc/s10052-015-3500-z. arXiv: 1412.7086 [hep-ex]
61. ATLAS Collaboration, Jet calibration and systematic uncertainties for jets reconstructed in the ATLAS detector at √s= 13 TeV ATL-PHYS-PUB-2015-15 (2015). https://cds.cern.ch/record/2037613
62. ATLAS Collaboration, Jet energy measurement and its systematic uncertainty in proton–proton collisions at $\sqrt{s}$ = 7 TeV with the ATLAS detector. Eur. Phys. J. C **75**, 17 (2015). https://doi.org/10.1140/epjc/s10052-014-3190-y. arXiv: 1406.0076 [hep-ex]
63. ATLAS Collaboration, Properties of jets and inputs to jet reconstruction and calibration with the ATLAS detector using proton-proton collisions at $\sqrt{s}$ = 13 TeV. ATL-PHYS-PUB-2015-036 (2015). https://cds.cern.ch/record/2044564
64. ATLAS Collaboration, Expected performance of the ATLAS b-tagging algorithms in Run-2. ATL-PHYS-PUB-2015-22 (2015). https://cds.cern.ch/record/2037697
65. ATLAS Collaboration, Measurements of Higgs boson production and couplings in the four-lepton channel in *pp* collisions at center-of-mass energies of 7 and 8 TeV with the ATLAS detector. Phys. Rev. D **91**, 12006 (2015). https://doi.org/10.1103/PhysRevD.91.012006. arXiv: 1408.5191 [hep-ex]
66. ATLAS Collaboration, Expected performance of missing transverse momentum reconstruction for the ATLAS detector at $\sqrt{s}$ = 13 TeV (2015). ATL-PHYS-PUB-2015-023. https://cds.cern.ch/record/2037700

67. ATLAS Collaboration, Selection of jets produced in 13 TeV proton-proton collisions with the ATLAS detector (2015). ATLAS-CONF-2015029. https://cds.cern.ch/record/2037702
68. ATLAS Collaboration, Performance of missing transverse momentum reconstruction with the ATLAS detector in the first proton-proton collisions at $\sqrt{s} = 13$ TeV (2015). ATL-PHYS-PUB-2015-027. https://cds.cern.ch/record/2037904
69. ATLAS Collaboration, Vertex reconstruction performance of the ATLAS detector at $\sqrt{s} = 13$ TeV (2015). ATL-PHYS-PUB-2015-026. https://cds.cern.ch/record/2037717
70. ATLAS Collaboration, Search for pair production of gluinos decaying via stop and sbottom in events with b-jets and large missing transverse momentum in pp collisions at $\sqrt{s} = 13$ TeV with the ATLAS detector. Phys. Rev. D **94**(3), 32003 (2016). https://doi.org/10.1103/PhysRevD.94.032003. arXiv: 1605.09318 [hep-ex]
71. ATLAS Collaboration, Electron efficiency measurements with the ATLAS detector using the 2015 LHC proton-proton collision data (2016). ATLAS-CONF-2016-024. https://cds.cern.ch/record/2157687
72. ATLAS Collaboration, Luminosity determination in pp collisions at $\sqrt{s} = 8$ TeV using the ATLAS detector at the LHC. Eur. Phys. J. C **76**, 653 (2016). https://doi.org/10.1140/epjc/s10052-016-4466-1. arXiv: 1608.03953 [hep-ex]
73. ATLAS Collaboration, The performance of the jet trigger for the ATLAS detector during 2011 data taking. Eur. Phys. J. C 76, 526 (2016). https://doi.org/10.1140/epjc/s10052-016-4325-0. arXiv: 1606.07759 [hep-ex]
74. ATLAS Collaboration, Modelling of the $t\bar{t}H$ and $t\bar{t}V (V = WZ)$ processes for $\sqrt{s} = 13$ TeV ATLAS analyses (2016). ATL-PHYS-PUB-2016-005. https://cds.cern.ch/record/2120826
75. ATLAS Collaboration, Multi-boson simulation for 13 TeV ATLAS analyses (2016). ATL-PHYS-PUB-2016-002. https://cds.cern.ch/record/2119986
76. ATLAS Collaboration, Performance of b-jet identification in the ATLAS experiment. J. Instrum. **11**, P04008 (2016). https://doi.org/10.1088/1748-0221/11/04/P04008. arXiv: 1512.01094 [hep-ex]
77. ATLAS Collaboration, Muon reconstruction performance of the ATLAS detector in proton-proton collision data at $\sqrt{s} = 13$ TeV. Eur. Phys. J. C **76**, 292 (2016). https://doi.org/10.1140/epjc/s10052-016-4120-y. arXiv: 1603.05598 [hep-ex]
78. ATLAS Collaboration, Performance of pile-up mitigation techniques for jets in pp collisions at $\sqrt{s} = 8$ TeV using the ATLAS detector. Eur. Phys. J. C **76**, 581 (2016). https://doi.org/10.1140/epjc/s10052-016-4395-z. arXiv: 1510.03823 [hep-ex]
79. ATLAS Collaboration, Search for pair production of gluinos decaying via stop and sbottom in events with b-jets and large missing transverse momentum in pp collisions at $\sqrt{s} = 13$ TeV with the ATLAS detector. Phys. Rev. D **94**, 032003 (2016). https://doi.org/10.1103/PhysRevD.94.032003. arXiv: 1605.09318 [hep-ex]
80. ATLAS Collaboration, Measurements of fiducial cross-sections for $t\bar{t}$ production with one or two additional b-jets in pp collisions at $\sqrt{s} = 8$ TeV using the ATLAS detector. Eur. Phys. J. C **76**, 11 (2016). https://doi.org/10.1140/epjc/s10052-015-3852-4. arXiv: 1508.06868 [hep-ex]
81. ATLAS Collaboration, Performance of b -jet identification in the ATLAS experiment. J. Instrum. **11**(4), P04008 (2016). http://stacks.iop.org/1748-0221/11/i=04/a=P04008
82. ATLAS Collaboration, Search for supersymmetry in final states with missing transverse momentum and multiple b-jets in proton–proton collisions at $\sqrt{s} = 13$ TeV with the ATLAS detector (2017). arXiv: 1711.01901 [hep-ex]
83. ATLAS Collaboration, Performance of top quark and W Boson tagging in run 2 with ATLAS (2017). ATLAS-CONF-2017-064. https://cds.cern.ch/record/2281054
84. ATLAS Collaboration, Jet reclustering and close-by effects in ATLAS Run 2 (2017). ATLAS-CONF-2017-062. https://cds.cern.ch/record/2275649
85. ATLAS Collaboration, Performance of the ATLAS trigger system in 2015. Eur. Phys. J. C **77**, 317 (2017). https://doi.org/10.1140/epjc/s10052-017-4852-3. arXiv: 1611.09661 [hep-ex]
86. ATLAS Collaboration, Topological cell clustering in the ATLAS calorimeters and its performance in LHC Run 1. Eur. Phys. J. C **77**, 490 (2017). https://doi.org/10.1140/epjc/s10052-017-5004-5. arXiv: 1603.02934 [hep-ex]

87. ATLAS Collaboration, Jet energy scale measurements and their systematic uncertainties in proton-proton collisions at $\sqrt{s}$ = 13 TeV with the ATLAS detector. Phys. Rev. D **96**, 72002 (2017). https://doi.org/10.1103/PhysRevD.96.072002. arXiv: 1703.09665 [hep-ex]
88. ATLAS Collaboration, Proton tagging with the one arm AFP detector (2017). ATL-PHYS-PUB-2017-012. https://cds.cern.ch/record/2265187
89. ATLAS Collaboration, Optimisation and performance studies of the ATLAS b-tagging algorithms for the 2017–18 LHC run (2017). ATL-PHYS-PUB-2017-013. https://cds.cern.ch/record/2273281
90. ATLAS Collaboration, Jet energy scale measurements and their systematic uncertainties in proton-proton collisions at $\sqrt{s}$ = 13 TeV with the ATLAS detector. Phys. Rev. **D96**(7), 072002 (2017). arXiv: 1703.09665 [hep-ex]
91. *ATLAS Inner Detector: Technical Design Report, 1* (CERN, Geneva, 1997). Technical Design Report ATLAS. http://cds.cern.ch/record/331063
92. ATLAS Level-1 Calorimeter Trigger Collaboration, The ATLAS level-1 calorimeter trigger: preprocessor implementation and performance. J. Instrum. **7**(12), P12008 (2012). http://stacks.iop.org/1748-0221/7/i=12/a=P12008
93. *ATLAS Liquid-Argon Calorimeter: Technical Design Report* (CERN, Geneva, 1996). Technical Design Report ATLAS. https://cds.cern.ch/record/331061
94. *ATLAS Muon Spectrometer: Technical Design Report* (CERN, Geneva, 1997). Technical Design Report ATLAS. https://cds.cern.ch/record/331068
95. *ATLAS Tile Calorimeter: Technical Design Report* (CERN, Geneva, 1996). Technical Design Report ATLAS. https://cds.cern.ch/record/331062
96. J.J. Aubert et al., Experimental observation of a heavy particle J, Phys. Rev. Lett. **33**, 1404–1406 (1974). https://doi.org/10.1103/PhysRevLett.33.1404
97. J.J. Aubert et al., The ratio of the nucleon structure functions F2N for iron and deuterium. Phys. Lett. B **123**(3), 275–278 (1983). ISSN: 0370-2693. https://doi.org/10.1016/0370-2693(83)90437-9
98. J.-E. Augustin et al., Discovery of a narrow resonance in electron-positron annihilation. Phys. Rev. Lett. **33**, , 1406–1408 (1974). https://doi.org/10.1103/PhysRevLett.33.1406
99. M. Baak et al., HistFitter software framework for statistical data analysis. Eur. Phys. J. C **75**, 153 (2015). https://doi.org/10.1140/epjc/s10052-015-3327-7. arXiv: 1410.1280 [hep-ex]
100. H. Baer et al., Naturalness, supersymmetry and light Higgsinos: a snowmass whitepaper, in *Proceedings, 2013 Community Summer Study on the Future of U.S. Particle Physics: Snowmass on the Mississippi (CSS2013): Minneapolis, July 29–August 6, 2013* (2013). arXiv: 1306.2926 [hep-ph]. https://inspirehep.net/record/1238302/files/arXiv:1306.2926.pdf
101. M. Bahr et al., Herwig++ physics and manual. Eur. Phys. J. C **58**, 639 (2008). https://doi.org/10.1140/epjc/s10052-008-0798-9. arXiv: 0803.0883 [hep-ph]
102. R.D. Ball et al., Parton distributions with LHC data. Nucl. Phys. B **867**, 244–289 (2013). https://doi.org/10.1016/j.nuclphysb.2012.10.003. arXiv: 1207.1303 [hep-ph]
103. R.D. Ball et al., Parton distributions for the LHC run II. J. High Energ. Phys. **2015**, 40 (2015). https://doi.org/10.1007/JHEP04(2015)040. arXiv: 1410.8849 [hep-ph]
104. R.D. Ball, Global parton distributions for the LHC run II. Nuovo Cim. **C38**(4), 127 (2016). https://doi.org/10.1393/ncc/i2015-15127-9. arXiv: 1507.07891 [hep-ph]
105. D.P. Barber et al., Discovery of three-jet events and a test of quantum chromodynamics at PETRA. Phys. Rev. Lett. **43**, 830–833 (1979).doi: https://doi.org/10.1103/PhysRevLett.43.830
106. R. Barbier et al., R-parity violating supersymmetry. Phys. Rept. **420**, 1–202 (2005). https://doi.org/10.1016/j.physrep.2005.08.006. arXiv: hep-ph/0406039 [hep-ph]
107. R. Barbieri, G.F. Giudice, Upper bounds on supersymmetric particle masses. Nucl. Phys. B **306**, 63 (1988). https://doi.org/10.1016/0550-3213(88)90171-X
108. J. Barranco García et al., Long term dynamics of the high luminosity Large Hadron Collider with crab cavities. Phys. Rev. Accel. Beams **19**, 101003 (2016). https://doi.org/10.1103/PhysRevAccelBeams.19.101003

109. F. Bauer et al., Construction and test of MDT chambers for the ATLAS muon spectrometer. Nucl. Instrum. Methods Phys. Res., Sect. A **461**(1), 17–20 (2001). ISSN: 0168-9002. https://doi.org/10.1016/S0168-9002(00)01156-6. 8th Pisa Meeting on Advanced Detectors
110. L. Beck, Search for the production of four top quarks at the CMS experiment at $\sqrt{s} = 13$ TeV. In: *Proceedings, 9th International Workshop on Top Quark Physics (TOP 2016): Olomouc, September 19–23, 2016* (2016). arXiv: 1611.09607 [hep-ex]. https://inspirehep.net/record/1500694/files/arXiv:1611.09607.pdf
111. W. Beenakker et al., Squark and gluino production at hadron colliders. Nucl. Phys. B **492**, 51–103 (1997). https://doi.org/10.1016/S0550-3213(97)00084-9. arXiv: hep-ph/9610490
112. W. Beenakker et al., Soft-gluon resummation for squark and gluino hadroproduction. J. High Energy Phys. **12**, 41 (2009). https://doi.org/10.1088/1126-6708/2009/12/041. arXiv: 0909.4418 [hep-ph]
113. W. Beenakker et al., Squark and gluino hadroproduction. Int. J. Mod. Phys. A **26**, 2637–2664 (2011). https://doi.org/10.1142/S0217751X11053560. arXiv: 1105.1110 [hep-ph]
114. M. Begel et al., Global feature extractor of the level-1 calorimeter trigger: ATLAS TDAQ phase-I upgrade gFEX final design report. Technical Report ATL-COM-DAQ-2016-184. Geneva: CERN (2016). https://cds.cern.ch/record/2233958
115. W.R. Bennett, Spectra of quantized signals. Bell Syst. Tech. J. **27**(3), 446–472 (1948). ISSN: 0005-8580. https://doi.org/10.1002/j.1538-7305.1948.tb01340.x
116. G. Bertone, D. Hooper, J. Silk, Particle dark matter: evidence, candidates and constraints. Phys. Rep. **405**(5–6), 279–390 (2005)
117. J.D. Bjorken, E.A. Paschos, Inelastic electron-proton and γ-proton scattering and the structure of the nucleon. Phys. Rev. **185**, 1975–1982 (1969). https://doi.org/10.1103/PhysRev.185.1975
118. E. D. Bloom et al., High-energy inelastic $e - p$ scattering at 6° and 10°. Phys. Rev. Lett. **23**, 930–934 (1969). https://doi.org/10.1103/PhysRevLett.23.930
119. C. Borschensky et al., Squark and gluino production cross sections in pp collisions at $\sqrt{s} = 13, 14, 33$ and 100 TeV. Eur. Phys. J. **C74**(12), 3174 (2014). https://doi.org/10.1140/epjc/s10052-014-3174-y. arXiv: 1407.5066 [hep-ph]
120. M. Breidenbach et al., Observed behavior of highly inelastic electron-proton scattering. Phys. Rev. Lett. **23**(16), 935–939 (1969). https://doi.org/10.1103/PhysRevLett.23.935
121. J.M. Butterworth et al. Jet substructure as a new Higgs search channel at the LHC. Phys. Rev. Lett. **100**, 242001 (2008). https://doi.org/10.1103/PhysRevLett.100.242001. arXiv: 0802.2470 [hep-ph]
122. J.M. Butterworth, I. Ochoa, T. Scanlon, Boosted Higgs $\rightarrow b\bar{b}$ in vector-boson associated production at 14 TeV. Eur. Phys. J. **C75**(8), 366 (2015). https://doi.org/10.1140/epjc/s10052-015-3592-5. arXiv: 1506.04973 [hep-ph]
123. M. Cacciari, G.P. Salam, Dispelling the N^3 myth for the k_t jet-finder. Phys. Lett. **B641**, 57–61 (2006). https://doi.org/10.1016/j.physletb.2006.08.037. arXiv: hep-ph/0512210 [hep-ph]
124. M. Cacciari, G.P. Salam, Pileup subtraction using jet areas. Phys. Lett. **B659**, 119–126 (2008). https://doi.org/10.1016/j.physletb.2007.09.077. arXiv: 0707.1378 [hep-ph]
125. M. Cacciari, G.P. Salam, G. Soyez, The anti-k_t jet clustering algorithm. J. High Energy Phys. **4**, 63 (2008). https://doi.org/10.1088/1126-6708/2008/04/063. arXiv: 0802.1189 [hep-ph]
126. M. Cacciari, G.P. Salam, G. Soyez, FastJet user manual. Eur. Phys. J. C **72**, 1896 (2012). https://doi.org/10.1140/epjc/s10052-012-1896-2. arXiv: 1111.6097 [hep-ph]
127. L. Canetti, M. Drewes, M. Shaposhnikov, Matter and antimatter in the universe. New J. Phys. **14**(9), 095012 (2012)
128. F. Cascioli, P. Maierhofer, S. Pozzorini, Scattering amplitudes with open loops. Phys. Rev. Lett. **108**, 111601 (2012). https://doi.org/10.1103/PhysRevLett.108.111601. arXiv: 1111.5206 [hep-ph]
129. S. Catani et al., Longitudinally invariant K_t clustering algorithms for hadron–hadron collisions. Nucl. Phys. **B406**, 187–224 (1993). https://doi.org/10.1016/0550-3213(93)90166-M
130. S. Catani et al., Vector boson production at hadron colliders: a fully exclusive QCD calculation at NNLO. Phys. Rev. Lett.**103**, 082001 (2009). https://doi.org/10.1103/PhysRevLett.103.082001. arXiv: 0903.2120 [hep-ph]

131. CERN SPS Collaboration, Experimental observation of lepton pairs of invariant mass around 95 GeV/c2 at the CERN SPS collider. Phys. Lett. B**126**(5), 398–410 (1983). ISSN: 0370-2693. https://doi.org/10.1016/0370-2693(83)90188-0
132. CERN Service graphique, Overall view of the LHC. Vue d'ensemble du LHC (2014). General photo. https://cds.cern.ch/record/1708849
133. J. Chadwick, Possible existence of a Neutron. Nature **129**, 312 (1932). http://dx.doi.org/10.1038/129312a0
134. S. Chatrchyan et al., Observation of a new boson at a mass of 125 GeV with the CMS experiment at the LHC. Phys. Lett. **B716**, 30–61 (2012). https://doi.org/10.1016/j.physletb.2012.08.021. arXiv: 1207.7235 [hep-ex]
135. F.E. Close. *An Introduction to Quarks and Partons* (Academic Press, London, 1979). ISBN: 9780121751524
136. CMS Collaboration, Summary plots from the CMS Supersymmetry physics group (2015). https://twiki.cern.ch/twiki/bin/view/CMSPublic/PhysicsResultsSUS. Accessed 08 April 2018
137. CMS Collaboration, Search for supersymmetry in multijet events with missing transverse momentum in proton-proton collisions at 13 TeV. Phys. Rev. D **96**, 32003 (2017). https://doi.org/10.1103/PhysRevD.96.032003. arXiv: 1704.07781 [hep-ex]
138. CMS Collaboration, Search for new phenomena with the M_{T2} variable in the all-hadronic final state produced in proton–proton collisions at $\sqrt{s} = 13$ TeV (2017). arXiv: 1705.04650 [hep-ex]
139. CMS Collaboration, Search for supersymmetry in pp collisions at $\sqrt{s} = 13$ TeV in the single-lepton final state using the sum of masses of large-radius jets. Phys. Rev. Lett. **119**, 151802 (2017). https://doi.org/10.1103/PhysRevLett.119.151802. arXiv: 1705.04673 [hep-ex]
140. CMS Collaboration, Search for supersymmetry in events with one lepton and multiple jets exploiting the angular correlation between the lepton and the missing transverse momentum in proton–proton collisions at $\sqrt{s} = 13$ TeV (2017). arXiv: 1709.09814 [hep-ex]
141. CMS Collaboration, Search for physics beyond the standard model in events with two leptons of same sign, missing transverse momentum, and jets in proton-proton collisions at $\sqrt{s} = 13$ TeV. Eur. Phys. J. C **77**, 578 (2017). https://doi.org/10.1140/epjc/s10052-017-5079-z. arXiv: 1704.07323 [hep-ex]
142. CMS Collaboration, Search for supersymmetry in events with at least three electrons or muons, jets, and missing transverse momentum in proton–proton collisions at $\sqrt{s} = 13$ TeV (2017). arXiv: 1710.09154 [hep-ex]
143. T. Cohen et al., Dissecting jets and missing energy searches using n-body extended simplified models. J. High Energy Phys. **8**, 38 (2016). https://doi.org/10.1007/JHEP08(2016)038. arXiv: 1605.01416 [hep-ph]
144. J.C. Collins, D.E. Soper, The theorems of perturbative QCD Annu. Rev. Nucl. Part. Sci. **37**, 383–409 (1987). http://arxiv.org/abs/https://doi.org/10.1146/annurev.ns.37.120187.002123. https://doi.org/10.1146/annurev.ns.37.120187.002123
145. J.C. Collins, D.E. Soper, G.F. Sterman, Factorization of hard processes in QCD. Adv. Ser. Direct. High Energy Phys. **5**, 1–91 (1989). https://doi.org/10.1142/9789814503266_0001. arXiv: hep-ph/0409313 [hep-ph]
146. Contemporary Physics Education Project, Contemporary Physics Education Project (2017). http://cpepweb.org/. Accessed 25 March 2018
147. W.N. Cottingham, D.A. Greenwood, *An Introduction to the Standard Model of Particle Physics* (Cambridge University Press, Cambridge, 2001)
148. G. Cowan et al., Asymptotic formulae for likelihood-based tests of new physics. Eur. Phys. J. C **71**, 1554 (2011). https://doi.org/10.1140/epjc/s10052-011-1554-0. arXiv: 1007.1727 [physics.data-an]. Erratum: Eur. Phys. J. C **73**, 2501 (2013)
149. K. Cranmer, I. Yavin, RECAST: extending the impact of existing analyses. J. High Energy Phys. **4**, 38 (2010). https://doi.org/10.1007/JHEP04(2011)038. arXiv: 1010.2506 [hep-ex]
150. K. Cranmer, I. Yavin, RECAST: extending the impact of existing analyses. J. High Energy Phys. **4**, 38 (2011). https://doi.org/10.1007/JHEP04(2011)038. arXiv: 1010.2506 [hep-ex]

151. K. Cranmer et al., HistFactory: a tool for creating statistical models for use with RooFit and RooStats (2012)
152. C. Csaki, Y. Grossman, B. Heidenreich, MFV SUSY: a natural theory for R-parity violation. Phys. Rev. **D85**, 095009 (2012). https://doi.org/10.1103/PhysRevD.85.095009. arXiv: 1111.1239 [hep-ph]
153. M. Czakon, A. Mitov, Top++: a program for the calculation of the top-pair cross-section at hadron colliders. Comput. Phys. Commun. **185**, 2930 (2014). https://doi.org/10.1016/j.cpc.2014.06.021. arXiv: 1112.5675 [hep-ph]
154. G. Danby et al., Observation of high-energy neutrino reactions and the existence of two kinds of neutrinos. Phys. Rev. Lett. **9**, 36–44 (1962). https://doi.org/10.1103/PhysRevLett.9.36
155. D. de Florian et al., Handbook of LHC higgs cross sections: 4. Deciphering the nature of the Higgs sector (2016). https://doi.org/10.23731/CYRM-2017-002. arXiv: 1610.07922 [hep-ph]
156. S. Dimopoulos, H. Georgi, Softly broken supersymmetry and SU(5). Nucl. Phys. B **193**(1), 150–162 (1981). https://doi.org/10.1016/0550-3213(81)90522-8
157. S. Dimopoulos, S. Raby, F. Wilczek, Supersymmetry and the scale of unification. Phys. Rev. D **24**, 1681–1683 (1981). https://doi.org/10.1103/PhysRevD.24.1681
158. Y.L. Dokshitzer et al., Better jet clustering algorithms. J. High Energy Phys. **8**, 1 (1997). https://doi.org/10.1088/1126-6708/1997/08/001. arXiv: hep-ph/9707323 [hep-ph]
159. H. Dreiner, An introduction to explicit R-parity violation, in *Perspectives on Supersymmetry II* (2010), pp. 565–583. https://doi.org/10.1142/9789814307505_0017. https://www.worldscientific.com/doi/pdf/10.1142/9789814307505_0017
160. D.W. Duke, J.F. Owens, Q^2-dependent parametrizations of parton distribution functions. Phys. Rev. D **30**, 49–54 (1984). https://doi.org/10.1103/PhysRevD.30.49
161. E.F. Eisenhandler, ATLAS level-1 calorimeter trigger algorithms. Technical Report. ATL-DAQ-2004-011. CERN-ATL-DAQ-2004-011. Geneva: CERN, 2004. http://cds.cern.ch/record/792528
162. J. Ellis, TikZ-Feynman: Feynman diagrams with TikZ. Comput. Phys. Commun. **210**, pp. 103–123 (2017). https://doi.org/10.1016/j.cpc.2016.08.019. arXiv: 1601.05437 [hep-ph]
163. J.R. Ellis, S. Rudaz, Search for supersymmetry in toponium decays. Phys. Lett. B **128**, 248 (1983). https://doi.org/10.1016/0370-2693(83)90402-1
164. L. Evans, P. Bryant, LHC machine. J. Instrum. **3**(8), S08001 (2008). http://stacks.iop.org/1748-0221/3/i=08/a=S08001
165. C.W. Fabjan, F. Gianotti, Calorimetry for particle physics. Rev. Mod. Phys. **75**, 1243–1286 (2003).CERN-EP-2003-075. 96 p. https://cds.cern.ch/record/692252
166. G.R. Farrar, P. Fayet, Phenomenology of the production, decay, and detection of new hadronic states associated with supersymmetry. Phys. Lett. B **76**(5), 575–579 (1978). ISSN: 0370-2693. https://doi.org/10.1016/0370-2693(78)90858-4
167. P. Fayet, Supersymmetry and weak, electromagnetic and strong interactions. Phys. Lett. B **64**(2), 159–162 (1976). ISSN: 0370-2693. https://doi.org/10.1016/0370-2693(76)90319-1
168. S. Ferrara, B. Zumino, Supergauge invariant Yang-Mills theories. Nucl. Phys. B **79**, 413 (1974). https://doi.org/10.1016/0550-3213(74)90559-8
169. S. Gieseke, C. Rohr, A. Siodmok, Colour reconnections in Herwig++. Eur. Phys. J. C **72**, 2225 (2012). https://doi.org/10.1140/epjc/s10052-012-2225-5. arXiv: 1206.0041 [hep-ph]
170. S.L. Glashow, Partial-symmetries of weak interactions. Nucl. Phys. **22**(4), 579–588 (1961). ISSN: 0029-5582. https://doi.org/10.1016/0029-5582(61)90469-2
171. T. Gleisberg, S. Höche, Comix, a new matrix element generator. J. High Energy Phys. **12**, 39 (2008). https://doi.org/10.1088/1126-6708/2008/12/039. arXiv: 0808.3674 [hep-ph]
172. T. Gleisberg et al., Event generation with SHERPA 1.1. J. High Energy Phys. **2**, 7 (2009). https://doi.org/10.1088/1126-6708/2009/02/007. arXiv: 0811.4622 [hep-ph]
173. Y.A. Golfand, E.P. Likhtman, Extension of the algebra of Poincare group generators and violation of p invariance. JETP Lett. **13**, 323–326 (1971). [Pisma Zh. Eksp. Teor. Fiz. **13**, 452 (1971)]
174. D.J. Gross, The discovery of asymptotic freedom and the emergence of QCD. Proc. Natl. Acad. Sci.**102**(26), 9099–9108 (2005). ISSN: 0027-8424. https://doi.org/10.1073/pnas.0503831102

175. D.J. Gross, F. Wilczek, Asymptotically free gauge theories. I. Phys. Rev. D **8**, 3633–3652 (1973). https://doi.org/10.1103/PhysRevD.8.3633
176. D.J. Gross, F. Wilczek, Ultraviolet behavior of non-abelian gauge theories. Phys. Rev. Lett. **30**, 1343–1346 (1973). https://doi.org/10.1103/PhysRevLett.30.1343
177. L.A. Harland-Lang et al., Parton distributions in the LHC era: MMHT 2014 PDFs. Eur. Phys. J. **C75**(5), 204 (2015). https://doi.org/10.1140/epjc/s10052-015-3397-6. arXiv: 1412.3989 [hep-ph]
178. S. Haywood et al., *ATLAS Inner Detector: Technical Design Report, 2* (CERN, Geneva, 1997). Technical Design Report ATLAS. https://cds.cern.ch/record/331064
179. S.W. Herb et al., Observation of a dimuon resonance at 9.5 GeV in 400-GeV proton-nucleus collisions. Phys. Rev. Lett. **39**, 252–255 (1977). https://doi.org/10.1103/PhysRevLett.39.252
180. HL-LHC High Luminosity Large Hadron Collider: The HL-LHC project. http://hilumilhc.web.cern.ch/about/hl-lhc-project
181. S. Hoche, Introduction to parton-shower event generators, in *Proceedings, The Oretical Advanced Study Institute in Elementary Particle Physics: Journeys Through the Precision Frontier: Amplitudes for Colliders (TASI 2014): Boulder, Colorado, June 2–27, 2014* (2015), pp. 235–295. https://doi.org/10.1142/9789814678766_0005. arXiv: 1411.4085 [hep-ph]. http://inspirehep.net/record/1328513/files/arXiv:1411.4085.pdf
182. S. Höche et al., QCD matrix elements + parton showers: the NLO case. J. High Energy Phys. **4**, 27 (2013). https://doi.org/10.1007/JHEP04(2013)027. arXiv: 1207.5030 [hep-ph]
183. A. Hocker et al., TMVA – toolkit for multivariate data analysis. PoS ACAT, 40 (2007). arXiv: physics/0703039 [PHYSICS]
184. T.-J. Hou et al., CT14 intrinsic charm parton distribution functions from CTEQ-TEA global analysis. J. High Energ. Phys. **2**, 59 (2018). https://doi.org/10.1007/JHEP02(2018)059. arXiv: 1707.00657 [hep-ph]
185. J.E. Huth et al., Toward a standardization of jet definitions, in *1990 DPF Summer Study on High-energy Physics: Research Directions for the Decade (Snowmass 90) Snowmass, Colorado, June 25–July 13* (1990), pp. 0134–136. http://lss.fnal.gov/cgi-bin/find_paper.pl?conf-90-249
186. L.E. Ibanez, G.G. Ross, Low-energy predictions in supersymmetric grand unified theories. Phys. Lett. B **105**, 439 (1981). https://doi.org/10.1016/0370-2693(81)91200-4
187. K. Inoue et al., Aspects of grand unified models with softly broken supersymmetry. Prog. Theor. Phys. **68**, 927 (1982). [Erratum: Prog. Theor. Phys.70,330(1983)]. https://doi.org/10.1143/PTP.68.927
188. B. Isildak, Measurement of the differential dijet production cross section in proton-proton collisions at $\sqrt{s} = 7$ tev. Ph.D. Thesis. Bogazici U. (2011). arXiv: 1308.6064 [hep-ex]. http://inspirehep.net/record/1251416/files/arXiv:1308.6064.pdf
189. P. Kant et al., HatHor for single top-quark production: updated predictions and uncertainty estimates for single top-quark production in hadronic collisions. Comput. Phys. Commun. **191**, 74–89 (2015). https://doi.org/10.1016/j.cpc.2015.02.001. arXiv: 1406.4403 [hep-ph]
190. R.K. Kaul, P. Majumdar, Cancellation of quadratically divergent mass corrections in globally supersymmetric spontaneously broken gauge theories. Nucl. Phys. B **199**(1), 36–58 (1982). ISSN: 0550-3213. https://doi.org/10.1016/0550-3213(82)90565-X
191. N. Kidonakis, Two-loop soft anomalous dimensions for single top quark associated production with a W^- or H^-. Phys. Rev. D **82**, 54018 (2010). https://doi.org/10.1103/PhysRevD.82.054018. arXiv: 1005.4451 [hep-ph]
192. N. Kidonakis, NNLL resummation for s-channel single top quark production. Phys. Rev. D **81**, 54028 (2010). https://doi.org/10.1103/PhysRevD.81.054028. arXiv: 1001.5034 [hep-ph]
193. N. Kidonakis, Next-to-next-to-leading-order collinear and soft gluon corrections for t-channel single top quark production. Phys. Rev. D **83**, 091503 (2011). https://doi.org/10.1103/PhysRevD.83.091503. arXiv: 1103.2792 [hep-ph]
194. K. Kodama et al., Observation of tau neutrino interactions. Phys. Lett. B **504**(3), 218–224 (2001). ISSN: 0370-2693. https://doi.org/10.1016/S0370-2693(01)00307-0

195. M. Kramer et al., Supersymmetry production cross sections in pp collisions at $\sqrt{s} = 7$ TeV (2012). arXiv: 1206.2892 [hep-ph]
196. D. Krohn, J. Thaler, L.-T. Wang, Jet trimming. J. High Energy Phys. **2**, 84 (2010). https://doi.org/10.1007/JHEP02(2010)084. arXiv: 0912.1342 [hep-ph]
197. A. Kulesza, L. Motyka, Threshold resummation for squark-antisquark and gluino-pair production at the LHC. Phys. Rev. Lett. **102**, 111802 (2009). https://doi.org/10.1103/PhysRevLett.102.111802. arXiv: 0807.2405 [hep-ph]
198. A. Kulesza, L. Motyka, Soft gluon resummation for the production of gluino-gluino and squark-antisquark pairs at the LHC. Phys. Rev. D **80**, 095004 (2009). https://doi.org/10.1103/PhysRevD.80.095004. arXiv: 0905.4749 [hep-ph]
199. H.-L. Lai et al., New parton distributions for collider physics. Phys. Rev. D **82**, 74024 (2010). https://doi.org/10.1103/PhysRevD.82.074024. arXiv: 1007.2241 [hep-ph]
200. D.J. Lange, The EvtGen particle decay simulation package. Nucl. Instrum. Meth. A **462**, 152 (2001). https://doi.org/10.1016/S0168-9002(01)00089-4
201. A.J. Larkoski, I. Moult, B. Nachman, Jet substructure at the large hadron collider: a review of recent advances in theory and machine learning (2017). arXiv: 1709.04464 [hep-ph]
202. C.G. Larrea et al., IPbus: a flexible Ethernet-based control system for xTCA hardware. J. Instrum. **10**(2), p. C02019 (2015). http://stacks.iop.org/1748-0221/10/i=02/a=C02019
203. Letter of intent for the phase-I upgrade of the ATLAS experiment. Technical Report CERN-LHCC-2011-012. LHCC-I-020. Geneva: CERN (2011). https://cds.cern.ch/record/1402470
204. P. Loch et al., Topological cell clustering in the ATLAS calorimeters and its performance in LHC Run I (2014). Technical Report ATL-COM-PHYS-2014-1439. Geneva: CERN. https://cds.cern.ch/record/1967028
205. E. Maguire, L. Heinrich, G. Watt, HEPData: a repository for high energy physics data. J. Phys. Conf. Ser. **898**(10), 102006 (2017). https://doi.org/10.1088/1742-6596/898/10/102006. arXiv: 1704.05473 [hep-ex]
206. S. Majewski et al., A thin multiwire chamber operating in the high multiplication mode. Nucl. Instrum. Meth. **217**, 265–271 (1983). https://doi.org/10.1016/0167-5087(83)90146-1
207. S.P. Martin, A supersymmetry primer. Adv. Ser. Direct. High Energy Phys. **18**(1), 1–98 (1998). https://doi.org/10.1142/9789812839657_0001, https://doi.org/10.1142/9789814307505_0001. arXiv: hep-ph/9709356 [hep-ph]
208. A.D. Martin et al., Parton distributions for the LHC. Eur. Phys. J. C **63**, 189 (2009). https://doi.org/10.1140/epjc/s10052-009-1072-5. arXiv: 0901.0002 [hep-ph]
209. D.W. Miller, Measurement of Hadronic Event Shapes and Jet Substructure in Proton-Proton Collisions at 7.0 TeV Center-of-Mass Energy with the ATLAS Detector at the Large Hadron Collider. Ph.D. Thesis. SLAC (2011). http://inspirehep.net/record/1088035/files/CERN-THESIS-2011-144.pdf
210. H. Miyazawa, Baryon number changing currents. Prog. Theor. Phys. **36**(6), 1266–1276 (1966). https://doi.org/10.1143/PTP.36.1266. oup/backfile/content_public/journal/ptp/36/6/10.1143/ptp.36.1266/2/36-6-1266.pdf
211. E. Mobs, The CERN accelerator complex (2016). General Photo. https://cds.cern.ch/record/2197559
212. P.J. Mohr, D.B. Newell, B.N. Taylor, CODATA recommended values of the fundamental physical constants: 2014. Rev. Mod. Phys. **88**, 035009 (2016). https://doi.org/10.1103/RevModPhys.88.035009
213. B. Nachman et al., Jets from jets: re-clustering as a tool for large radius jet reconstruction and grooming at the LHC. J. High Energy Phys. **2**, 75 (2015). https://doi.org/10.1007/JHEP02(2015)075. arXiv: 1407.2922 [hep-ph]
214. Y. Nambu, G. Jona-Lasinio, Dynamical model of elementary particles based on an analogy with superconductivity. I. Phys. Rev. 122, 345–358 (1961). https://doi.org/10.1103/PhysRev.122.345
215. S.H. Neddermeyer, C.D. Anderson, Note on the nature of cosmic-ray particles. Phys. Rev. **51**, 884–886 (1937). https://doi.org/10.1103/PhysRev.51.884

216. J. Neyman, E.S. Pearson, IX. On the problem of the most efficient tests of statistical hypotheses. Philos. Trans. R. Soc. Lond. A **231**, 694–706 (1933). ISSN: 0264-3952. https://doi.org/10.1098/rsta.1933.0009. http://rsta.royalsocietypublishing.org/content/231/694-706/289.full.pdf.
217. On a new kind of rays. Nature **53**, 274–276 (1896). http://dx.doi.org/10.1038/053274b0
218. C. Patrignani et al., (Particle Data Group). Review of particle physics. Chin. Phys. **C40**(10), 100001 (2016). https://doi.org/10.1088/1674-1137/40/10/100001
219. M.L. Perl et al., Evidence for anomalous lepton production in electron-positron annihilation. Phys. Rev. Lett. **35**, 1489–1492 (1975). https://doi.org/10.1103/PhysRevLett.35.1489
220. G. Piacquadio, C. Weiser, A new inclusive secondary vertex algorithm for b-jet tagging in ATLAS. J. Phys. Conf. Ser. **119**, 032032 (2008). https://doi.org/10.1088/1742-6596/119/3/032032
221. T. Plehn, Lectures on LHC physics. Lect. Notes Phys. **844**, 1–193 (2012). https://doi.org/10.1007/978-3-642-24040-9. arXiv: 0910.4182 [hep-ph]
222. H.D. Politzer, Reliable perturbative results for strong interactions. Phys. Rev. Lett. **30**, 1346–1349 (1973). https://doi.org/10.1103/PhysRevLett.30.1346
223. H.D. Politzer, Asymptotic freedom: an approach to strong interactions. Phys. Rep. **14**(4), 129–180 (1974)
224. J. Pumplin et al., New generation of parton distributions with uncertainties from global QCD analysis. J. High Energy Phys. **7**, 12 (2002). https://doi.org/10.1088/1126-6708/2002/07/012. arXiv: hep-ph/0201195
225. A. Purcell, *Go on a Particle Quest at the First CERN Webfest* (CERN Publications, Meyrin, 2012), p. 10. https://cds.cern.ch/record/1473657. BUL-NA-2012-269.35/2012.
226. A. Quadt, Top quark physics at hadron colliders. Eur. Phys. J. C **48**(3), 835–1000 (2006). ISSN: 1434-6052. https://doi.org/10.1140/epjc/s2006-02631-6
227. L. Randall, R. Sundrum, A large mass hierarchy from a small extra dimension. Phys. Rev. Lett. **83**, 3370–3373 (1999). https://doi.org/10.1103/PhysRevLett.83.3370. arXiv: hep--ph/9905221 [hep-ph]
228. A.L. Read, Presentation of search results: the CL(s) technique. J. Phys. G **28**, 2693 (2002). https://doi.org/10.1088/0954-3899/28/10/313
229. F. Reines, C.L. Cowan Jr., The neutrino. Nature **178**, 446–449 (1956). http://dx.doi.org/10.1038/178446a0
230. W. Riegler et al., Resolution limits of drift tubes. Nucl. Instrum. Methods Phys. Res., Sect. A **443**(1), 156-163 (2000). ISSN: 0168-9002. https://doi.org/10.1016/S0168-9002(99)01014-1
231. W.B. Rolnick, *The Fundamental Particles and Their Interactions* (Addison-Wesley, Reading, 1994)
232. L. Rossi, The LHC superconducting magnets, in *Proceedings of the Particle Accelerator Conference, 2003. PAC 2003*, vol. 1 (IEEE, Piscataway, 2003), pp. 141–145
233. E. Rutherford, Collision of alpha particles with light atoms. Lond. Edinb. Dubl. Philos. Mag. J. Sci. **37**(222), 581–587 (1919). https://doi.org/10.1080/14786440608635919
234. N. Sakai, Naturalness in supersymmetric guts. Z. Phys. C **11**, 153 (1981). https://doi.org/10.1007/BF01573998
235. A. Salam, Weak and electromagnetic interactions. Conf. Proc. **C680519**, 367–377 (1968)
236. G.P. Salam, Towards jetography. Eur. Phys. J. **C67**, 637–686 (2010). https://doi.org/10.1140/epjc/s10052-010-1314-6. arXiv: 0906.1833 [hep-ph]
237. A. Salam, J.A. Strathdee, Supersymmetry and nonabelian gauges. Phys. Lett. B **51**, 353–355 (1974). https://doi.org/10.1016/0370-2693(74)90226-3
238. S. Schumann, F. Krauss, A parton shower algorithm based on Catani-Seymour dipole factorisation. J. High Energy Phys. **3**, 38 (2008). https://doi.org/10.1088/1126-6708/2008/03/038. arXiv: 0709.1027 [hep-ph]
239. J. Shelton, Jet substructure, in *Proceedings, Theoretical Advanced Study Institute in Elementary Particle Physics: Searching for New Physics at Small and Large Scales (TASI 2012): Boulder, Colorado, June 4–29, 2012* (2013), pp. 303–340. https://doi.org/10.1142/9789814525220_0007. arXiv: 1302.0260 [hep-ph]. http://inspirehep.net/record/1217434/files/arXiv:1302.0260.pdf

240. M. Shochet et al., Fast TracKer (FTK) technical design report. Technical Report CERN-LHCC-2013-007. ATLAS-TDR-021. ATLAS Fast Tracker Technical Design Report (2013). http://cds.cern.ch/record/1552953
241. T. Sjöstrand, S. Mrenna, P.Z. Skands, PYTHIA 6.4 physics and manual. J. High Energy Phys. **5**, 26 (2006). https://doi.org/10.1088/1126-6708/2006/05/026. arXiv: hep-ph/0603175
242. T. Sjöstrand, S. Mrenna, P.Z. Skands, A brief introduction to PYTHIA 8.1. Comput. Phys. Commun. **178**, 852 (2008). doi: 10.1016/j.cpc.2008.01.036. arXiv: 0710.3820 [hep-ph]
243. P.Z. Skands, Tuning Monte Carlo generators: the Perugia tunes. Phys. Rev. D **82**, 074018 (2010). https://doi.org/10.1103/PhysRevD.82.074018. arXiv: 1005.3457 [hep-ph]
244. M.J. Swiatlowski, A. Schwartzman, Measuring the standard model and searching for new physics using the ATLAS detector. Presented 2015 (2015). https://cds.cern.ch/record/2040684
245. The ALICE Collaboration, The ALICE experiment at the CERN LHC. J. Instrum. **3**(8), S08002 (2008). https://doi.org/10.1088/1748-0221/3/08/S08002
246. The ATLAS TRT Collaboration, The ATLAS transition radiation tracker (TRT) proportional drift tube: design and performance. J. Instrum. **3**(2), P02013 (2008). http://stacks.iop.org/1748-0221/3/i=02/a=P02013
247. The ATLAS TRT Collaboration, The ATLAS TRT barrel detector. J. Instrum. **3**(2), P02014 (2008). http://stacks.iop.org/1748-0221/3/i=02/a=P02014
248. The CMS Collaboration, The CMS experiment at the CERN LHC. J. Instrum. **3**(8), S08004 (2008). https://doi.org/10.1088/1748-0221/3/08/S08004
249. The LHCb Collaboration, The LHCb detector at the LHC. J. Instrum. **3**(8), S08005 (2008). https://doi.org/10.1088/1748-0221/3/08/S08005
250. J.J. Thomson, XL. Cathode rays. Lond. Edinb. Dubl. Philos. Mag. J. Sci. **44**(269), 293–316 (1897). https://doi.org/10.1080/14786449708621070
251. G. 't Hooft, Dimensional regularization and the renormalization group. Nucl. Phys. **B61**, 455–468 (1973). https://doi.org/10.1016/0550-3213(73)90376-3
252. G. 't Hooft et al., Recent developments in gauge theories. Proceedings, Nato Advanced Study Institute, Cargese, France, August 26–September 8, 1979. NATO Sci. Ser. B **59**, 1–438 (1980). https://doi.org/10.1007/978-1-4684-7571-5
253. D.V. Volkov, V.P Akulov, Is the neutrino a goldstone particle? Phys. Lett. B **46**(1), 109–110 (1973). ISSN: 0370-2693. https://doi.org/10.1016/0370-2693(73)90490-5
254. B.R. Webber, Fragmentation and hadronization. Int. J. Mod. Phys. **A15S1**, 577–606 (2000). eConfC990809,577(2000). https://doi.org/10.1142/S0217751X00005334. arXiv: hep-ph/9912292 [hep-ph]
255. P. Weber, ATLAS calorimetry: trigger, simulation and jet calibration. Ph.D. Thesis. Kirchhoff Inst. Phys. (2008). http://www.ub.uni-heidelberg.de/archiv/8170/
256. S. Weinberg, A model of leptons. Phys. Rev. Lett. **19**, 1264–1266 (1967). https://doi.org/10.1103/PhysRevLett.19.1264
257. S. Weinberg, New approach to the renormalization group. Phys. Rev. D **8**, 3497–3509 (1973). https://doi.org/10.1103/PhysRevD.8.3497
258. N. Wermes, G. Hallewel, *ATLAS Pixel Detector: Technical Design Report* (CERN, Geneva, 1998). Technical Design Report ATLAS. http://cds.cern.ch/record/381263
259. J. Wess, B. Zumino, Supergauge transformations in four-dimensions. Nucl. Phys. B **70**, 39–50 (1974). https://doi.org/10.1016/0550-3213(74)90355-1
260. J. Wess, B. Zumino, Supergauge invariant extension of quantum electrodynamics. Nucl. Phys. B **78**, 1 (1974). https://doi.org/10.1016/0550-3213(74)90112-6
261. E. Witten, Dynamical breaking of supersymmetry. Nucl. Phys. B **188**(3), 513–554 (1981). ISSN: 0550-3213. https://doi.org/10.1016/0550-3213(81)90006-7
262. M. Wu, D. Miller, Measurement of Collinear *W* Boson Emission from High Transverse Momenta Jets Using the ATLAS Detector (2017). Presented 31 Jan 2017. https://cds.cern.ch/record/2298601
263. C.N. Yang, R.L. Mills, Conservation of isotopic spin and isotopic gauge invariance. Phys. Rev. **96**, 191–195 (1954). https://doi.org/10.1103/PhysRev.96.191

The manufacturer's authorised representative in the EU is Springer Nature Customer Service Centre GmbH, Europaplatz 3, 69115 Heidelberg, Germany. If you have any concerns regarding our products, please contact ProductSafety@springernature.com

Printed and bound by CPI Group (UK) Ltd, Croydon, CR0 4YY
05/10/2026
02228339-0005